Remote Sensing for Environmental Monitoring

Vara Saritha · Chaitanya Baliram Pande ·
Raj Singh · Mohammad Shahid
Editors

Remote Sensing for Environmental Monitoring

Springer

Editors
Vara Saritha
Department of Environmental Science
GITAM Deemed to be University
Visakhapatnam, Andhra Pradesh, India

Raj Singh
Department of Environmental Science
GITAM Deemed to be University
Visakhapatnam, Andhra Pradesh, India

Chaitanya Baliram Pande
Department of Civil Engineering, School of Core Engineering, Faculty of Science, Technology and Architecture (FoSTA)
Manipal University Jaipur
Jaipur, Rajasthan, India

Mohammad Shahid
Department of Chemistry
Koneru Lakshmaiah Education Foundation
Vaddeswaram, Andhra Pradesh, India

ISBN 978-981-96-5545-8 ISBN 978-981-96-5546-5 (eBook)
https://doi.org/10.1007/978-981-96-5546-5

This Springer imprint is published by the registered company Springer Nature Singapore Pte Ltd.
The registered company address is: 152 Beach Road, #21-01/04 Gateway East, Singapore 189721, Singapore

Preface

The book *Remote Sensing for Environmental Monitoring* explores the diverse applications of remote sensing in monitoring and managing environmental resources. It covers key areas such as water ecosystems, agriculture, forests, land use and land cover analysis, atmospheric and climate studies, coastal and marine environments, urban landscapes, and wildlife habitat assessment. Moreover, the book highlights its significance and wide-ranging applications in environmental assessment. It explains fundamental principles of image interpretation and analysis, details various satellite platforms and sensors, and examines aerial and ground-based remote sensing technologies through real-world case studies. Additionally, this book's chapters discuss advancements in remote sensing for Earth resource monitoring, including data acquisition, preprocessing techniques, and specialized imaging methods such as multispectral, hyperspectral, and thermal infrared imaging. This volume contains a total of 15 characters. Chapter 1 explores multispectral, hyperspectral, and thermal infrared imaging for environmental monitoring. It covers data acquisition methods, image processing techniques, and applications in vegetation health, water quality, and urban heat analysis. Chapter 2 examines spatiotemporal changes in snow cover within the Gori Ganga watershed, a key tributary of the Kali River in the Uttarakhand Himalayas. Using NDSI analysis, it maps snow cover and glacial lake dynamics from 1990 to 2022. Chapter 3 focuses on monitoring forest changes in the Western Ghats using Synthetic Aperture Radar (SAR) imagery. It applies preprocessing techniques and the Radar Vegetation Index (RVI) with Sentinel-1C data to detect deforestation trends. Chapter 4 discusses data acquisition, preprocessing, and analytical methods, highlighting applications in forest health, water and soil assessment, and urban heat island analysis. Case studies validate these technologies, while future challenges and advancements, including AI integration and sensor improvements, are explored. Chapter 5 examines the role of remote sensing in geriatric nursing, focusing on vital sign monitoring, fall detection, cognitive health assessment, and environmental monitoring. It also addresses technical, ethical, and operational challenges, along with future advancements in sensor technology, AI, and telehealth. Chapter 6 applies a geospatial approach to mapping groundwater potential and identifying suitable sites for artificial recharge in the Narmada River Basin, India. Using eight thematic factors and the Analytical Hierarchy Process (AHP), five groundwater potential zones were classified and validated with well-yield data. Chapter 7 explores remote sensing applications in detecting freshwater algal blooms, highlighting key spectral indices like NDCI, FAI, and ABDI. It discusses their strengths, limitations, and challenges in species-level detection. Chapter 8 reviews impact of climate change on agriculture, particularly in mountainous regions and developing nations. It highlights shifts in monsoons, gendered effects, and the need for adaptation strategies. Chapter 9 evaluates the water quality of Lake Llanquihue by estimating chlorophyll-a concentrations using Sentinel-2 satellite data and three atmospheric correction methods. The Acolite B4/B5 algorithm provided the best results, though accuracy was limited by sparse monitoring stations. Chapter 10 examines groundwater quality in Karnataka's Dakshina Kannada district, focusing on the hydrogeochemistry of the Netravathi and Gurapura catchments. ArcGIS mapping highlights spatial variations, emphasizing the need for improved monitoring and sustainable groundwater management. Chapter 11 assesses groundwater depletion in Uttar Pradesh's NCR region using primary and secondary data. Geospatial tools aid in analysis, providing insights for sustainable water management and policy recommendations. Chapter 12 evaluates soil erosion in Dehradun using Google Earth Engine and RUSLE. Sustainable land management, conservation measures, and reforestation are recommended to mitigate risks and support environmental sustainability. Chapter 13 documents eleven *Fissidens* moss taxa across multiple locations in the Eastern Ghats, with *Fissidens pulchellus* newly reported. These species exhibit diverse ecological preferences. Chapter 14 discusses the Bisalpur Wetland in Rajasthan, which supports biodiversity, including migratory and vulnerable bird species. It provides essential ecosystem services but faces threats from human activities like fuelwood collection.

This volume includes research conducted by scientists, researchers, professors, and planners from universities, research labs, and organizations in India and other countries. It provides an in-depth analysis of remote sensing applications in natural resource mapping while exploring emerging technologies and future trends in environmental monitoring. The book offers valuable insights into potential applications, challenges, and future directions in the field.

Visakhapatnam, India — Vara Saritha
Pune, India — Chaitanya Baliram Pande
Visakhapatnam, India — Raj Singh
Vaddeswaram, India — Mohammad Shahid

Contents

About the Editors

Dr. Vara Saritha M.Sc., M.Tech, Ph.D., is working as Associate Professor in the Department of Life Science, Environmental Science Division, GITAM Deemed to be University, Visakhapatnam, Andhra Pradesh. She has 20 years of teaching and research experience. Specializations of the author include environmental planning and sustainable development, addressing challenges of potable water, and remote sensing technology. The author has published 80 research papers and is also active in addressing issues of public and occupational health and corrosion studies. She authored four textbooks. She presented around 50 research papers in several national and international seminars and conferences. She has completed three research projects funded by University Grants Commission, Department of Science and Technology and GITAM. Three students have been awarded Ph.D. under her guidance, and currently, she is mentoring four students. She is Reviewer to 12 international journals in the field of Environmental Science and also is Associate Editor, Editorial Board Member for three journals published by IGI Global in the field of Environmental Science.

Dr. Chaitanya Baliram Pande Ph.D., completed his Ph.D. in environment science at Sant Gadge Baba Amravati University, India, and an M.Sc. in geoinformatics at Amravati University in 2011. He has more than 13+ years of teaching, research and industry experience. He is included in the 2% Stanford scientist list, 2024. He is Reviewer for several scientific journals of international repute, and he is Editorial Board Member in the American Journal of Agricultural and Biological Sciences. He has published 120+ research papers, two textbooks, 10 edited books, 19 conference papers and 25+ book chapters. Altogether, they have received more than 4761 citations as per Google Scholar. His research interests include remote sensing, GIS, air pollution, Google Earth Engine, machine learning, watershed management, hydrogeology, hydrological modelling, drought monitoring, land use and land cover analysis, groundwater quality, urban planning, hydro-geochemistry, groundwater modelling, geology, hyperspectral remote sensing, and GIS applications in natural resources and watershed management, and environmental monitoring and assessment.

Raj Singh M.Sc., PGD, is Research Scholar in the Department of Environmental Science GITAM Deemed to be University, Visakhapatnam, Andhra Pradesh, India. He served as Assistant Professor at Dr. K. N. Modi University, Newai, Rajasthan, and Tula's Institute, Dehradun, India. He is Alumnus of the Indian Space Research Organization (ISRO), Department of Space, Government of India, and also worked on a France-funded IDDRI project at BITS Pilani, K. K. Birla Campus, Goa, India. He is Active Member of Wetland International, Wageningen, Netherlands, and the British Ecological Society London, UK. His expertise is in Space Science, Environmental Science, Wetland Ecology, Remote sensing, and GIS. He has published over 20 scientific articles in reputed national and international peer-reviewed journals, review, book chapters, presented six papers at international conferences, and edited three books with Jenny Stanford and Springer Nature publisher.

Dr. Mohammad Shahid Ph.D., is Assistant Professor in the Department of Chemistry at Koneru Lakshmaiah Education Foundation, India. He earned his Master's in Chemistry from Shibli National College, Azamgarh (2006), and Ph.D. from Jamia Millia Islamia, New Delhi (2014). He pursued postdoctoral research at Soochow University, China, the University of Glasgow, UK (Marie Curie Fellow), and the Institute of Chemical Technology, Mumbai (Dr. D. S. Kothari Fellow). He also served as Assistant Professor and Head of the Department of Applied Science at Dr. K. N. Modi University and the Department of Chemistry at Mewar University, Rajasthan. Dr. Shahid has received several prestigious fellowships to support his research pursuits, including grants from UGC, India, China Postdoctoral Fellowship, EU Marie Curie Fellowship, and Dr. D. S. Kothari Fellowship. He has contributed to over 50 scientific publications with ~4000 citations and h-index of 24. In addition, has also edited and authored several books with renowned international publishing houses.

Chapter 1
Remote Sensing for Environment Assessment: Multispectral, Hyperspectral, and Thermal Imaging Applications

Aman Srivastava and Shubham Jain

Abstract Environmental challenges are increasingly multifaceted, necessitating sophisticated tools for accurate and holistic assessments. Traditional monitoring methods commonly fail to capture the intricate details required for effective environmental management. This study addresses these limitations by investigating the capabilities of multispectral, hyperspectral, and thermal infrared imaging technologies. These remote sensing techniques utilize a broad spectrum of wavelengths to detect and analyze environmental data beyond human perception, providing a critical understanding of key environmental factors such as vegetation health, water quality, urban heat patterns, and soil composition. The study introduces the fundamental principles of these imaging technologies, highlighting their physical interactions with environmental features. It further explores data acquisition methodologies, ranging from satellite platforms and uncrewed aerial vehicles to ground-based sensors. It focuses on advanced image processing techniques, including machine learning (ML) algorithms like the maximum likelihood classifier, K-means clustering, and the ISODATA algorithm. The practical applications of these techniques are illustrated through case studies addressing diverse environmental contexts, such as vegetation monitoring in the Amazon, urban heat island mitigation in New York City, and water quality assessment in the Great Lakes. Additionally, the study emphasizes the integration of thermal infrared imaging with multispectral and hyperspectral data, providing temperature-related understandings crucial for environmental studies. The synergistic combination of these technologies, alongside geographic information systems (GIS), improves decision-making and facilitates more sustainable environmental management practices. Lastly, the study discusses emerging challenges and future research directions in the field, aiming to equip researchers

A. Srivastava
Formerly, Centre for Technology Alternatives for Rural Areas (CTARA), Indian Institute of Technology Bombay, Mumbai, India

S. Jain (✉)
Department of Civil Engineering, Indian Institute of Technology Bombay, Mumbai, India
e-mail: shubhamjain752@gmail.com

V. Saritha et al. (eds.), *Remote Sensing for Environmental Monitoring*,
https://doi.org/10.1007/978-981-96-5546-5_1

and policymakers with advanced methodologies for addressing ongoing and future environmental challenges.

Keywords Environmental monitoring · Geospatial technologies · UAV-based sensing · Machine learning in remote sensing · Urban heat islands · Disaster risk management

1.1 Introduction

Remote sensing has become an indispensable tool in environmental assessment, delivering unparalleled capabilities for monitoring and analyzing Earth's ecosystems across various spatial and temporal scales. It involves acquiring information about objects or phenomena without physical contact. It typically uses satellite or aircraft-based sensors to detect and record reflected or emitted energy from the Earth's surface and atmosphere (Gao et al. 2019; Lillesand et al. 2015). By providing a unique vantage point for observing complex environmental processes and changes, remote sensing supports various applications, including land cover mapping, vegetation monitoring, water quality assessment, and urban planning (Mashala et al. 2023). The fundamental principle of remote sensing lies in measuring and analyzing electromagnetic radiation (EMR) reflected or emitted from the Earth's surface. Different materials interact uniquely with EMR due to their physical and chemical properties, allowing sensors to capture distinct spectral signatures that can be analyzed to identify and monitor various environmental features (Abdulraheem et al. 2023). For example, healthy vegetation reflects strongly in the near-infrared region while absorbing visible red light, enabling the assessment of plant health and biomass. Remote sensing provides critical data that underpin environmental research and management, and its importance continues to grow with technological advancements.

Advancements in sensor technologies and data processing capabilities have significantly expanded the application of remote sensing in environmental assessment over the past few decades (Toth and Jóźków 2016). Remote sensing provides wide spatial coverage, as satellite-based sensors can capture data over vast areas, facilitating regional to global-scale assessments (Chuvieco, 2020). The temporal frequency of many platforms provides regular revisit times, enabling monitoring of environmental changes over time (Manfreda et al. 2018). Additionally, remote sensing can be more cost-effective than extensive field surveys for large-scale monitoring (Fascista 2022). It allows for observing areas that are difficult or dangerous to access on the ground, minimizing risks of disturbance or damage (Pettorelli et al. 2014). Furthermore, sensors capture data across various parts of the electromagnetic spectrum, revealing information invisible to the human eye, such as thermal properties and chemical compositions (Reddy 2018; Lei et al. 2022). These capabilities have made remote sensing an essential tool in numerous environmental applications, including land use and land cover mapping, vegetation monitoring, water resource management, climate change studies, and disaster risk assessment (Thenkabail et al. 2018; Rane et al. 2023; Liladhar Rane et al. 2024; Pande et al. 2024; Srivastava and Chinnasamy 2021, 2024; Jain et al. 2024a, b).

Remote sensing technologies are categorized into passive and active systems based on energy sources. Passive remote sensing relies on natural energy, primarily sunlight, to capture reflected or emitted energy from the Earth's surface using sensors on satellites or aircraft. This method collects data in the electromagnetic spectrum's visible, near-infrared, and thermal infrared regions, with examples including satellites like Landsat and Sentinel-2 (Drusch et al. 2012). While effective for large-scale, continuous observations of vegetation health, land cover changes, and surface temperature variations, passive remote sensing is limited by factors such as cloud cover, atmospheric conditions, and the availability of sunlight (Li and Roy 2017). Active remote sensing involves systems emitting energy to scan objects and areas. Technologies like RADAR and LiDAR send signals reflected to the sensor, creating detailed images or data models. Active systems can operate day and night and penetrate atmospheric obstructions like clouds and rain, providing reliable data under varied environmental conditions (National Academy of Science 2015). Radar systems, such as Synthetic Aperture Radar (SAR), emit microwave signals to gather detailed information about surface characteristics. At the same time, LiDAR uses laser pulses for high-resolution topographic mapping and vegetation structure analysis (Pirotti 2011). The selection between passive and active remote sensing depends on the specific environmental variables being monitored, the required spatial and temporal resolution, and the project's overall goals. By combining these technologies, researchers can holistically understand environmental changes and their impacts on ecosystems and human populations.

Multispectral, hyperspectral, and thermal imaging are advanced remote sensing technologies that provide unique capabilities for environmental assessment. Multispectral imaging captures data in multiple broad spectral bands, typically ranging from visible to near-infrared regions. This allows for the differentiation of land cover types and assessment of vegetation health using indices like the Normalized Difference Vegetation Index (NDVI) (Xue and Su 2017). Satellites such as Landsat 8 and Sentinel-2 employ multispectral sensors for agricultural monitoring, forest management, and land use classification (Soni et al. 2022). Hyperspectral imaging collects data across hundreds of narrow, contiguous spectral bands, enabling the detection of subtle differences in the reflectance properties of Earth's surface features (Homolová et al. 2013). This high spectral resolution provides detailed information about material composition, facilitating the identification of specific minerals, assessing plant species diversity, detecting plant stress, and monitoring water quality (Wambugu et al. 2021). Thermal imaging measures the heat emitted by objects on Earth's surface, providing essential understandings of energy balance, heat fluxes, and temperature variations. Thermal sensors on satellites like Landsat 8 Thermal Infrared Sensor (TIRS) and the ECOsystem Spaceborne Thermal Radiometer Experiment on Space Station (ECOSTRESS) are used to monitor urban heat islands, detect forest fires, assess water body temperatures, and evaluate vegetation stress and soil moisture content (Fisher et al. 2020). Combining multispectral or hyperspectral data with thermal imagery reveals relationships between land cover, vegetation health, and surface temperature dynamics, improving environmental assessments (Guha et al. 2020). These advanced imaging technologies enable more precise and all-rounded environmental monitoring, addressing complex ecological questions and management needs.

Advancements in data processing techniques, particularly machine learning algorithms, have revolutionized remote sensing data analysis. Machine learning enables handling large datasets and extracting complex patterns, improving classification accuracy and predictive capabilities (Resch et al. 2018; Maxwell et al. 2018). Algorithms such as Support Vector Machines (SVM), Random Forests, and Convolutional Neural Networks (CNNs) are increasingly employed for land cover classification, vegetation monitoring, and change detection tasks (Yuan et al. 2020). Integrating machine learning with remote sensing facilitates automated feature extraction, improves the interpretation of multispectral and hyperspectral data, and supports the development of sophisticated environmental models (Vali et al. 2024). Deep learning techniques have improved the accuracy of urban land cover mapping and detecting subtle environmental changes, contributing to better urban planning and resource management (Li et al. 2024). Combining advanced algorithms and high-quality data refines the ability to monitor environmental conditions and predict future changes.

Remote sensing is essential in environmental studies, providing critical data for understanding natural and human-induced changes. Its vast applications encompass climate change monitoring, biodiversity conservation, water resources management, urban planning, and disaster management. In climate change studies, remote sensing enables monitoring of glacier retreats, sea-level rise, alterations in weather patterns, and the distribution of carbon stocks (Lenton et al. 2024). For biodiversity conservation, it assists in mapping habitat distribution, tracking species migration, and identifying threatened ecosystems (Cavender-Bares et al. 2022). In water resources management, remote sensing assesses water quality, monitors changes in water bodies, and aids in managing agricultural water usage (Kumar et al. 2022). Urban planning utilizes remote sensing to monitor urban sprawl, land use changes, and infrastructure development (Abdi 2020). Disaster management provides crucial data for responding to natural disasters like floods, wildfires, and hurricanes, enabling rapid assessment and resource allocation (Lakshmi 2016). As environmental challenges grow more complex, these advanced imaging technologies are essential for providing accurate, timely, and detailed information and supporting evidence-based decision-making in environmental management and policy formulation (Turner et al. 2015). Integrating remote sensing data with other geospatial technologies uplifts the effectiveness of environmental monitoring and contributes to sustainable development goals.

In light of the growing complexities of environmental challenges, this study aims to holistically examine advanced remote sensing technologies, focusing on multispectral, hyperspectral, and thermal infrared imaging. The primary objective is to explore how these technologies, integrated with machine learning algorithms, can improve environmental monitoring by capturing and analyzing data beyond the capabilities of conventional methods. This study will further explore key principles, data acquisition methodologies, and advanced processing techniques, emphasizing the use of machine learning algorithms for classification and analysis. Additionally, the study will investigate various applications, from vegetation and water quality monitoring to urban heat assessment and disaster management. Through the integration of remote sensing with GIS, the scope of this study extends to evaluating how these combined approaches can boost decision-making in environmental management. Case studies will provide real-world context, while discussions on the limitations of current technologies, emerging trends, and future research directions will contribute to a holistic understanding of the evolving field of remote sensing. Ultimately, this work seeks to equip researchers and practitioners with the knowledge to capitalize on these advanced technologies for more informed and sustainable environmental assessments.

1.2 Remote Sensing Technologies and Concepts

Remote sensing technologies have evolved significantly, providing advanced tools for environmental assessment by capturing and analyzing electromagnetic radiation (EMR) interactions with Earth's surface materials. Understanding the principles and applications of multispectral, hyperspectral, and thermal infrared imaging is essential for utilizing these technologies effectively.

1.2.1 Principles of Multispectral and Hyperspectral Imaging

Multispectral and hyperspectral imaging are grounded in the principle that different materials on Earth's surface interact uniquely with EMR, resulting in distinct spectral signatures. These signatures are patterns of reflected, absorbed, and emitted radiation across varying wavelengths, characteristic of particular materials such as vegetation, water, soil, or urban structures (Lillesand et al. 2015). The unique spectral behavior arises from materials' physical and chemical properties, including molecular composition, surface texture, and moisture content (Jensen 2009). For example, due to cellular structure, healthy vegetation exhibits strong reflectance in the near-infrared region. It absorbs red light because of chlorophyll, while water bodies absorb most near-infrared radiation, appearing dark in these bands.

The interaction of EMR with Earth materials involves processes like reflection, absorption, transmission, scattering, and emission, forming the basis of remote sensing techniques that allow the differentiation of various surface features (Richards 2013). Multispectral imaging captures data across several broad spectral bands, typically ranging from visible to near-infrared regions, providing general information about the reflectance characteristics of materials. Common multispectral sensors include Landsat Operational Land Imager (OLI), Sentinel-2 MultiSpectral Instrument (MSI), and Moderate Resolution Imaging Spectroradiometer (MODIS), which are widely used for applications in agricultural monitoring, forest management, and land use classification (Roy et al. 2014; Drusch et al. 2012).

In contrast, hyperspectral imaging collects data across hundreds of narrow, contiguous spectral bands, providing continuous spectral signatures that enable the detection of subtle differences in material properties. This high spectral resolution allows for detailed material composition analysis, facilitating tasks such as identifying specific minerals, assessing plant species diversity, detecting vegetation stress, and monitoring water quality (Govender et al. 2007; Adam et al. 2010). Examples of hyperspectral sensors include NASA's Airborne Visible/Infrared Imaging Spectrometer (AVIRIS) and the spaceborne Hyperion sensor on the EO-1 satellite (Goetz 2009; Pearlman et al. 2003).

The choice between multispectral and hyperspectral imaging depends on the specific application, required spectral resolution, and considerations of practical data handling and processing complexity. While hyperspectral data provide more detailed spectral information, they result in increased data volume and require sophisticated processing techniques. Multispectral data, with fewer bands, are less demanding computationally and are suitable for broader applications where detailed spectral information is not critical (Shaw and Burke 2003). Selecting the appropriate imaging technology involves balancing the need for spectral detail with considerations of data volume, processing capabilities, and the specific environmental features of interest.

1.2.2 Basics of Thermal Infrared Imaging

Thermal infrared (TIR) imaging differs from multispectral and hyperspectral imaging, focusing on detecting emitted rather than reflected radiation. All objects with a temperature above absolute zero emit thermal radiation, with the amount and wavelength determined by the object's temperature according to Planck's Law (Vollmer and Möllmann 2010). For Earth observation, the thermal infrared region of interest lies between 3 and 14 μm, encompassing the wavelengths where Earth's surface emits most of its thermal energy (Kuenzer and Dech 2013).

Key thermal properties relevant to TIR remote sensing include emissivity, thermal inertia, and heat capacity. Emissivity measures how efficiently a surface emits thermal radiation compared to a perfect blackbody, affecting the apparent temperature recorded in thermal imagery (Dash et al. 2002). Different materials have varying emissivities; for example, vegetation typically has high emissivity values, while bare soils and urban materials may vary widely. Thermal inertia describes a material's resistance to temperature change, influencing how quickly it heats up or cools down, which is critical in assessing diurnal temperature variations (Zhang et al. 2016). Heat capacity, the amount required to raise a material's temperature by a certain amount, also affects temperature dynamics and is essential in interpreting thermal imagery.

Thermal infrared imaging has numerous applications in environmental studies. Urban climatology uses it to analyze urban heat islands by mapping surface temperature variations and identifying heat hotspots (Voogt and Oke 2003). In agriculture, thermal imaging assists in detecting vegetation stress by identifying changes in canopy temperature, which can indicate water stress or disease before visible symptoms appear (Jones et al. 2009). Thermal data are also employed in soil moisture assessment based on thermal inertia differences between wet and dry soils (Zhang and Zhou 2016; Petropoulos et al. 2015). Additionally, thermal imaging is vital in monitoring water temperatures of lakes, rivers, and coastal areas, contributing to studies on aquatic ecosystem health and thermal pollution (Handcock et al. 2012). In geologic applications, thermal infrared sensors detect volcanic activity by identifying thermal anomalies associated with lava flows or geothermal areas (Blackett, 2017).

Combining thermal infrared data with multispectral and hyperspectral data provides comprehensive perspectives into environmental conditions. While multispectral or hyperspectral data can reveal the composition and health of vegetation, thermal data add information about energy fluxes and stress levels. For instance, integrating these datasets allows for a nuanced understanding of how vegetation health correlates with surface temperature variations, aiding in drought assessment and resource management (Karnieli et al. 2010). This integrated approach improves the ability to monitor and model complex environmental processes by providing multidimensional datasets that capture both spectral and thermal properties.

1.2.3 Overview of Data Acquisition Platforms

Data acquisition platforms play a crucial role in remote sensing, determining the spatial resolution, temporal frequency, and spectral capabilities of the collected data. The primary platforms include satellites, airborne systems, Unmanned Aerial Vehicles (UAVs), and ground-based sensors, each yielding distinct advantages and limitations (Colomina and Molina 2014) (see Table 1.1).

Satellites provide extensive spatial coverage and regular revisit times, making them ideal for large-scale and long-term environmental monitoring. They carry a variety of sensors, including multispectral (e.g., Landsat, Sentinel-2), hyperspectral (e.g., Hyperion), and thermal infrared instruments (e.g., Landsat Thermal Infrared Sensor, MODIS) (Roy et al. 2014; Drusch et al. 2012). Satellites provide consistent data acquisition but are limited by fixed orbits, atmospheric conditions, and relatively coarse spatial resolutions compared to airborne or UAV platforms. Airborne systems involve sensors mounted on human-crewed aircraft, providing higher spatial resolution data due to lower altitudes and flexible flight planning. Airborne platforms are often used for hyperspectral imaging (e.g., AVIRIS) and LiDAR surveys, enabling detailed mapping of small to medium-sized areas (Goetz 2009). They yield flexibility in data acquisition timing and sensor configurations but are costlier and have limited temporal frequency compared to satellites. UAVs have emerged as

Table 1.1 Comparison of data acquisition platforms

Platform	Altitude	Spatial resolution	Temporal resolution	Application areas	References
Satellites	600–36,000 km	10 m to 1 km	Days to weeks (fixed orbits)	Global monitoring, land cover, climate studies	Roy et al. (2014), Drusch et al. (2012)
Airborne systems	1–10 km	Sub-meter to 5 m	On-demand (mission-specific)	Detailed mapping, hyperspectral imaging, LiDAR	Goetz (2009)
UAVs	< 500 m	Centimeter-level	On-demand (flexible missions)	Agriculture, forestry, infrastructure inspection	Manfreda et al. (2018)
Ground-based	Ground level	Millimeter-level	Continuous or periodic	Calibration, validation, detailed site studies	Milton et al. (2009)

important platforms for remote sensing, providing high spatial resolution, flexibility, and cost-effectiveness for local-scale studies (Aasen et al. 2018). UAVs can carry various sensors, including multispectral, hyperspectral, thermal infrared, and LiDAR, allowing data collection for specific applications (Manfreda et al. 2018). They are ideal for monitoring small areas, conducting frequent surveys, and accessing hard-to-reach locations. However, UAVs have limitations in flight time, payload capacity, and regulatory constraints regarding airspace usage. Ground-based sensors include portable spectroradiometers and thermal cameras for field measurements and calibration (Milton et al. 2009). These instruments provide the highest spatial resolution and are essential for validating and calibrating data from airborne and satellite platforms. Ground-based observations are critical for detailed studies of specific sites but lack spatial coverage and are labor-intensive.

The selection of an appropriate data acquisition platform depends on the study's objectives, required spatial and temporal resolutions, area of interest, and resource availability. Combining data from multiple platforms can refine analysis by utilizing their strengths. For example, satellite data provide broad coverage and temporal consistency, while UAVs yield high-resolution insights into specific areas of interest (Pricope et al. 2022). Integrating datasets from different platforms facilitates multi-scale analyses, improving the understanding of environmental processes across various spatial and temporal scales.

1.3 Data Processing and Machine Learning Techniques

The transformation of raw remote sensing data into actionable environmental strategies involves several critical steps, including image pre-processing, feature extraction, classification using machine learning algorithms, and accuracy assessment. These processes refine the quality of the data, extract meaningful features, and classify environmental features with high accuracy, ultimately supporting effective environmental monitoring and decision-making.

1.3.1 Image Pre-processing and Feature Extraction

Image pre-processing is a fundamental step that aims to correct distortions or degradations in the raw data resulting from sensor limitations, atmospheric conditions, and geometric discrepancies. Pre-processing ensures that the images accurately represent the Earth's surface, essential for reliable analysis and interpretation (Richards 2013). Radiometric correction addresses variations in sensor response and atmospheric effects that can alter the true radiance values of the Earth's surface. Techniques such as dark object subtraction and atmospheric correction models like Fast Line-of-sight Atmospheric Analysis of Spectral Hypercubes (FLAASH) adjust for atmospheric scattering and absorption, improving the radiometric integrity of the data (Chavez 1996). Geometric correction rectifies spatial distortions caused by sensor motion, Earth rotation, and terrain relief, aligning the images to a coordinate system through image-to-map rectification and image-to-image registration using ground control points (Jensen 2015).

Feature extraction involves transforming the pre-processed data into variables more directly related to the environmental parameters of interest. Vegetation indices are among the most widely used feature extraction techniques, providing quantitative measures of vegetation health, biomass, and productivity. The Normalized Difference Vegetation Index (NDVI), calculated using near-infrared (NIR) and red-light reflectance, is instrumental in monitoring vegetation dynamics, drought conditions, and deforestation (Tucker 1979). For instance, NDVI has been used to assess the impact of drought on vegetation in the Sahel region, revealing significant correlations between NDVI values and rainfall patterns (Anyamba and Tucker 2005).

The Enhanced Vegetation Index (EVI) improves upon NDVI by optimizing sensitivity in high-biomass regions and reducing atmospheric influences, making it suitable for monitoring dense vegetation canopies (Huete et al. 2002). Other indices like the Soil-Adjusted Vegetation Index (SAVI) incorporate soil brightness correction factors, refining vegetation monitoring in arid and semi-arid regions where soil reflectance can affect vegetation signal (Xu et al. 2006; Huete 1988). The Normalized Difference Water Index (NDWI) monitors water bodies, assisting in flood mapping, wetland studies, and irrigation management (McFeeters 1996). These indices enable researchers to derive meaningful environmental information from spectral data, facilitating better understanding and management of natural resources.

Spectral unmixing is another critical feature extraction technique that is instrumental in analyzing hyperspectral data where pixels may contain mixtures of different materials. Linear spectral unmixing decomposes mixed pixels into fractional abundances of pure spectral endmembers, allowing for the quantification of specific materials within a pixel (Keshava and Mustard 2002). This technique has been effectively applied in urban environments to map impervious surfaces, vegetation, and soil, aiding urban planning and management (Roberts et al. 1998). In mineral exploration, spectral unmixing assists in identifying mineral compositions, improving the detection of ore deposits (Van der Meer et al. 2012).

1.3.2 Machine Learning Algorithms for Classification

Machine learning algorithms have revolutionized the classification of remote sensing data by enabling handling large datasets and capturing complex patterns within the data. These algorithms can be broadly categorized into supervised and unsupervised learning methods, each with advantages and environmental monitoring applications.

Supervised classification methods rely on labeled training data to learn the relationships between spectral signatures and land cover classes. The Maximum Likelihood Classifier (MLC) is a traditional statistical method that assumes a normal data distribution within each class. It calculates the probability of a pixel belonging to a particular class based on its spectral values (Otukei and Blaschke 2010). While MLC is straightforward, it may not perform well with non-normally distributed data or complex class boundaries. Support Vector Machines (SVM) are robust classifiers that can handle high-dimensional data and complex class separations by finding the optimal hyperplane that maximizes the margin between classes (Mountrakis et al. 2011; Elbeltagi et al. 2022). SVMs are particularly effective in hyperspectral image classification due to their ability to manage the high dimensionality and limited training samples often associated with such data (Melgani and Bruzzone 2004). For example, SVMs have been successfully applied to accurately classify urban land cover types, aiding in urban growth monitoring and planning (Zhu and Blumberg 2002). Random Forests (RF) are ensemble learning methods that construct multiple decision trees during training and output the mode of the classes for classification tasks (Breiman 2001). RF algorithms are known for their high accuracy, resistance to overfitting, and ability to handle datasets with a large number of input variables. They have been widely used in land cover classification, vegetation mapping, and change detection studies (Belgiu and Drăguţ 2016; Elbeltagi et al. 2023a; Masood et al. 2023; Vishwakarma et al. 2024). For instance, RF has been utilized to map forest types and health conditions, contributing to forest management and conservation efforts (Gislason et al. 2006).

Unsupervised classification methods do not require labeled training data; instead, natural groupings within the data are identified based on spectral similarity. K-means clustering partitions the data into a predefined number of clusters by minimizing the variance within each cluster (MacQueen 1967). The Iterative Self-Organizing Data Analysis Technique Algorithm (ISODATA) extends K-means by allowing the number of clusters to change during the iteration process based on specific criteria (Ball and Hall 1965). These methods are practical for preliminary analysis and exploratory data assessment, especially in areas with limited prior knowledge.

Deep learning approaches have emerged as powerful tools for remote sensing image classification. While CNNs excel at automatically learning hierarchical features from raw image data and capturing spatial and spectral patterns that traditional methods may miss (Zhang et al. 2019; Zhang et al. 2016), Deep Neural Networks (DNNs) provide a broader framework for learning complex data representations. DNNs have been widely utilized in remote sensing for tasks requiring multisource data integration, such as fusing spectral, spatial, and temporal information for land cover classification and allied applications. Their ability to model intricate relationships between input features makes them particularly effective in detecting subtle changes in land use or vegetation dynamics (Elbeltagi et al. 2023b). CNNs have been applied to classify hyperspectral images for crop type identification, improving precision agriculture practices (Li et al. 2019; Elbeltagi et al. 2024).

Object-Based Image Analysis (OBIA) integrates spectral information with spatial context by segmenting images into meaningful objects rather than individual pixels (Blaschke 2010). OBIA considers shape, texture, and contextual relationships, improving classification accuracy in high-resolution imagery where spatial patterns are significant. This approach has been practical in urban mapping, where features like buildings, roads, and vegetation patches exhibit distinct spatial characteristics (Blaschke et al. 2014).

The selection of an appropriate machine learning algorithm depends on factors such as the classification task's complexity, the input data's nature, computational resources, and the desired level of accuracy. Table 1.2 summarizes key machine learning classification techniques, their applications, strengths, weaknesses, and suitable data types.

1.3.3 Accuracy Assessment and Validation Techniques

Assessing the accuracy of classification results is essential to ensure the reliability and validity of remote sensing analyses. Accuracy assessment involves comparing the classified data with reference information that is assumed to be true, typically derived from ground truth data, higher-resolution imagery, or expert interpretation (Congalton and Green 2019). The confusion matrix, also known as the error matrix, is a standard tool for accuracy assessment. It tabulates the relationship between known reference data and the corresponding classified data, allowing for the calculation of

Table 1.2 Machine learning-based classification techniques

Algorithm	Type	Applications	Strengths	Weaknesses	Suitable data	References
Maximum likelihood classifier	Supervised	General land cover mapping	Simple implementation	Assumes normal distribution; may misclassify complex data	Multispectral, Hyperspectral	Otukei and Blaschke (2010)
Support vector machines	Supervised	Hyperspectral classification	Handles high-dimensional data; robust	Computationally intensive	Hyperspectral	Mountrakis et al. (2011)
Random forests	Supervised	Land cover, vegetation mapping	High accuracy; resistant to overfitting	Requires tuning; less interpretable	Multispectral, Hyperspectral	Belgiu and Drăguţ (2016)
K-means clustering	Unsupervised	Preliminary data analysis	Simple; no training data required	Requires several clusters; may converge to local minima	Multispectral	MacQueen (1967)
ISODATA	Unsupervised	Exploratory analysis	Adapts number of clusters; flexible	Computationally intensive	Multispectral	Ball and Hall (1965)
Convolutional neural networks	Supervised	Complex pattern recognition	Automatically extracts features; high accuracy	Requires large datasets; computationally demanding	Multispectral, Hyperspectral	Zhang et al. (2016)
Object-based image analysis	Supervised/ Unsupervised	High-resolution imagery	Incorporates spatial context; accurate	Complex implementation requires segmentation	High-resolution imagery	Blaschke (2010)

various accuracy metrics. Overall accuracy represents the proportion of correctly classified samples over the total number of samples. Producer's accuracy measures the probability that a reference sample is correctly classified, indicating errors of omission. User's accuracy reflects the probability that a classified pixel represents the actual class on the ground, indicating errors of commission (Foody 2020). The Kappa coefficient is a statistical measure that accounts for the possibility of agreement occurring by chance. It provides a more rigorous classification accuracy assessment, particularly when the dataset has an unequal distribution of classes (Cohen 1960). However, some studies suggest caution in interpreting Kappa due to its sensitivity to class prevalence and the number of classes (Foody 2020). Spatially explicit accuracy assessment methods consider the spatial distribution of classification errors, identifying patterns or clusters of misclassifications that may be associated with specific land cover types or environmental conditions (Olofsson et al. 2014). Additionally, cross-validation techniques, such as k-fold cross-validation, can be used to evaluate the generalizability of the classification model by partitioning the data into training and testing subsets multiple times. Validation of remote sensing products is further improved by integrating field measurements and high-resolution reference data. For example, in vegetation studies, in-situ leaf area index (LAI) or biomass measurements can validate remote sensing-derived estimates, improving model calibration and accuracy (Weiss et al. 2020). In water quality assessments, ground-based measurements of parameters like chlorophyll concentration or turbidity provide essential validation data for satellite-derived indices (Rolim et al. 2023).

1.4 Environmental Monitoring Applications

Advanced imaging technologies such as multispectral, hyperspectral, and thermal imaging have revolutionized environmental monitoring by providing detailed, timely, and spatially extensive data. These technologies enable the assessment of vegetation health, land cover dynamics, biodiversity, water quality, coastal zone management, urban heat islands, and air quality. This section explores these applications, highlighting how remote sensing contributes to understanding and managing environmental systems.

1.4.1 Vegetation, Land Cover, and Biodiversity

Remote sensing technologies have significantly boosted the ability to monitor vegetation health, assess biodiversity, and track land cover changes at various scales. Multispectral and hyperspectral imaging are extensively used in precision agriculture to monitor crop health and predict yields. Vegetation indices such as the Normalized Difference Vegetation Index (NDVI) and the Enhanced Vegetation Index (EVI) detect stress in crops before it becomes visible to the naked eye, facilitating timely interventions (Mulla 2013). NDVI, calculated using near-infrared and red reflectance, ranges from – 1 to 1, with higher values indicating healthier vegetation. EVI improves upon NDVI by reducing atmospheric influences and improving sensitivity in high-biomass regions (Huete et al. 2002).

In forestry, these technologies enable the assessment of forest health, deforestation detection, and forest regeneration monitoring. Combining LiDAR with multispectral data has effectively estimated biomass and carbon stocks in tropical forests, contributing to climate change studies and conservation efforts (Asner et al. 2012). Hyperspectral imaging detects subtle changes in plant spectral signatures associated with disease or pest infestations, allowing for early intervention and management (Mahlein et al. 2012). For biodiversity assessment, hyperspectral data can distinguish between plant species based on their unique spectral signatures, aiding in mapping species distribution and detecting invasive species (Fassnacht et al. 2016). This detailed spectral information supports conservation planning and ecosystem management. Multispectral imagery also assesses habitat fragmentation and connectivity, which are crucial for wildlife conservation efforts by identifying corridors and barriers affecting species movement (Vos et al. 2001).

Remote sensing plays a vital role in land cover change detection, enabling the monitoring of urban expansion, agricultural development, and natural habitat loss. Time series of multispectral images track changes over time, informing land use planning and environmental impact assessments (Song et al. 2018). Satellite-based monitoring is essential for programs like REDD + (Reducing Emissions from Deforestation and Forest Degradation), which rely on remote sensing to track forest cover changes and support climate change mitigation efforts (Reiche et al. 2016). Additionally, remote sensing assists in mapping cropland extent and crop type identification, monitoring agricultural intensification, and informing food security assessments and agricultural policy (Teluguntla et al. 2015).

1.4.2 Water Quality Monitoring and Coastal Zone Management

Remote sensing yields a synoptic view of water bodies, enabling large-scale assessment of water quality parameters such as surface temperature, turbidity, and chlorophyll concentration. Thermal infrared and multispectral sensors on satellites and UAVs monitor these parameters, indicating pollutants and algal blooms (Kutser 2009). High chlorophyll levels, often indicative of harmful algal blooms, are detected by analyzing reflectance in the blue and green bands using multispectral and hyperspectral data, aiding in water quality management and public health protection. Turbidity and sediment load increase reflectance in the visible and near-infrared bands due to suspended sediments; monitoring these changes helps manage erosion processes and maintain aquatic ecosystem health (Doxaran et al. 2002). Thermal infrared imaging maps surface water temperatures, which is crucial for understanding aquatic ecosystem dynamics and the impacts of thermal pollution (Handcock et al. 2012). Hyperspectral imaging can detect chemical pollutants and oil spills on water surfaces, improving environmental monitoring and disaster response (Leifer et al. 2012).

In coastal zone management, remote sensing is crucial in monitoring shoreline changes, assessing coastal erosion, and managing resources. Multispectral imagery tracks shoreline dynamics over time, informing erosion management strategies and infrastructure planning (Boak and Turner 2005). Hyperspectral data assess coral reef health and detect bleaching events, contributing to marine conservation efforts (Mumby et al. 2004). Combining multispectral imagery with digital elevation models helps assess the potential impacts of sea-level rise on coastal communities, supporting adaptation strategies (Nicholls and Cazenave 2010).

1.4.3 Urban Heat Island and Air Quality Assessment

Thermal infrared imaging is essential for studying urban climatology and the urban heat island (UHI) effect, where urban structures absorb and re-radiate more heat than surrounding rural areas. By analyzing thermal infrared images, researchers identify areas with significant temperature variations, indicating heat hotspots within cities (Voogt and Oke 2003). Combining thermal data with multispectral land cover classifications helps understand the relationship between urban land use and temperature patterns, informing urban planning strategies to mitigate heat accumulation (Weng et al. 2004). Time series of thermal imagery reveals how urban heat patterns change diurnally and seasonally, guiding the implementation of mitigation measures such as increasing vegetation cover and using reflective materials (Imhoff et al. 2010). Remote sensing also contributes to air quality assessment, which is traditionally reliant on ground-based sensors. Multispectral data from sensors like MODIS estimate aerosol optical depth, providing observations into aerosol concentrations and distribution (Levy et al. 2007). Hyperspectral sensors detect and quantify atmospheric gases, including nitrogen dioxide and sulphur dioxide, refining understanding of air pollution patterns and supporting regulatory compliance (Streets et al. 2013). These capabilities contribute to public health assessments and inform policies to improve air quality.

1.5 Case Studies

Several real-world case studies exemplify the practical applications of multispectral, hyperspectral, and thermal imaging, highlighting their impact on environmental monitoring and management.

1.5.1 Vegetation Monitoring in the Amazon Rainforest

The Amazon Rainforest, a critical global ecosystem, has been extensively monitored using remote sensing technologies. Multispectral and hyperspectral imagery have been employed to assess deforestation rates, forest health, and biodiversity. Vegetation indices such as NDVI provide an understanding of the extent of deforestation and the health of remaining forests, facilitating the tracking of illegal logging activities and informing conservation strategies (Hansen et al. 2013). Integrating satellite data with ground observations supports initiatives to preserve the rainforest and mitigate climate change.

1.5.2 Agricultural Drought Assessment in India

India frequently faces agricultural droughts that significantly affect crop yields and food security. Remote sensing technologies have been instrumental in assessing and monitoring drought conditions nationwide. Multispectral imagery from satellites like Landsat and Sentinel-2 has been used to calculate vegetation indices such as NDVI and the Vegetation Condition Index (VCI) to monitor crop health and soil moisture levels (Kogan 2001; Rhee et al. 2010). These indices help identify drought-affected areas by detecting anomalies in vegetation vigor, enabling timely interventions by policymakers and farmers to mitigate impacts. For example, the Indian government utilizes remote sensing data to implement drought relief measures and optimize water resource allocation, contributing to more resilient agricultural practices (Singh et al. 2003; Chinnasamy and Srivastava 2021).

1.5.3 Urban Heat Island Mitigation in New York City

New York City has been the focus of studies on urban heat islands using thermal infrared remote sensing. Data from satellites like Landsat and MODIS have been used to map surface temperatures and identify heat hotspots within the city (Rosenzweig et al. 2009). The findings have guided the implementation of mitigation strategies, including the installation of green roofs, the creation of urban parks, and the use of reflective materials, all aimed at reducing the UHI effect and improving urban liveability. These measures contribute to energy savings, reduce heat-related health risks, and boost urban sustainability.

1.5.4 Water Quality Monitoring in the Great Lakes

The Great Lakes in North America have been monitored for water quality using remote sensing technologies. Multispectral sensors on satellites such as Sentinel-2 have been used to track algal blooms, sediment plumes, and changes in water temperature (Binding et al. 2018). This information is crucial for managing water resources, protecting aquatic ecosystems, and ensuring safe drinking water supplies for millions of people. Remote sensing data support early warning systems for harmful algal blooms, inform regulatory actions and guide conservation efforts in the region.

1.5.5 Desertification Assessment in the Sahel Region, Africa

The Sahel region in Africa faces severe desertification due to climatic variability and human activities such as overgrazing and deforestation. Remote sensing provides significant data for assessing land degradation and vegetation dynamics in this vulnerable region. Using multispectral data from MODIS and Landsat, researchers calculate vegetation indices to monitor changes in vegetation cover over time, identifying areas undergoing desertification (Mayaux et al. 2013; Anyamba and Tucker 2005). These assessments inform land management practices, support reforestation efforts, and contribute to programs like the Great Green Wall initiative to combat desertification and restore degraded lands (Goffner et al. 2019).

Table 1.3 demonstrates the versatility and global applicability of multispectral, hyperspectral, and thermal imaging in environmental monitoring. From agricultural management in India to coral reef conservation in Australia and wildfire monitoring in the United States, remote sensing technologies provide critical data that inform environmental policies, resource management, and disaster response strategies. As sensor technologies continue to advance and new analysis techniques are developed, the scope of these applications is likely to expand, delivering increasingly detailed and timely information for environmental management and policy-making worldwide.

1.6 Integrating Remote Sensing with Geographic Information Systems

Integrating remote sensing data with Geographic Information Systems (GIS) has revolutionized environmental analysis by empowering all-rounded, spatially explicit, and temporally dynamic assessments. Remote sensing provides detailed imagery and spectral data, while GIS provides powerful spatial analysis and visualization capabilities. This synergy improves the ability to understand, model, and manage complex environmental processes, supporting informed decision-making in environmental management and policy formulation (Rane et al. 2023; Liladhar Rane et al. 2024).

Table 1.3 Environmental case studies from across the globe

Case study	Objectives	Applied Techniques	Imaging Technology Used	References
Agricultural drought assessment in India	Monitor drought conditions, assess crop health	NDVI, VCI analysis, anomaly detection	Landsat, Sentinel-2	Kogan (2001), Singh et al. (2003)
Urban heat island analysis in Delhi, India	Identify heat hotspots, guide mitigation strategies	Thermal infrared imaging, land use analysis	Landsat 8	Grover and Singh (2015)
Deforestation monitoring in the Congo Basin, Africa	Track forest cover changes, detect illegal logging	Time-series analysis, change detection	Landsat, MODIS	Potapov et al. (2012)
Desertification assessment in the Sahel Region, Africa	Assess land degradation, monitor vegetation dynamics	Vegetation indices (NDVI), trend analysis	MODIS, Landsat	Anyamba and Tucker (2005)
Coral reef monitoring in the Great Barrier Reef, Australia	Map coral health, detect bleaching events	Hyperspectral imaging, spectral analysis	Airborne and satellite hyperspectral sensors	Hedley et al. (2016)
Glacial retreat monitoring in Greenland	Assess glacial melting, model sea-level rise	Thermal infrared imaging, temporal analysis	MODIS, ASTER	Cooper and Smith (2019)
Forest fire monitoring in California, USA	Detect active fires, assess burn severity	Thermal imaging, fire mapping	MODIS, VIIRS	Giglio et al. (2008)
Algal bloom monitoring in Lake Victoria, East Africa	Detect and monitor harmful algal blooms	Chlorophyll mapping, spectral analysis	Sentinel-2, multispectral sensors	Ogembo and Mohamed (2023)
Vegetation monitoring in the Amazon Rainforest	Assess deforestation rates, forest health, biodiversity	NDVI analysis, change detection	Multispectral and hyperspectral imagery	Hansen et al. (2013)
Urban heat island mitigation in New York City, USA	Identify heat hotspots, guide mitigation strategies	Thermal infrared imaging, land cover classification	Landsat, MODIS	Rosenzweig et al. (2009)
Water quality monitoring in the Great Lakes, North America	Monitor algal blooms, sediment plumes, water temperature	Chlorophyll concentration mapping, thermal analysis	Sentinel-2, multispectral sensors	Binding et al. (2018)

1.6.1 Remote Sensing and GIS Synergy

Remote sensing and GIS are complementary technologies that, when integrated, form a powerful toolkit for environmental analysis. Remote sensing supplies up-to-date spatial data that can be incorporated into GIS platforms, combining various data layers—such as elevation, land cover, and hydrology—with remote sensing imagery to create a multidimensional spatial analysis environment (Ehlers et al. 1989). The synergy between these technologies refines understanding, modeling, and managing complex environmental processes by allowing holistic, spatially explicit, and temporally dynamic assessments (Goodchild 2018).

Key processes in integrating remote sensing data with GIS include georeferencing and orthorectification and aligning remote sensing images with GIS coordinate systems for accurate overlay with other spatial data (Toutin 2004). Remote sensing data are often converted into GIS-compatible formats, such as raster datasets or vector layers, facilitating seamless integration. Additionally, GIS tools can extract attributes from remote sensing data and associate them with spatial features, enriching existing GIS datasets (Congalton and Green 2019). The combination of GIS and time-series remote sensing data enables the analysis of environmental changes over time, helping identify trends and patterns such as deforestation progression, urban expansion, or changes in water quality (Singh 1989; Singh et al. 2024). Thermal data can be combined with optical and spectral data to understand environmental conditions comprehensively; integrating temperature data with vegetation indices helps assess plant health under heat stress (Karnieli et al. 2010). The combination of GIS and time-series remote sensing data enables the analysis of environmental changes over time, which is crucial for understanding long-term environmental trends and assessing the effectiveness of conservation efforts (Lu and Weng 2007; Sahu et al. 2024; Singh et al. 2025).

1.6.2 Applications of Remote Sensing and GIS in Land Use, Urban Planning, and Water Resources

1.6.2.1 Land Use and Land Cover Mapping

Integrating remote sensing and GIS significantly improves land use and land cover mapping. GIS data, such as topography and soil types, improve the accuracy of land cover classifications derived from remote sensing (Lu and Weng 2007). Time-series remote sensing data integrated with GIS enable detailed analysis of land use and land cover changes over time, supporting sustainable land management practices and policymaking. For example, India's National Remote Sensing Centre (NRSC) uses satellite imagery integrated with GIS to monitor agricultural lands, forests, urban areas, and water bodies, producing detailed maps that inform land management decisions (https://www.nrsc.gov.in/, accessed 21 May 2025).

1.6.2.2 Urban Planning and Management

When combined with GIS layers of infrastructure and zoning information, remote sensing data on urban extent enable predictive modeling of urban growth patterns (Herold et al. 2003). Thermal remote sensing data integrated with GIS layers of population density and land use inform strategic planning of urban green spaces to mitigate urban heat islands (Voogt and Oke 2003; Norton et al. 2015). Cities like Singapore employ high-resolution satellite imagery and GIS to support urban planning and development, using remote sensing data to monitor land use changes, assess green cover, and manage infrastructure development (Chow and Roth 2006; Peijun et al. 2010).

1.6.2.3 Water Resource Management

Integrating remote sensing data on land cover and topography with GIS hydrological modeling tools boosts watershed-scale analysis of water resources (Jenson and Domingue 1988). Combining remote sensing-derived flood extent data with GIS layers of elevation, infrastructure, and population allows for detailed flood risk assessment and management (Sanyal and Lu 2004). In the Colorado River Basin, remote sensing data from satellites like Landsat and MODIS are integrated into GIS to monitor water quality and quantity, aiding in managing water resources by providing real-time data on water levels, temperature, and sediment load (Mishra and Singh 2010).

1.7 Improving Environmental Decision-Making Through GIS Integration

Integrating remote sensing data into GIS platforms forms the backbone of environmental decision support systems (DSS), providing policymakers and environmental managers with tools to simulate scenarios and predict outcomes of various management decisions (Malczewski 2006). GIS-based DSS enables visualization of the consequences of actions such as deforestation, urban expansion, or new conservation policies. This capability is crucial for making informed, data-driven decisions that balance short-term needs with long-term environmental sustainability.

Integrating real-time remote sensing data into GIS allows for prompt decision-making in situations requiring immediate response, such as wildfire management or flood prevention (Goodchild 2007). Real-time monitoring and alert systems can generate notifications based on predefined criteria, facilitating rapid mobilization of resources and efficient response strategies (Goodchild and Glennon 2010). Web-based platforms like Google Earth Engine and ArcGIS Online have democratized remote sensing and GIS integration access, enabling collaborative environmental monitoring and decision-making (Gorelick et al. 2017). Integrating machine learning and artificial intelligence (AI) with GIS and remote sensing data further refines decision-making capabilities, enabling advanced predictive modeling and automated feature extraction (Ma et al. 2019).

1.8 Advanced Integration Techniques

1.8.1 Machine Learning and AI in GIS-Remote Sensing Integration

The fusion of machine learning and artificial intelligence (AI) with GIS and remote sensing data has led to significant advancements in environmental analysis (Table 1.4). Object-Based Image Analysis (OBIA) combines remote sensing image segmentation with GIS data to improve classification accuracy, particularly in heterogeneous landscapes (Blaschke 2010). Deep learning approaches, such as Convolutional Neural Networks (CNNs), are increasingly utilized to analyze integrated remote sensing and GIS data for tasks like land cover classification and change detection (Zhang et al. 2016). These models automatically learn hierarchical features from raw data, improving classification performance and enabling analysis of large, high-dimensional datasets.

1.8.2 Web-Based GIS and Remote Sensing Platforms

Advancements in cloud computing and big data technologies address the challenges of processing and analyzing vast volumes of remote-sensing data (Yang et al. 2017). Distributed processing platforms like Google Earth Engine enable planetary-scale analysis of remote sensing data, providing researchers with the computational resources necessary to handle big data without extensive local infrastructure (Gorelick et al. 2017). Similarly, ArcGIS Online facilitates integrating and analyzing remote sensing data with other GIS layers, supporting collaborative efforts in environmental management (Fu and Sun 2010).

1.9 Conclusion

Multispectral, hyperspectral, and thermal imaging technologies play a pivotal role in environmental monitoring and assessment by revealing critical information about the Earth's surface and atmosphere that is inaccessible through conventional means. By interlinking different portions of the electromagnetic spectrum, these remote sensing

Table 1.4 Challenges, emerging trends, and future research directions in remote sensing and GIS technologies

Challenges	Emerging trends	Future research directions	References
Spatial and temporal resolution trade-offs	Development of CubeSats and SmallSats providing high temporal resolution at lower cost	Multi-scale integration across UAVs, airborne, and satellite platforms	Whitcraft et al. (2015), Transon et al. (2018)
Atmospheric interference	Advanced atmospheric correction algorithms	Development of robust algorithms for atmospheric correction in diverse conditions	Gao et al. (2009)
Data volume and processing challenges	Cloud computing and big data technologies	Distributed processing platforms and data cubes	Yang et al. (2017)
Calibration and validation difficulties	Harmonization of sensor data	Consistent long-term datasets for environmental studies	Justice et al. (2002), Wulder et al. (2015)
Classification and interpretation complexities	Machine learning and AI applications	Improved algorithms for feature extraction and classification	Ma et al. (2019)
Data quality and accuracy concerns	Advanced statistical models for uncertainty quantification	Methods to quantify and communicate uncertainty in remote sensing products	Atkinson and Foody (2002)
Interoperability issues	Adoption of open standards for geospatial data	Improved data sharing and integration capabilities across platforms	Yang et al. (2014)
Real-time integration challenges	Internet of Things (IoT) and sensor networks integration	Development of real-time monitoring and early warning systems	Hart and Martinez (2006)
Capacity building needs	User-friendly tools and platforms	Training programs and infrastructure development; improving the accessibility of remote sensing data to non-experts	Nedovic-Budic et al. (2011)

technologies allow thorough analyses of vegetation health, water quality, urban heat dynamics, and soil composition. Integrated with GIS, they boost the capacity to analyze, visualize, and interpret complex environmental data, facilitating spatially explicit and temporally dynamic assessments.

Sophisticated data acquisition platforms—from satellites to UAVs—collect vast amounts of environmental data and transform them into implementable actions through advanced processing techniques, including machine learning algorithms. These advancements have expanded the range of applications, such as land cover analysis, water quality assessment, urban heat island studies, and emerging areas like air quality monitoring and disaster management. The synergy between remote sensing and GIS amplifies the utility of both technologies, supporting evidence-based decision-making, global monitoring capabilities, rapid responses to environmental changes, and cost-effective assessments. Looking ahead, advancements in sensor technology, artificial intelligence, and big data analytics promise to further advance the capabilities of remote sensing in environmental monitoring. Developing more sophisticated sensors, improved data processing algorithms, and integration with emerging technologies like the IoT will lead to more detailed, accurate, and timely environmental visions. However, realizing the full potential of these technologies requires addressing challenges related to data accessibility, interoperability, capacity building, and ethical considerations in data use. As remote sensing technologies continue to evolve, they will play an increasingly crucial role in sustainable resource management, climate change mitigation and adaptation, and biodiversity conservation. It is imperative for researchers, practitioners, and policymakers to stay abreast of these advancements and effectively integrate the wealth of information provided by remote sensing into decision-making processes, thereby contributing to more effective and sustainable environmental management practices.

Competing Interests The authors declare no potential conflicts of interest.

References

Aasen H, Burkart A, Bolten A, Bareth G (2018) Generating 3D hyperspectral information with lightweight UAV snapshot cameras for vegetation monitoring: from camera calibration to quality assurance. ISPRS J Photogramm Remote Sens 145:85–97

Abdi AM (2020) Land cover and land use classification performance of machine learning algorithms in a boreal landscape using Sentinel-2 data. Giscience & Rem Sens 57(1):1–20

Abdulraheem MI, Zhang W, Li S, Moshayedi AJ, Farooque AA, Hu J (2023) Advancement of remote sensing for soil measurements and applications: a comprehensive review. Sustainability 15(21):15444

Adam E, Mutanga O, Rugege D (2010) Multispectral and hyperspectral remote sensing for identification and mapping of wetland vegetation: a review. Wetlands Ecol Manage 18(3):281–296

Anyamba A, Tucker CJ (2005) Analysis of Sahelian vegetation dynamics using NOAA-AVHRR NDVI data from 1981–2003. J Arid Environ 63(3):596–614

Asner GP, Clark JK, Mascaro J, Galindo García GA, Chadwick KD, Navarrete Encinales DA, Paez-Acosta G, Cabrera Montenegro E, Kennedy-Bowdoin T, Duque Á, Balaji A (2012) High-resolution mapping of forest carbon stocks in the Colombian Amazon. Biogeosciences 9(7):2683–2696

Atkinson PM, Foody GM (2002) Uncertainty in remote sensing and GIS: fundamentals. In: Uncertainty in remote sensing and GIS, pp 1–18

Ball GH, Hall DJ (1965) ISODATA: A novel method of data analysis and pattern classification. Proc IEEE 57(8):697–698

Belgiu M, Drăguţ L (2016) Random forest in remote sensing: a review of applications and future directions. ISPRS J Photogramm Remote Sens 114:24–31

Binding CE, Greenberg TA, Bukata RP (2018) The MERIS MCI and its potential for satellite detection of winter diatom blooms on partly ice-covered Lake Erie. J Geophys Res: Oceans 123(3):1665–1682

Blackett M (2017) An overview of infrared remote sensing of volcanic activity. J Imaging 3(2):13

Blaschke T (2010) Object based image analysis for remote sensing. ISPRS J Photogramm Remote Sens 65(1):2–16

Blaschke T, Hay GJ, Kelly M, Lang S, Hofmann P, Addink E, Feitosa RQ, Van der Meer F, Van der Werff H, Van Coillie F, Tiede D (2014) Geographic object-based image analysis–towards a new paradigm. ISPRS J Photogramm Remote Sens 87:180–191

Boak EH, Turner IL (2005) Shoreline definition and detection: a review. J Coastal Res:688–703

Breiman L (2001) Random forests machine learning. 45(1):5–32

Cavender-Bares J, Schneider FD, Santos MJ, Armstrong A, Carnaval A, Dahlin KM, Fatoyinbo L, Hurtt GC, Schimel D, Townsend PA, Ustin SL (2022) Integrating remote sensing with ecology and evolution to advance biodiversity conservation. Nat Ecol Evol 6(5):506–519

Chavez PS (1996) Image-based atmospheric corrections—revisited and improved. Photogramm Eng Remote Sens 62(9):1025–1036

Chinnasamy P, Srivastava A (2021) Revival of traditional cascade tanks for achieving climate resilience in drylands of South India. Front Water 3:639637

Chow WT, Roth M (2006) Temporal dynamics of the urban heat island of Singapore. Int J Climatol 26(15):2243

Chuvieco E (2020) Fundamentals of satellite remote sensing: an environmental approach. CRC Press

Cohen J (1960) A coefficient of agreement for nominal scales. Educ Psychol Measur 20(1):37–46

Colomina I, Molina P (2014) Unmanned aerial systems for photogrammetry and remote sensing: a review. ISPRS J Photogramm Remote Sens 92:79–97

Congalton RG, Green K (2019) Assessing the accuracy of remotely sensed data: principles and practices. CRC Press

Cooper MG, Smith LC (2019) Satellite remote sensing of the Greenland ice sheet ablation zone: a review. Remote Sens 11(20):2405

Dash P, Göttsche F-M, Olesen F-S, Fischer H (2002) Land surface temperature and emissivity estimation from passive sensor data: theory and practice—current trends. Int J Remote Sens 23(13):2563–2594

Doxaran D, Froidefond JM, Lavender S, Castaing P (2002) Spectral signature of highly turbid waters: application with SPOT data to quantify suspended particulate matter concentrations. Remote Sens Environ 81(1):149–161

Drusch M, Del Bello U, Carlier S, Colin O, Fernandez V, Gascon F, Hoersch B, Isola C, Laberinti P, Martimort P, Meygret A (2012) Sentinel-2: ESA's optical high-resolution mission for GMES operational services. Remote Sens Environ 120:25–36

Ehlers M, Edwards G, Bédard Y (1989) Integration of remote sensing with geographic information systems: a necessary evolution. Photogramm Eng Remote Sens 55(11):1619–1627

Elbeltagi A, Srivastava A, Kushwaha NL, Juhász C, Tamás J, Nagy A (2022) Meteorological data fusion approach for modeling crop water productivity based on ensemble machine learning. Water 15(1):30

Elbeltagi A, Srivastava A, Deng J, Li Z, Raza A, Khadke L et al (2023) Forecasting vapor pressure deficit for agricultural water management using machine learning in semi-arid environments. Agric Water Manag 283:108302

Elbeltagi A, Srivastava A, Li P, Jiang J, Jinsong D, Rajput J et al (2023) Forecasting actual evapotranspiration without climate data based on stacked integration of DNN and meta-heuristic models across China from 1958 to 2021. J Environ Manage 345:118697

Elbeltagi A, Srivastava A, Ehsan M, Sharma G, Yu J, Khadke L et al (2024) Advanced stacked integration method for forecasting long-term drought severity: CNN with machine learning models. J Hydrol: Region Stud 53:101759

Fascista A (2022) Toward integrated large-scale environmental monitoring using WSN/UAV/Crowdsensing: a review of applications, signal processing, and future perspectives. Sensors 22(5):1824

Fassnacht FE, Latifi H, Stereńczak K, Modzelewska A, Lefsky M, Waser LT, Straub C, Ghosh A (2016) Review of studies on tree species classification from remotely sensed data. Remote Sens Environ 186:64–87

Fisher JB, Melton F, Middleton E et al (2020) ECOSTRESS: NASA's next generation mission to measure evapotranspiration from the international space station. Water Resour Res 56(4):e2019WR026058

Foody GM (2020) Explaining the unsuitability of the kappa coefficient in the assessment and comparison of the accuracy of thematic maps obtained by image classification. Remote Sens Environ 239:111630

Fu P, Sun J (2010) Web GIS: principles and applications. Esri Press

Gao BC, Montes MJ, Davis CO, Goetz AF (2009) Atmospheric correction algorithms for hyperspectral remote sensing data of land and ocean. Remote Sens Environ 113:S17–S24

Gao B-C, Xiong X, Li R-R, Zhang H (2019) Overview of remote sensing of the Earth from space. Appl Sci 9(9):1966

Giglio L, Csiszar I, Restás Á, Morisette JT, Schroeder W, Morton D, Justice CO (2008) Active fire detection and characterization with the advanced spaceborne thermal emission and reflection radiometer (ASTER). Remote Sens Environ 112(6):3055–3063

Gislason PO, Benediktsson JA, Sveinsson JR (2006) Random forests for land cover classification. Pattern Recogn Lett 27(4):294–300

Goetz AFH (2009) Three decades of hyperspectral remote sensing of the Earth: a personal view. Remote Sens Environ 113:S5–S16

Goffner D, Sinare H, Gordon LJ (2019) The Great Green Wall for the Sahara and the Sahel Initiative as an opportunity to enhance resilience in Sahelian landscapes and livelihoods. Reg Environ Change 19(5):1417–1428

Goodchild MF (2007) Citizens as sensors: the world of volunteered geography. GeoJournal 69(4):211–221

Goodchild MF (2018) Reimagining the history of GIS. Ann GIS 24(1):1–8

Goodchild MF, Glennon JA (2010) Crowdsourcing geographic information for disaster response: a research frontier. Int J Digit Earth 3(3):231–241

Gorelick N, Hancher M, Dixon M, Ilyushchenko S, Thau D, Moore R (2017) Google Earth engine: planetary-scale geospatial analysis for everyone. Remote Sens Environ 202:18–27

Govender M, Chetty K, Bulcock H (2007) A review of hyperspectral remote sensing and its application in vegetation and water resource studies. Water SA 33(2):145–151

Grover A, Singh RB (2015) Analysis of urban heat island (UHI) in relation to normalized difference vegetation index (NDVI): A comparative study of Delhi and Mumbai. Environments 2(2):125–138

Guha S, Govil H, Mukherjee S, Ganguly D (2020) A review of multispectral and hyperspectral remote sensing data for wildfire fuel modeling. ISPRS J Photogramm Remote Sens 168:236–247

Handcock RN, Torgersen CE, Cherkauer KA, Gillespie AR, Tockner K, Faux RN, Tan J (2012) Thermal infrared remote sensing of water temperature in riverine landscapes. In: Fluvial remote sensing for science and management, pp 85–113

Hansen MC, Potapov PV, Moore R, Hancher M, Turubanova SA, Tyukavina A, Townshend JRG et al (2013) High-resolution global maps of 21st-century forest cover change. Science 342(6160):850–853

Hart JK, Martinez K (2006) Environmental sensor networks: a revolution in the earth system science? Earth Sci Rev 78(3–4):177–191

Hedley JD, Roelfsema CM, Chollett I, Harborne AR (2016) Coral reef applications of Sentinel-2: coverage, characteristics, bathymetry and benthic mapping with comparison to Landsat 8. Remote Sens Environ 204:308–326

Herold M, Goldstein NC, Clarke KC (2003) The spatiotemporal form of urban growth: measurement, analysis and modeling. Remote Sens Environ 86(3):286–302

Homolová L, Malenovský Z, Gallé A, Rascher U, Mohammed G (2013) Review of optical-based remote sensing for plant trait mapping. Ecol Complex 15:1–16

Huete AR (1988) A soil-adjusted vegetation index (SAVI). Remote Sens Environ 25(3):295–309

Huete A, Didan K, Miura T, Rodriguez EP, Gao X, Ferreira LG (2002) Overview of the radiometric and biophysical performance of the MODIS vegetation indices. Remote Sens Environ 83(1–2):195–213

Imhoff ML, Zhang P, Wolfe RE, Bounoua L (2010) Remote sensing of the urban heat island effect across biomes in the continental USA. Remote Sens Environ 114(3):504–513

Jain S, Srivastava A, Khadke L, Chatterjee U, Elbeltagi A (2024a) Global-scale water security and desertification management amidst climate change. Environ Sci Poll Res 31:58720–58744

Jain S, Srivastava A, Vishwakarma DK, Rajput J, Rane NL, Salem A, Elbeltagi A (2024b) Protecting ancient water harvesting technologies in India: strategies for climate adaptation and sustainable development with global lessons. Front Water 6:1441365

Jensen JR (2009) Remote sensing of the environment: an earth resource perspective 2/e. Pearson Education India

Jensen JR (2015) Introductory digital image processing: a remote sensing perspective, 4th ed. Pearson

Jenson SK, Domingue JO (1988) Extracting topographic structure from digital elevation data for geographic information system analysis. Photogramm Eng Remote Sens 54(11):1593–1600

Jones HG, Serraj R, Loveys BR, Xiong L, Wheaton A, Price AH (2009) Thermal infrared imaging of crop canopies for the remote diagnosis and quantification of plant responses to water stress in the field. Funct Plant Biol 36(11):978–989

Justice CO, Townshend JRG, Vermote EF, Masuoka E, Wolfe RE, Saleous N, Roy DP, Morisette JT (2002) An overview of MODIS Land data processing and product status. Remote Sens Environ 83(1–2):3–15

Karnieli A et al (2010) Use of NDVI and land surface temperature for drought assessment: merits and limitations. J Clim 23(3):618–633

Keshava N, Mustard JF (2002) Spectral unmixing. IEEE Signal Process Mag 19(1):44–57

Kogan FN (2001) Operational space technology for global vegetation assessment. Bull Am Meteor Soc 82(9):1949–1964

Kuenzer C, Dech S (eds) (2013) Thermal infrared remote sensing: sensors, methods, applications. Springer

Kumar S, Meena RS, Sheoran S, Jangir CK, Jhariya MK, Banerjee A, Raj A (2022) Remote sensing for agriculture and resource management. In: Natural resources conservation and advances for sustainability. Elsevier, pp 91–135

Kutser T (2009) Passive optical remote sensing of cyanobacteria and other intense phytoplankton blooms in coastal and inland waters. Int J Remote Sens 30(17):4401–4425

Lakshmi V (ed) (2016) Remote sensing of hydrological extremes. Springer

Lei T, Wang J, Li X, Wang W, Shao C, Liu B (2022) Flood disaster monitoring and emergency assessment based on multi-source remote sensing observations. Water 14(14):2207

Leifer I, Lehr WJ, Simecek-Beatty D, Bradley E, Clark R, Dennison P, Hu Y, Matheson S, Jones CE, Holt B, Reif M (2012) State of the art satellite and airborne marine oil spill remote sensing: application to the BP Deepwater Horizon oil spill. Remote Sens Environ 124:185–209

Lenton TM, Abrams JF, Bartsch A, Bathiany S, Boulton CA, Buxton JE, Conversi A, Cunliffe AM, Hebden S, Lavergne T, Poulter B (2024) Remotely sensing potential climate change tipping points across scales. Nat Commun 15(1):343

Levy RC, Remer LA, Mattoo S, Vermote EF, Kaufman YJ (2007) Second-generation operational algorithm: Retrieval of aerosol properties over land from inversion of moderate resolution imaging spectroradiometer spectral reflectance. J Geophys Res: Atmos 112(D13)

Li J, Roy DP (2017) A global analysis of Sentinel-2A, Sentinel-2B and Landsat-8 data revisit intervals and implications for terrestrial monitoring. Rem Sens 9(9):902

Li W, Fu H, Yu L, Cracknell A (2019) Deep learning based oil palm tree detection and counting for high-resolution remote sensing images. Rem Sens 11(1):5

Li Z, Chen B, Wu S, Su M, Chen JM, Xu B (2024) Deep learning for urban land use category classification: a review and experimental assessment. Remote Sens Environ 311:114290

Liladhar Rane N, Purushottam Choudhary S, Saha A, Srivastava A, Pande CB, Alshehri F et al (2024) Delineation of environmentally sustainable urban settlement using GIS-based MIF and AHP techniques. Geocarto Int 39(1):2335249

Lillesand TM, Kiefer RW, Chipman JW (2015) Remote sensing and image interpretation, 7th ed. Wiley

Lu D, Weng Q (2007) A survey of image classification methods and techniques for improving classification performance. Int J Remote Sens 28(5):823–870

Ma L, Liu Y, Zhang X, Ye Y, Yin G, Johnson BA (2019) Deep learning in remote sensing applications: a meta-analysis and review. ISPRS J Photogramm Remote Sens 152:166–177

MacQueen J (1967). Some methods for classification and analysis of multivariate observations. In: Proceedings of the fifth Berkeley symposium on mathematical statistics and probability, vol 1, pp 281–297. University of California Press

Mahlein A-K, Oerke E-C, Steiner U, Dehne H-W (2012) Recent advances in sensing plant diseases for precision crop protection. Eur J Plant Pathol 133(1):197–209

Manfreda S, McCabe MF, Miller PE, Lucas R, Pajuelo Madrigal V, Mallinis G, Ben Dor E, Helman D, Estes L, Ciraolo G, Müllerová J (2018) On the use of unmanned aerial systems for environmental monitoring. Rem Sens 10(4):641

Masood A, Hameed MM, Srivastava A, Pham QB, Ahmad K, Razali SFM, Baowidan SA (2023) Improving PM2.5 prediction in New Delhi using a hybrid extreme learning machine coupled with snake optimization algorithm. Scient Rep 13(1):21057

Maxwell AE, Warner TA, Fang F (2018) Implementation of machine-learning classification in remote sensing: an applied review. Int J Remote Sens 39(9):2784–2817

Mayaux P, Pekel JF, Desclée B, Donnay F, Lupi A, Achard F, Clerici M, Bodart C, Brink A, Nasi R, Belward A (2013) State and evolution of the African rainforests between 1990 and 2010. Philos Trans R Soc B: Biol Sci 368(1625):20120300

Milton EJ, Schaepman ME, Anderson K, Kneubühler M, Fox N (2009) Progress in field spectroscopy. Remote Sens Environ 113:S92–S109

Malczewski J (2006) GIS-based multicriteria decision analysis: a survey of the literature. Int J Geogr Inf Sci 20(7):703–726

Mashala MJ, Dube T, Mudereri BT, Ayisi KK, Ramudzuli MR (2023) A systematic review on advancements in remote sensing for assessing and monitoring land use and land cover changes impacts on surface water resources in semi-arid tropical environments. Rem Sens 15(16):3926

McFeeters SK (1996) The use of the Normalized Difference Water Index (NDWI) in the delineation of open water features. Int J Remote Sens 17(7):1425–1432

Melgani F, Bruzzone L (2004) Classification of hyperspectral remote sensing images with support vector machines. IEEE Trans Geosci Remote Sens 42(8):1778–1790

Mishra AK, Singh VP (2010) Changes in extreme precipitation in Texas. J Geophys Res: Atmospheres 115(D14)

Mountrakis G, Im J, Ogole C (2011) Support vector machines in remote sensing: a review. ISPRS J Photogramm Remote Sens 66(3):247–259

Mulla DJ (2013) Twenty-five years of remote sensing in precision agriculture: key advances and remaining knowledge gaps. Biosys Eng 114(4):358–371

Mumby PJ, Skirving W, Strong AE, Hardy JT, LeDrew EF, Hochberg EJ, Stumpf RP, David LT (2004) Remote sensing of coral reefs and their physical environment. Mar Pollut Bull 48(3–4):219–228

National Academies of Sciences, Division on Engineering, Physical Sciences, Board on Physics and Committee on a Survey of the Active Sensing Uses of the Radio Spectrum (2015) A strategy for active remote sensing amid increased demand for radio spectrum. National Academies Press

Nedovic-Budic Z, Pinto JK, Warnecke L (2011) GIS database development and exchange: interaction mechanisms and motivations. In: Spatial data infrastructures in context: North and south, p 69

Nicholls RJ, Cazenave A (2010) Sea-level rise and its impact on coastal zones. Science 328(5985):1517–1520

Norton BA, Coutts AM, Livesley SJ, Harris RJ, Hunter AM, Williams NS (2015) Planning for cooler cities: a framework to prioritise green infrastructure to mitigate high temperatures in urban landscapes. Landsc Urban Plan 134:127–138

Ogembo V, Mohamed GA (2023) Spectral reflectance and algal bloom monitoring of Lake Victoria using remote sensing techniques, Winum Gulf of Kenya. Earth 12(4):90–98

Olofsson P, Foody GM, Herold M, Stehman SV, Woodcock CE, Wulder MA (2014) Good practices for estimating area and assessing accuracy of land change. Remote Sens Environ 148:42–57

Otukei JR, Blaschke T (2010) Land cover change assessment using decision trees, support vector machines and maximum likelihood classification algorithms. Int J Appl Earth Obs Geoinf 12:S27–S31

Pande CB, Srivastava A, Moharir KN, Radwan N, Mohd Sidek L, Alshehri F et al (2024) Characterizing land use/land cover change dynamics by an enhanced random forest machine learning model: a Google Earth Engine implementation. Environ Sci Eur 36(1):84

Pearlman JS, Carman SL, Segal C (2003) Hyperion imaging spectrometer on the New Millennium Program Earth Observer-1 system. IEEE Trans Geosci Remote Sens 41(6):1160–1173

Peijun DU, Xingli LI, Wen CAO, Yan LUO, Zhang H (2010) Monitoring urban land cover and vegetation change by multi-temporal remote sensing information. Min Sci Technology (China) 20(6):922–932

Petropoulos GP, Griffiths HM, Tarantola S (2015) Surface soil moisture retrievals from remote sensing: current status, products and future trends. Phys Chem Earth, Parts a/b/c 83–84:36–56

Pettorelli N, Laurance WF, O'Brien TG, Wegmann M, Nagendra H, Turner W (2014) Satellite remote sensing for applied ecologists: opportunities and challenges. J Appl Ecol 51(4):839–848

Pirotti F (2011) Analysis of full-waveform LiDAR data for forestry applications: a review of investigations and methods. iForest-Biogeosciences and Forestry 4(3):100

Potapov PV, Turubanova SA, Hansen MC, Adusei B, Broich M, Altstatt A, Mane L, Justice CO (2012) Quantifying forest cover loss in Democratic Republic of the Congo, 2000–2010, with Landsat ETM+ data. Remote Sens Environ 122:106–116

Pricope NG, Minei A, Halls JN, Chen C, Wang Y (2022) UAS Hyperspatial LiDAR data performance in delineation and classification across a gradient of wetland types. Drones 6(10):268

Rane NL, Achari A, Choudhary SP, Mallick SK, Pande CB, Srivastava A, Moharir KN (2023) A decision framework for potential dam site selection using GIS, MIF and TOPSIS in Ulhas river basin, India. J Clean Prod 423:138890

Reddy GO (2018) Satellite remote sensing sensors: principles and applications. In: Geospatial technologies in land resources mapping, monitoring and management, pp 21–43

Reiche J, Lucas R, Mitchell AL, Verbesselt J, Hoekman DH, Haarpaintner J, Kellndorfer JM, Rosenqvist A, Lehmann EA, Woodcock CE, Seifert FM (2016) Combining satellite data for better tropical forest monitoring. Nat Clim Chang 6(2):120–122

Resch B, Usländer F, Havas C (2018) Combining machine-learning topic models and spatiotemporal analysis of social media data for disaster footprint and damage assessment. Cartogr Geogr Inf Sci 45(4):362–376

Rhee J, Im J, Carbone GJ (2010) Monitoring agricultural drought for arid and humid regions using multi-sensor remote sensing data. Remote Sens Environ 114(12):2875–2887

Richards JA (2013) Remote sensing digital image analysis, 5th ed. Springer

Roberts DA, Gardner M, Church R, Ustin S, Scheer G, Green RO (1998) Mapping chaparral in the Santa Monica Mountains using multiple endmember spectral mixture models. Remote Sens Environ 65(3):267–279

Rolim SBA, Veettil BK, Vieiro AP, Kessler AB, Gonzatti C (2023) Remote sensing for mapping algal blooms in freshwater lakes: a review. Environ Sci Pollut Res 30(8):19602–19616

Rosenzweig C, Solecki WD, Parshall L, Lynn B, Cox J, Goldberg R, Hodges S, Gaffin S, Slosberg RB, Savio P, Dunstan F (2009) Mitigating New York City's heat island: integrating stakeholder perspectives and scientific evaluation. Bull Am Meteor Soc 90(9):1297–1312

Roy DP, Wulder MA, Loveland TR et al (2014) Landsat-8: Science and product vision for terrestrial global change research. Remote Sens Environ 145:154–172

Sanyal J, Lu XX (2004) Application of remote sensing in flood management with special reference to monsoon Asia: a review. Nat Hazards 33(2):283–301

Sahu H, Purohit P, Srivastava A, Singh R, Mishra AP, Arunachalam K, Kumar U (2024) Comparative assessment of soil parameters and ecological dynamics in the Western Himalayan wetland and its surrounding periphery. Environ Qual Manage 34(1):e22283

Shaw GA, Burke HK (2003) Spectral imaging for remote sensing. Lincoln Lab J 14(1):3–28

Singh A (1989) Review article digital change detection techniques using remotely-sensed data. Int J Remote Sens 10(6):989–1003

Singh RP, Roy S, Kogan F (2003) Vegetation and temperature condition indices from NOAA AVHRR data for drought monitoring over India. Int J Remote Sens 24(22):4393–4402

Singh R, Saritha V, Pande CB (2024) Dynamics of LULC changes, LST, vegetation health and climate interactions in wetland buffer zone: a remote sensing perspective. Phys Chem Earth, Parts A/B/C:103660

Singh R, Saritha V, Mishra AP, Pande CB, Sahu H (2025) A comprehensive analysis of water quality index in a wetland ecosystem supporting drinking water to major cities in Rajasthan, India. J Clean Prod:144593

Song XP, Hansen MC, Stehman SV, Potapov PV, Tyukavina A, Vermote EF, Townshend JR (2018) Global land change from 1982 to 2016. Nature 560(7720):639–643

Soni PK, Rajpal N, Mehta R, Mishra VK (2022) Urban land cover and land use classification using multispectral sentinal-2 imagery. Multimedia Tools Appl 81(26):36853–36867

Srivastava A, Chinnasamy P (2021) Investigating impact of land-use and land cover changes on hydro-ecological balance using GIS: insights from IIT Bombay, India. SN Appl Sci 3(3):343

Srivastava A, Chinnasamy P (2024) Watershed development interventions for rural water safety, security, and sustainability in semi-arid region of Western-India. Environ Dev Sustain 26(7):18231–18265

Streets DG, Canty T, Carmichael GR, de Foy B, Dickerson RR, Duncan BN, Edwards DP, Haynes JA, Henze DK, Houyoux MR, Jacob DJ (2013) Emissions estimation from satellite retrievals: a review of current capability. Atmos Environ 77:1011–1042

Teluguntla P, Thenkabail PS, Xiong J, Gumma MK, Giri C, Milesi C, Ozdogan M, Congalton R, Tilton J, Sankey TT, Massey R (2015) Global cropland area database (GCAD) derived from remote sensing in support of food security in the twenty-first century: current achievements and future possibilities

Thenkabail PS, Lyon JG, Huete A (eds) (2018) Hyperspectral remote sensing of vegetation, 2nd ed. CRC Press

Toth CK, Jóźków G (2016) Remote sensing platforms and sensors: a survey. ISPRS J Photogramm Remote Sens 115:22–36

Toutin T (2004) Geometric processing of remote sensing images: models, algorithms and methods. Int J Remote Sens 25(10):1893–1924

Transon J, D'Andrimont R, Maugnard A, Defourny P (2018) Survey of hyperspectral earth observation applications from space in the Sentinel-2 context. Remote Sensing 10(2):157

Tucker CJ (1979) Red and photographic infrared linear combinations for monitoring vegetation. Remote Sens Environ 8(2):127–150

Turner W, Rondinini C, Pettorelli N, Mora B, Leidner AK, Szantoi Z, Buchanan G, Dech S, Dwyer J, Herold M, Koh LP (2015) Free and open-access satellite data are key to biodiversity conservation. Biol Cons 182:173–176

Vali A, Comai S, Matteucci M (2024) An automated machine learning framework for adaptive and optimized hyperspectral-based land cover and land-use segmentation. Rem Sens 16(14):2561

Van der Meer FD, Van der Werff HM, Van Ruitenbeek FJ, Hecker CA, Bakker WH, Noomen MF, Van Der Meijde M, Carranza EJM, De Smeth JB, Woldai T (2012) Multi-and hyperspectral geologic remote sensing: a review. Int J Appl Earth Obs Geoinf 14(1):112–128

Vishwakarma DK, Kumar P, Yadav KK, Ali R, Markuna S, Chauhan S et al (2024) Evaluation of CatBoost method for predicting weekly Pan evaporation in subtropical and sub-humid regions. Pure Appl Geophys 181(2):719–747

Vollmer M, Möllmann K-P (2010) Infrared thermal imaging: fundamentals, research and applications. Wiley-VCH

Voogt JA, Oke TR (2003) Thermal remote sensing of urban climates. Remote Sens Environ 86(3):370–384

Vos CC, Verboom J, Opdam PFM, Ter Braak CJF (2001) Toward ecologically scaled landscape indices. Am Nat 157(1):24–41

Wambugu N, Chen Y, Xiao Z, Tan K, Wei M, Liu X, Li J (2021) Hyperspectral image classification on insufficient-sample and feature learning using deep neural networks: a review. Int J Appl Earth Obs Geoinf 105:102603

Weiss M et al (2020) Review of methods for in situ leaf area index (LAI) determination: Part II. Estimating LAI from gap fraction or indirect measurements. Agric For Meteorol 291:108091

Weng Q, Lu D, Schubring J (2004) Estimation of land surface temperature–vegetation abundance relationship for urban heat island studies. Remote Sens Environ 89(4):467–483

Whitcraft AK, Vermote EF, Becker-Reshef I, Justice CO (2015) Cloud cover throughout the agricultural growing season: impacts on passive optical earth observations. Remote Sens Environ 156:438–447

Wulder MA, Hilker T, White JC, Coops NC, Masek JG, Pflugmacher D, Crevier Y (2015) Virtual constellations for global terrestrial monitoring. Remote Sens Environ 170:62–76

Xu H (2006) Modification of normalised difference water index (NDWI) to enhance open water features in remotely sensed imagery. Int J Remote Sens 27(14):3025–3033

Xue J, Su B (2017) Significant remote sensing vegetation indices: a review of developments and applications. J Sens 2017:1–17

Yang C, Huang Q, Li Z, Liu K, Hu F (2017) Big data and cloud computing: innovation opportunities and challenges. Int J Digit Earth 10(1):13–53

Zhang D, Zhou G (2016) Estimation of soil moisture from optical and thermal remote sensing: a review. Sensors 16(8):1308

Zhang L, Zhang L, Du B (2016) Deep learning for remote sensing data: a technical tutorial on the state of the art. IEEE Geosci Rem Sens Mag 4(2):22–40

Zhang S, Li Y, Jia X, Shen H (2019) Spectral–spatial classification of hyperspectral imagery using a dual-channel convolutional neural network. Remote Sensing Letters 10(5):445–454

Yang C, Liu K, Li Z, Li W, Wu H, Xia J, Huang Q, Li J, Sun M, Miao L, Zhou N (2014) GEOSS clearinghouse: integrating geospatial resources to support the global earth observation system of systems. In: Big Data: techniques and technologies in geoinformatics. CRC Press, pp 31–54

Yuan Q, Shen H, Li T, Li Z, Li S, Jiang Y, Xu H, Tan W, Yang Q, Wang J, Gao J (2020) Deep learning in environmental remote sensing: achievements and challenges. Remote Sens Environ 241:111716

Zhu G, Blumberg DG (2002) Classification using ASTER data and SVM algorithms; The case study of Beer Sheva, Israel. Remote Sens Environ 80(2):233–240

Chapter 2
Monitoring Snow Cover and Glacially Impounded Lakes in the Uttarakhand Himalayan Watershed Gori Ganga by Using Geospatial Applications

D. S. Parihar

Abstract The snow cover of the fragile and youngest folded mountain chain known to have the highest water tower on the earth, viz., the Himalayan is decreasing steadily due to worldwide warming. There is a necessary study the pattern of spatiotemporal dynamics in different Himalayan watersheds. The present objective and results of the study delivers a pattern and rate of spatiotemporal change of an Uttarakhand Himalayan watershed (ex. Gori Ganga) which is a major tributary river as well as watershed of the Kali River (Sharda River) that makes a boundary between India and Nepal. Present research analysed the NDSI for showing the snow cover delineation area. The study refers to mapping and monitoring the dynamics of snow cover from 1990 to 2022 and rapidly developed glacially impounded lakes over the study area. The study will help to determine snow and glacial lakes inventory. It is kinetic change rate and also useful to international level studies on SCCD and the risk of GLOF. The study reveals that in 1990, about 30.97% of the Gori Ganga basin was under snow cover, while in 2022, only 10.99% was found. If this depletion rate of snow cover continues, the study advocates that there shall be a finished snow cover area in the Gori Ganga basin before 2040. Therefore, there is an urgent need for mitigation of the depletion of snow cover area rapidly for the long-term survival of Himalayan glacial-fed rivers.

Keywords NDSI · Snow/ice cover and line · Glacial lakes · GLOF · Global warming · Geospatial application

Abbreviations

CORREL	Correlation coefficient
GEE	Google Earth engine pro

D. S. Parihar (✉)
Department of Geography, Kumaun University, D.S.B. Campus, Nainital, Uttarakhand, India
e-mail: devendraparihar@kunainital.ac.in

V. Saritha et al. (eds.), *Remote Sensing for Environmental Monitoring*,
https://doi.org/10.1007/978-981-96-5546-5_2

GIS	Geographical information system
GLOF	Glacial lake outburst floods
GPS	Global positioning system
IPCC	Intergovernmental panel on climate change
MoFE	Ministry of Forest and Environment
NDSI	Normalized difference snow index
NDVI	Normalized difference vegetation index
NSIDC	National snow and ice data center
OLI	Operational land images
RS	Remote sensing
SCCD	Snow cover change detection
TIRS	Thermal infrared sensors
TM	Thematic mapper
USGS	United States geological survey
UTM	Universal transverse mercator
WGM	World glacier monitoring

2.1 Introduction

Compared to much larger ice caps, glaciers are found in milder environments where most of their surface is subject to annual ablation, making glaciers particularly vulnerable to climate change and a visible indicator of global warming (Wouters et al. 2019). Observational records of IPCC (2013), suggest above-average warming of Himalayan Mountain (~ 0.9–1.6 °C), the global average warming (0.85 °C) over the last century (Bhutiyani 2016). The widespread increase in surface temperature (Diodato et al. 2012) and changes in precipitation regime (Negi et al. 2018) in the Himalayan region have caused the thinning and retreat of glaciers (Bolch et al. 2012; IPCC 2013). Kumar et al. 2008 study present that in the Alaknanda Valley (Western Himalaya) the mean annual temperature has increased by 0.150 °C from 1960 to 2000. Bhutiyani et al. (2007) suggest a temperature increase of about 1.60 °C/100 years in the North-western Himalayan region over the past century, which is in close agreement with the 2013 IPCC report prediction of the rise in global average temperature.

The generalized retreat of glaciers in the regions of the Himalayas, the Alps, etc. towards Antarctica and the Arctic is an unmistakable indicator of global warming (IPCC 2013). WGM (2018) service monitors glacier mass balance using a global reference glacier set with more than 30 years of observations in 19 mountain regions from 1970 to 2018, the results of which are based on a subset of glaciers, showing that the cumulative ice loss since 1970 is 21.1 m of water equivalent. NSIDC (2020), reports that snow cover covers an average of about 46 million km^2 of the Earth's surface, with about 45.32 million km^2 (98.52%) of the Earth's snow cover located in the northern hemisphere. The extent of snow in the northern hemisphere is divided into two geographical regions: the northern hemisphere has about 28.84 million km^2 (62.69%) and Eurasia has 16.48 million km^2 (35.83%). Suggestions after studied by using satellite images that nearly 67% of Himalayan glaciers have retreated (Ageta and Kadota 1992).

Revenga et al. (2003) studied the great rivers of Central Asia depend on the Himalayas as the source of water (for drinking, sanitation, irrigation etc.) for more than 1200 million people. The highest water tower in the world is very rich in glaciers and snow. The MoFE New Delhi (SGIM 2010) reveals that the three basins viz., the Indus, Ganga and Brahmaputra put together have 71,182.08 km^2 of glaciated area with 32,392 numbers of glaciers. Of these, 16,049 glaciers occupied 32,246.43 km^2 of the glacier-covered area in the Indus River basin; 18,392.90 km^2 of the glaciated area of 6237 glaciers must be occupied by the Ganga River basin; and 20,542.75 km^2 of the glaciated area must be occupied by 10,106 glaciers in the Brahmaputra River basin. Of the 2767 glaciers monitored over the three Himalayan basins, 2184 are retreating at an average rate of 3.75%, 435 are advancing, and 148 glaciers exhibit no change. For a period between 1990 and 2001, retreat is found to be higher in Spiti and Alaknanda basins, even though both basins are located in different climatic zones. 10% losses were observed for glaciers having an extent higher than 15 km^2 for the period 1962 and 2001/2002. Similar trend was observed for a period between 2001/2002 to 2007, indicating an influence of glacier size on retreat. According to Parihar and Rawat (2021), the distribution of snow cover in 1990 was about 30.97% (678.87 km^2), in 1999 about 25.77% (564.92 km^2) of the Gori Ganga watershed area was covered with snow while in 2016 snow cover was only detected at 15.08% (330.44 km^2) and data suggest that about 348.43 km^2 snow cover area has been converted into the non-snow cover area at an average rate 13.40 km^2/year from 1990 to 2016. At the regional scale and level of different Himalayan valley glaciers, snow sheets has an impact on the local climate and the availability of water resources for drinking, livelihood, domestic, agriculture, industry uses, etc. (Konig et al. 2001; Wang and Li 2003; Kargel et al. 2005).

According to Fairbridge (1968) all lakes originating from present or past glaciers, in continental ice sheets, mountain and valley glaciers are described as glacier lakes. Bolch et al. (2012), studied have shown the twentieth century has had a significant impact on glaciers as well as surrounding environments due to global warming and climate change in the Himalayas. Thus, in response to global warming and climate change, the volume and number of potentially hazardous moraine-dammed lakes and glacial lakes caused for GLOF in the different glaciers and parts of the Uttarakhand Himalaya have increased (Parihar 2022; Richardson and Reynolds 2000; Quincey et al. 2005; Gardelle et al. 2011; Raj et al. 2013).

This research will highlight the temporal changes and signature of global warming on snow cover areas and glacial lakes (Fig. 2.1) in the Gori watershed. Main aims of this survey are to study the Himalayan watershed, which encompass the following: Understanding the impact of global warming on snow cover area and glacier lakes. Status, dynamics and future projections of snow cover area and increasing glacial lakes in the glacier region of the study area which is an alarm for GLOF. Study and mapping with the help of RS and GIS of snow cover area delineation, snow

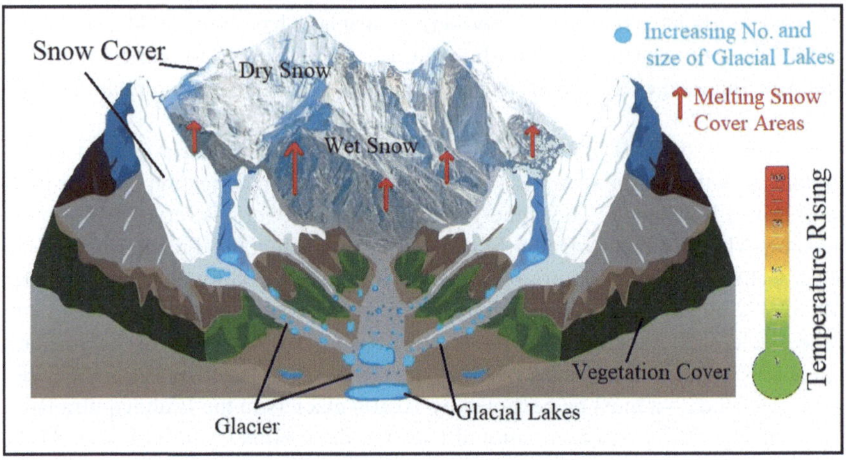

Fig. 2.1 Graphical abstract of the study

cover dynamics, increasing glacial lakes and other uses in the study area. Mapping of glaciers via manual digitization of glacier outlines (Hall et al. 1992), NDSI based snow cover estimation (Dozier, 1989; Hall et al. 1995; Winther and Hall 1999; Silverio and Jaquet 2005). Earlier studies con ducted to examine the signature of global warming in the study area concluded that the dynamics of vegetation lines (Parihar et al. 2021), timberlines (Parihar 2021a), snowlines (Parihar 2021b), and snow cover (Parihar and Rawat 2021) by using geospatial technology with NDVI and NDSI method in the Gori Ganga watershed were dynamic.

2.2 Extension: Study Area

The Central Himalayan basin, viz., Gori Ganga (29° 45′ 0″ N to 30° 35′ 47″ N latitudes and 79° 59′ 33″ E to 80° 29′ 25″ E longitude), encompasses a total area of 2191.93 km^2 elevation varies between 626 to 6639 m in Fig. 2.2 (Parihar 2021c). Total population living in the study area is about 40,616 (2011) in 168 different villages. In this basin is the internationally known Milam glacier (near Milam Village) from where the river Gori Ganga originates. It is a valley glacier that has a complex basin belonging to the southeast of the Trisul peak. The Milam Glacier is a major glacier in the Kumaon Himalaya, 16.7 km long. It gathers snow from Trishul Peak and seven different tributary glaciers and mountain snow-caps in the watershed. The town of Munsyari remains one of the last hill stations accessible by the region's road network. The inner villages of Munsyari and Madkote are currently the starting points for many inner Himalayan Mountain routes. The town of Munsyari, one of its main attractions, offers a stunning view of the high Himalayan Panchachuli. It is an impressive place surrounded by the support of unspoiled nature and high mountains.

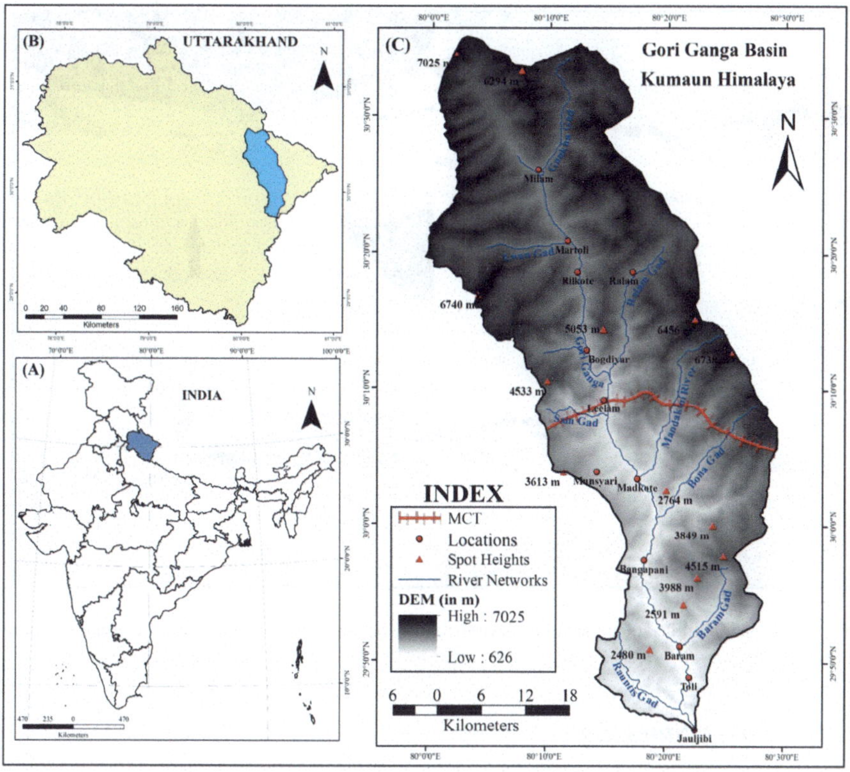

Fig. 2.2 Location and extension of the study area, the Gori Ganga basin, Central Himalaya

2.3 Data Acquaintance and Methodology

2.3.1 Field Data Collection

The Garmin GPS etrex20 handheld device was used in a field survey for the collection of ground coordinates and Elevation of snow cover line during October 2017, Chhipla Kedar Alpine, snow line position was recorded at 29° 56′ 14.94″ N and 80° 22′ 49.01″ E with Elevation recorded at 4069 m that had a mean accuracy of 11.38 m. In April 2018, Thalba Alpine, snow line position was recorded at 29° 58′ 15.58″ N and 80° 23′ 45.02″ E with Elevation recorded at 3780 m which had a mean accuracy of 13.3 m. In May 2019, Charthi Alpine, snow line position was recorded at 29° 59′ 39.37″ N and 80° 24′ 4.81″ E with Elevation recorded at 3702 m that had a mean accuracy of 12 m during the field survey, the ground photographic pieces of evidence of snout of glacier area (Fig. 2.3) were also recorded and used as a piece of reference information for further analysis.

Fig. 2.3 Charthi alpine at Central Himalaya with clearly visible **a**, **b** snow cover area and **c**, **d** alpine lakes

2.3.2 Layer Stacking

Satellite data (Landsat series) has 4–11 bands and every band has containeddifferent information about the land cover area. The layer stacking process (all the bands were computed in one layer) with the help of Q-GIS 3.4.3 software.

2.3.3 Data Used and Overview of the Study Design

The present work was started by downloading and processing Landsat-5 TM and Landsat-8 OLI and TIRS images (Table 2.1) for the years 1990, 1999, 2016 and 2022 (http://earthexplorer.usgs.gov/). For all 3 years, photos have been decided on for the November month is pretty cloudless and most snow cowl may be visible on this month. Application of ERDAS software is done to do geometric corrections to re-project the original sinusoidal file to geometric lat-long projection, where nearest neighbor re-sampling method and images required bands were layer stacked. Figure 2.4 presents methodological flow chart.

The study area, i.e., Gori Ganga basin was clipped using its shape file from satellite images and the images have been given the base map coordinates, i.e., UTM projection, 44 N zone for the purpose to identify the study area in the images, the NDSI raster calculated data of 1990 to 2022 (Fig. 2.5) was calculated in Arc

Table 2.1 Specification of data used

Satellite	Band	Wavelength (micrometers)	Resolution (in m)	Path/Row
LANDSAT-5	Green Band 2	0.52–0.60	30	145/039
	SWIR Band 5	1.55–1.75	30	145/039
LANDSAT-8	Green Band 3	0.53–0.59	30	145/039
	SWIR Band 6	0.57–1.65	30	145/039

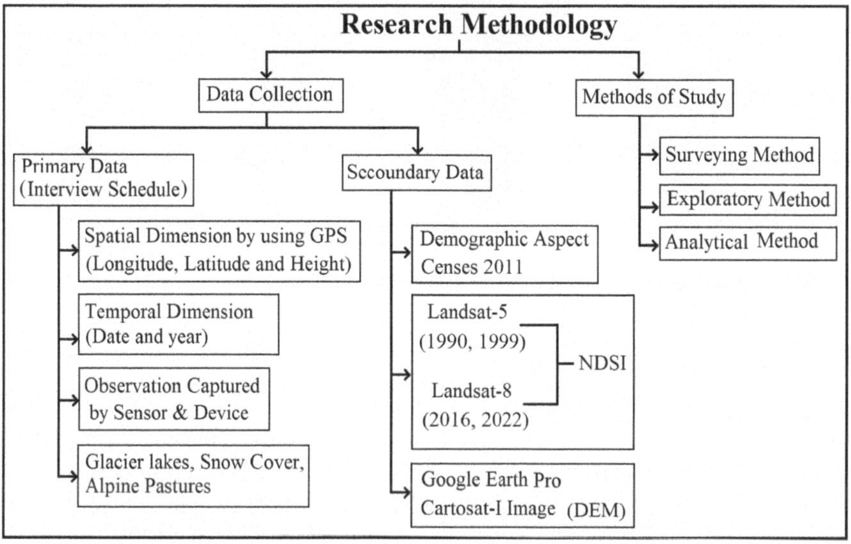

Fig. 2.4 Methodological flow chart

GIS 10.2.2 software using the NDSI index using the following equation (Hall et al. 2002):

$$\text{NDSI} = \frac{\text{Green} - \text{SWIR}}{\text{Green} + \text{SWIR}}$$

NDSI is useful for identifying snow and ice and for distinguishing snow from most mounds. This method is generally used for snowpack mapping using satellite data (Kulkarni et al. 2006; Gupta et al. 2005; Negi et al. 2008). An NDSI calculation threshold value of 0.4 is defined for satellite images from different sensors (Xiao et al. 2001) and to manage the mixed zone, the cutoff value was lowered from 0.4 to 0.1 (Klein et al. 1998).

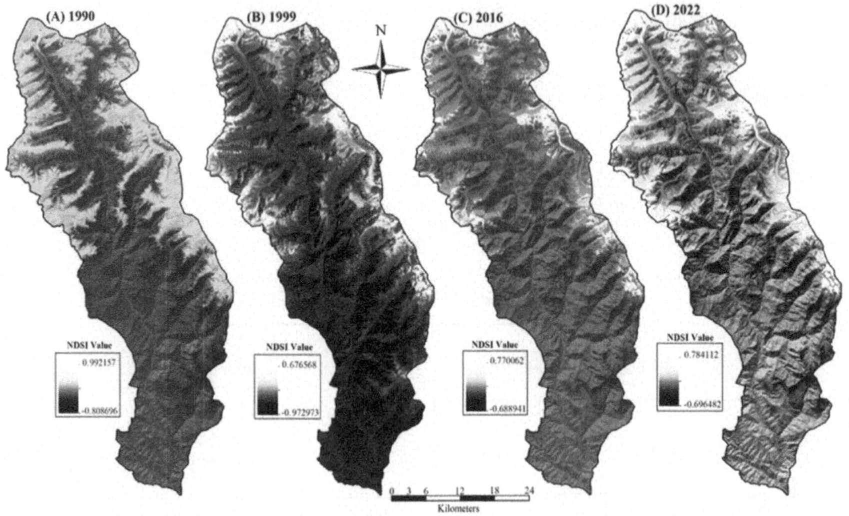

Fig. 2.5 Geographical distribution of NDSI values: 1990, 1999, 2016

2.3.4 Data Interpretation

The year 1990, 1999, 2016 and 2022 snow cover areas were calculated by using NDSI formula and snow cover positions were visually interpreted and digitized as polygon files in Arc GIS 10.8 software. Areas of each polygon were calculated and compared in Microsoft Excel 2019. Snow cover area statistics based on characteristics of the terrain were analyzed using average annual values.

2.4 Result and Discussion

The results obtained through the analysis of NDSI imagery are diagrammatically illustrated in Fig. 2.5 and data is registered in Table 2.2.

2.4.1 Impacts of Global Warming:

Thus it is necessary to study the different effects of global warming and climate change based on the available long-term satellite data over the region. The Central Himalayan basin, viz., Gori Ganga is facing many changes in the snow and glacier cover area, glacial lakes, etc. The present study on the impacts of global warming on snow and glacier cover area and the increase in the number of glacial lakes conducted in a Himalayan river basin has wide impacts.

Table 2.2 Measured and projected snow cover area depletion period, rate and area in the Gori Ganga basin (*Based on Landsat-5 and 8, satellite imageries*)

Data types	Year	Temporal changes (in km^2)	Period	Years	Change from snow to non-snow area		Reference
					in km^2	in km^2/year	
Measured	1990	678.87	–	–	–	–	Parihar and Rawat (2021)
	1999	564.92	1990–1999	9	104.64	11.63	
	2016	330.44	1999–2016	17	243.79	14.34	
	2022	240.92	2016–2022	6	89.52	14.92	
Projected	2030	116.40	2022–2030	8	124.52	15.56	On the basis of depletion rate
	2040	0	2030–2040	10	163	16.35	

(i) Understanding the impact of global warming in snow cover areas, dynamics on snow cover areas which affect glacial cover and help to increase the number of glacier lakes.
(ii) In forecasting the sifting of snow cover area through estimating projection. In forecasting the sifting of snow cover area through estimating projection.

A brief account of the impacts is presented in the following paragraphs.

2.4.1.1 Dynamics on Snow Cover Area

Gori Ganga basin was monitored using Indian RS Landsat sensor data for 32 years from 1990 to 2022. Results are presented in Fig. 2.6 and Table 2.2. There was 678.87 km^2 (30.97% of total basin area) snow cover in 1990 which was less 240.92 km^2 (10.99%) in 2022.

2.4.1.2 Snow Cover Area Changes Pattern

Based on Fig. 2.6 and Table 2.2, data reveal that overall during the last 32 years about 437.95 km^2 (19.98%) of the Gori Ganga basin has been converted from a snow cover area to a non-snow cover area.

2.4.1.3 Snow Cover Area Change Rate and Trend

The amount of shifted snow-covered area and the rate of dynamics trend in the study basin were worked out (Fig. 2.6 and Table 2.2). Data reveals that during 1990–2022 (32 years) about 19.98% area of the Gori Ganga basin area changed from snow cover area to non-snow cover area at the average rate of 13.63 km^2/year. The pattern of 32 years of snow cover studies in different periods from 1990 to 2022 reveals that

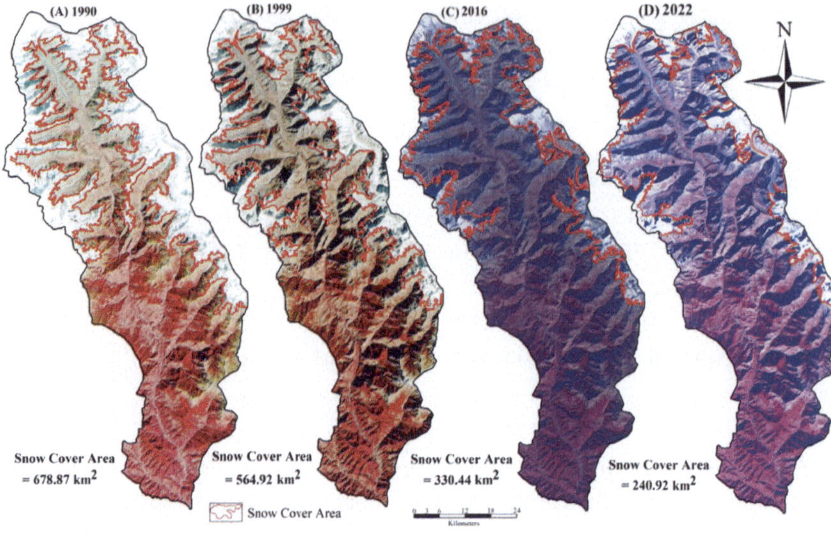

Fig. 2.6 Temporal changes in snow cover area in the Gori Ganga basin, Central Himalaya: 1990, 1999, 2016

there is a strong negative trend of snow cover retreating with global warming. This means that the temperature is rising due to global warming and the snow cover area is decreasing steadily.

2.4.2 *Predicted Snow Cover Area*

The present study based on 32 years of satellite data from four different years, i.e., 1990, 1999, 2016 and 2022 provides the present status and historical changes of the snow cover area because the present study reveals that the Gori Ganga basin experienced snow cover area is decreasing year by year. The future decreasing status of the snow cover area, snow cover depletion area and rates of depletion was projected for the years 2030 and 2040 based on these four years' data (Table 2.2) using linear regression. These data reveal that the snow cover area was depleting at the rate of 11.63 km^2/year in the 1990s which shall be increased at the rate of 16.35 km^2/year during 2030–2040. Based on these data it can be stated that the snow cover area shall be completely loss in 2040 in the Gori Ganga basin. Figure 2.7 is a graphically presented linear regression of measured and projected temporal data in the study area. The CORREL statistical method test showed that the trend of the upward shifting amount and rate of snow cover from 1990 to 2040 was negative relation (− 0.30) where slope is − 0.0077 and intercept is 15.74. Here, in the test, we found that the relation variables showed the snow cover area is shrinking.

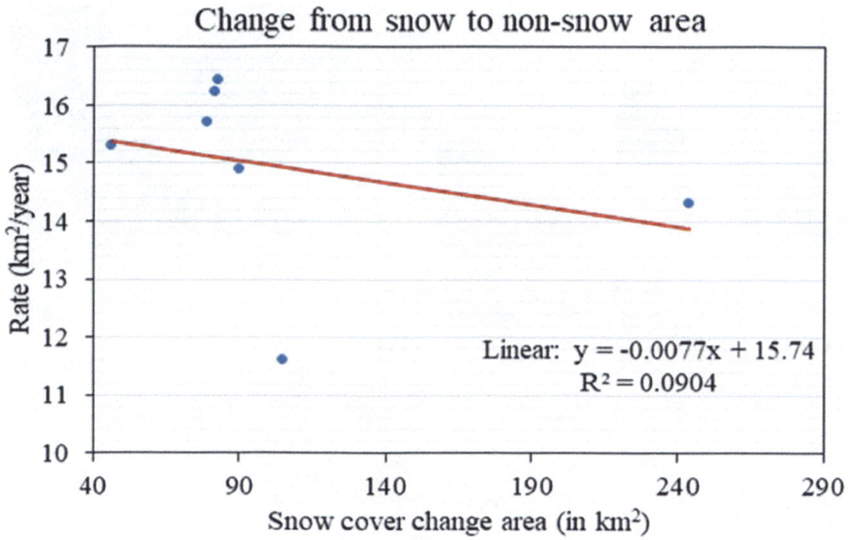

Fig. 2.7 Graphical linear regression representation of temporal dynamics from snow to non-snow area from 1990 to 2040

2.5 Signature of Changes in Snow and Glacier Cover

Long-term more than three decades (1990–2022) satellite data reveals that about 19.98% area of the study area was lost from snow cover area to non-snow cover area at the average rate of 13.63 km^2/year due to global warming. Changing climate and global warming is likely to increase the need for water and may reduce the supplying resources become the snow line has been shifted upward. These data suggest that the study area loss is being affected by global warming by which new glacial landforms such as ice caves, Moulin and glacial lakes are developing in the region. Figure 2.8 present some examples of such new developing landforms caused by global warming. Examples of newly developed glacial lakes after 2010 are presented in Fig. 2.9. The ground photographic pieces of evidence of the largest glacial lake of Gaukha glacier area (Fig. 2.10) were captured during the field survey 2019 and used as reference information for further analysis. Figure 2.10 showing the largest glacial lake (30° 33′ 48.61″ N 80° 10′ 33.74″ E, 4877 m) at the terminal point in the east Gaukha glacier. Figure 2.11 depicts the distribution of glacial lakes in the Gori Ganga watershed till 2010 and developed from 2010 to 2023 which is registered in Table 2.3.

Fig. 2.8 Icebergs falling on Milam glacier is a signature impact of global warming in Milam glacial areas in the Gori Ganga basin: **a** at terminal, **b** river path along glacier terminal **c** Moulin and **d** ice bores

2.6 Findings

The fundamental objective of the present study is to study snow cover dynamics and glacial lakes increased due to global warming in the Gori Ganga basin, which also includes the study of their patterns, velocities and trends using geospatial applications. The snow cover area of the basin is depleting steadily due to global warming. In 1990 the snow cover area in the basin area was about 30.97% which was found 10.99% in 2022. These data suggest that on average during the last 32 years (1990–2022) about 19.98% of snow cover has been converted into the non-snow cover area at the rate of 13.69 km^2/year. If their global warming continues, it can be extrapolated that the snow cover area in the Gori Ganga basin shall be about 5.31% in 2030 and completed loss of permanent snow cover in 2040. There was a total of 30 glacial lakes exist in the region where the largest glacial lakes about 21 found in the

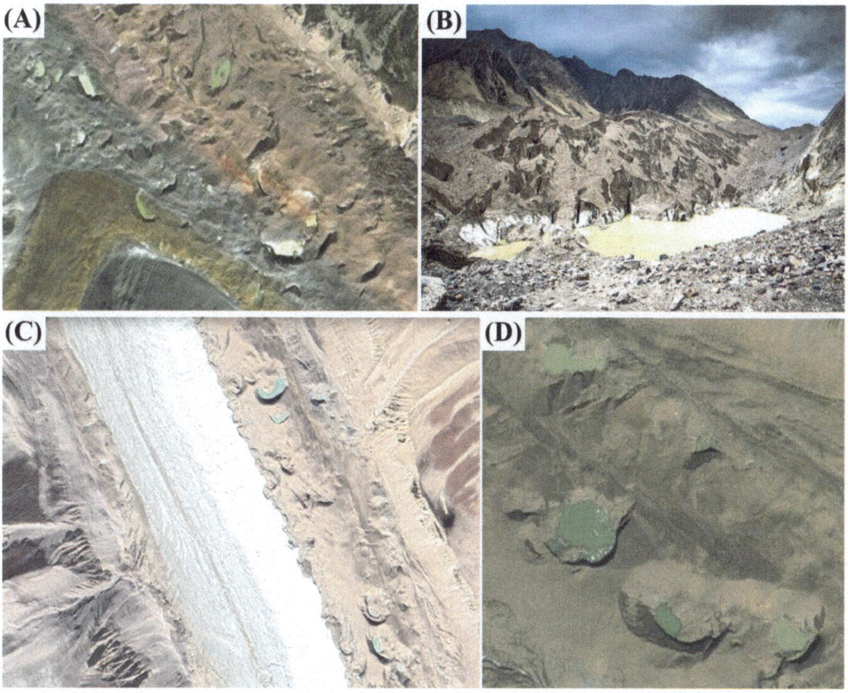

Fig. 2.9 Newly developed glacial lakes after 2010 in the Gori Ganga basin: **a** in the Milam glacier (in 2019), **b** at the terminal of the Milam glacier (in 2019), **c** near snout at Ralam glacier (GEP, 2020) and **d** in lower part of Ralam glacier area (2019)

Fig. 2.10 Gaukha Glacier at Central Himalaya ground truthing of largest glacial lake in the basin area

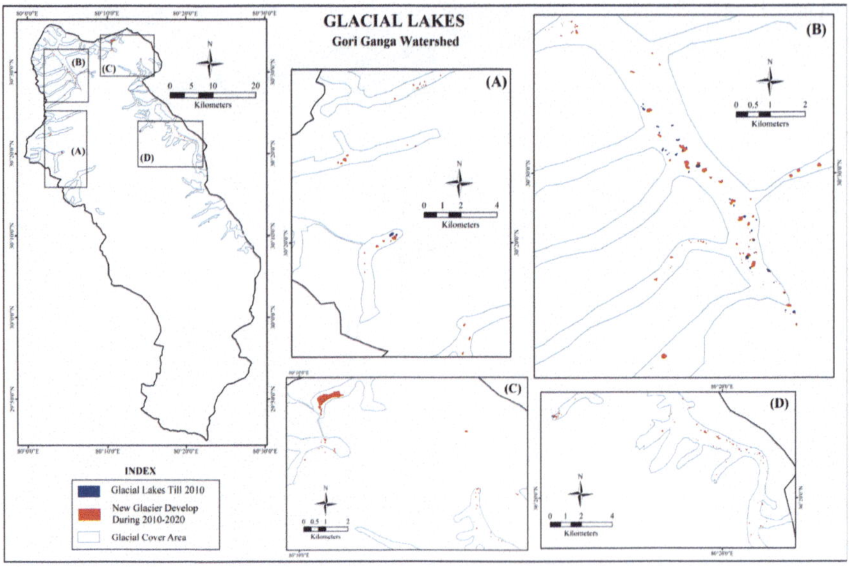

Fig. 2.11 Distribution of glacial lakes in the Gori Ganga basin till 2010 and developed from 2010 to 2023: **a** Lwan and Martoliya **b** Milam, **c** Gaukha and **d** Ralam

Table 2.3 Number of glacial lakes in the Gori Ganga basin till 2010 and developed during 2010 to 2023

S. N.	Years	Major glacial lakes on different glacier area					Total lakes
		Milam	Gaukha	Ralam	Lwan	Martoliya	
1	2010	21	Nil	16	Nil	2	39
2	2023	+ 15	+ 7	+ 9	+ 3	+ 4	+ 38
Increase no. of lakes		36	7	25	3	6	77

Milam glacier while two glaciers, viz., Gaukha and Lwan were found without any glacial lakes till 2010. There were 38 glacial lakes developed due to global warming in different glaciers between 2010 to 2023 where the highest lakes were developed about 15 in the Milam glacier and 3 glacier lakes found in the Lwan glacier. The melting of the permanent snow reservoir, a phenomenon that intensified in the twentieth century, is dragging our planet to iceless. Over the last few decades, steadily increasing global temperatures have caused ice sheets' rapid melting and changing snow volume in different spectacles. Due to increase in retreating rate leads to the formation of new lakes and the enlargement in diameter and volume of existing ones in the Himalayan region. These water bodies created by the melting of glaciers due to temperature changes and global warming are known as glacial lakes. This plays a crucial role as a freshwater source for rivers in this region. GLOF is types of outburst flood caused by the failure of a release of melt water from a moraine-dam due to dam failure. They cause for disastrous floods and create huge damage of natural environment as well loss of lives, livestock, connectivity and property in the downstream areas. Figure 2.10 is the dangerous for GLOF in this area. Sattar et al. (2020) studied by using satellite imageries of 50 years (1968 to 2018) about Gaukha Glacier lake which is largest glacial lake in Gori Ganga basin. A temporal analysis by using a hazard assessment of the lake using 1D and 2D hydrodynamic modeling of the lake surface shows that the lake has grown more than double its size from 0.10 km^2 to 0.23 km^2 and identified the potential for GLOF triggering factors.

This study is based on Geospatial application using satellite imageries, i.e., Landsat-5, Landsat-8 and Cartosat-1. Therefore the study demonstrates that the RS and GIS techniques are very useful for the study of the dynamics of snow cover area and increased glacial lakes in the Gori Ganga basin.

Acknowledgements I am highly thankful to the many regional and national newspaper and news channels, which are Times of India, Hindustan, Dainik Jagran, NavBharat Times, AajTk, Assam Times Post, SANDARP, news1hindustan and Uttarakhand Tak, etc. for published my article on newspapers and broadcast in Television channels ***"New glacial lakes formed due to melting of glaciers in Gori Ganga watershed Uttarakhand and any future geological activities can cause the lake to burst"*** in the month of February/March 2023. Author thanks to USGS for providing Landsat series data.

Competing Interests The authors declare no potential conflicts of interest.

Data Availability Statement The datasets analyzed during the current study are available from the corresponding author on reasonable request.

Funding Sources This research did not receive any specific grant from funding agencies in the public, commercial, or not-for-profit sectors.

References

Ageta Y, Kadota T (1992) Predictions of changes of glacier mass balance in the Nepal 36 K. Higuchi Himalaya and Tibetan plateau: a case study of air temperature increase for three glaciers. Ann Glaciol 16:89–94

Bhutiyani MR (2016) Spatial and temporal variability of climate change in high-altitude regions of NW Himalaya. In: Singh RB, Schickho AU, Mal S (eds) Climate change, glacier response, and vegetation dynamics in the Himalaya. Springer, pp 87–101

Bhutiyani MR, Kale VS, Pawar NJ (2007) Long-term trends in maximum, minimum and mean annual air temperatures across the northwestern Himalaya during the 20th Century. Clim Change 85(1–2):159–177

Bolch T, Kulkarni AV, Kaab A, Huggel C, Paul F, Cogley G, Frey H, Kargel JS, Fujita K, Scheel M, Stoffel M, Bajracharya S (2012) The state and fate of Himalayan glaciers. Sci 336:310–314

Diodato N, Bellocchi G, Tartari G (2012) How do Himalayan areas respond to global warming? Int J Climat 32:975–982

Dozier J (1989) Spectral signature of alpine snow covers from the Landsat Thematic Mapper. Remote Sens Environ 28(1):9–22

Fairbridge RW (1968) Glaciation glacierization. In: Geomorphology encyclopedia of earth science. Springer, Berlin, Heidelberg. https://doi.org/10.1007/3-540-31060-6_162

Gardelle J, Arnaud Y, Berthier E (2011) Contrasted evolution of glacial lakes along the Hindu Kush Himalaya mountain range between 1990 and 2009. Global Planet Change 75:47–55

Gupta RP, Haritashya UK, Singh P (2005) Mapping dry/wet snow cover in the Indian Himalayas using IRS multispectral imagery. Remote Sens Environ 97:458–469

Hall DK, Williams RS Jr, Bayr KJ (1992) Glacier recession in Iceland and Austria as observed from space. Eos 73(12):129–141. https://doi.org/10.1029/91EO00104

Hall DK, Riggs GA, Salomonson VV (1995) Developments of methods for mapping global snow cover using Moderate Resolution Imaging Spectroradiometer (MODIS) data. Remote Sens Environ 54(2):127–140. https://doi.org/10.1016/0034-4257(95)00137-P

Hall DK, Riggs GA, Salomanson VV, Di-Girolamo N, Bayr KJ (2002) MODIS snow cover products. Remote Sens Environ 83:181–194

http://news1hindustan.com/danger-77-lakes-formed-due-to-melting-of-glacier-in-uttarakhand-floods-can-create-havoc-in-this-district/. Accessed 22 July 2024

https://timesofindia.indiatimes.com/city/dehradun/77-new-glacial-lakes-add-to-kumaon-flood-risk/articleshow/98347815.cms. Accessed 22 July 2024

https://www.jagran.com/uttarakhand/nainital-global-warming-impact-in-high-himalayan-region-77-lakes-formed-due-to-glacier-melting-in-uttarakhand-23346000.html. Accessed 22 July 2024

https://www.livehindustan.com/uttarakhand/story-77-new-lakes-formed-in-uttarakhand-due-to-melting-of-glacier-experts-are-worried-about-this-7834478.html. Accessed 22 July 2024

https://sandrp.in/2023/03/06/drp-nb-060323-indias-regulators-blind-to-increasing-threats-in-himalayas/. Accessed 22 July 2024

https://assamtimespost.in/?p=96110. Accessed 22 July 2024

IPCC (2013) Climate change 2013: the physical science basis. Contribution of working group-I to the fifth assessment report of the IPCC. In: Stocker TF, Qin D, Plattner GK, Tignor M, Allen SK, Boschung J, Nauels A, Xia Y, Bex V, Midgley PM (eds) Cambridge University Press, Cambridge, United Kingdom and New York, USA. https://www.ipcc.ch/report/ar5/wg1/. Accessed 26 July 2024

Kargel JS, Abrams MJ, Bishop MP, Bush A, Hamilton G, Jiskoot H, Kaab A, Kieffer HH, Lee EM, Paul Rau FF, Raup B, Shroder JF, Soltesz D, Stainforth D, Stearns L, Wessels R (2005) Multispectral imaging contributions to global land ice measurements from space. Remote Sens Environ 99(1–2):187–219. https://doi.org/10.1016/j.rse.2005.07.004

Klein AG, Hall DK, Riggs GA (1998) Improving snow cover mapping in forests through the use of a canopy reflectance model. Hydrol Process 12:1723–1744

Konig M, Winther JG, Isaksson E (2001) Measuring snow and glacier ice properties from satellite. Rev Geophys 39(1):1–28. https://doi.org/10.1029/1999RG000076

Kulkarni AV, Dhar S, Rathor BP, Raj KBG, Kalia R (2006) Recession of Samundratapu glacier, Chandra river basin, Himachal Pradesh. JISRS 34(1):39–46

Kumar K, Joshi S, Joshi V (2008) Climate variability, vulnerability, and coping mechanism in Alaknanda catchment, Central Himalaya, India. Ambio 37:286–291

Negi H, Snehmani S, Thakur NK (2008) Operational snow cover monitoring in NW-Himalaya using terra and aqua MODIS imageries, proceedings. International workshop on snow, ice, glacier and avalanches at IIT Mumbai, India, 7–9 January, pp 11–25

Negi HS, Kanda N, Shekhar MS, Ganju A (2018) Recent wintertime climatic variability over the North West Himalayan Cryosphere. Curr Sci 114(4):760–770

NSIDC (2020) National snow and ice data center. https://nsidc.org/cryosphere/snow/climate.html. Accessed 28 July 2024

Parihar DS (2021) Dynamics of timberline due to Spatio-temporal changes using GIS and RS in the Gori Ganga watershed, Kumaun Himalaya, Uttarakhand. IJGGS 8(1):42–47. https://doi.org/10.14445/23939206/IJGGS-V8I1P105

Parihar DS (2021b) Due to global warming: snow line dynamics in the Gori Ganga watershed, Kumaun Himalaya by using RS and GIS. Int J Ecol Env Sci 3(1):226–233. https://www.ecologyjournal.in/archives/2021/vol3/issue1/3-1-41. Accessed 28 July 2024

Parihar DS (2021) Study of suitable sites for tourism development in the Gori Ganga watershed Kumaun Himalaya by using Remote Sensing and GIS. Int Multidiscip Res J 11(12):321–328

Parihar DS (2022) A review on century of global warming impacts as a global issue in the Kumaun Himalaya, India. In: Chitsimran, Sidhu A, Kaur M, Parasmehak, Kaur D, Sharma S, Prakash S (ed) Chapter-19 on Contemporary issues in business and economics Vol-1. Red Shine Publication, pp 194–222. ISBN-978–1–387–65048–4. https://doi.org/10.25215/1387650483

Parihar DS, Rawat JS (2021) Spatio-temporal change in snow cover area using RS & GIS in the Gori Ganga watershed, Kumaun Himalaya. Int J Adv Res 9(3):30–34. https://doi.org/10.21474/IJAR01/12550

Parihar DS, Rawat JS, Singh M and Pant NC (2021) Spatio-temporal change of vegetation line in the Gori Ganga watershed, Kumaun Himalaya by using Remote Sensing and GIS techniques. Int J Trend Sci Res Dev 6(1):409–416. https://www.ijtsrd.com/papers/ijtsrd47806.pdf. Accessed 28 July 2024

Quincey DJ, Lucas RM, Richardson SD, Glasser NF, Hambrey MJ, Reynolds JM (2005) Optical remote sensing techniques in high-mountain environments: application to glacial hazards. Prog in Phys Geogr 29(4):475–505

Raj KBG, Remya SN, Kumar V (2013) Remote sensing-based hazard assessment of glacial lakes in Sikkim Himalaya. Curr Sci 104(3):359–364

Revenga C, Nackoney J, Hoshino E, Kura Y, Maidens J (2003) Watersheds of the world CD, IUCN, International Water Management Institute (IWMI), Ramsar Convention Bureau, World Resources Institute (WRI), Washington, ISBN: 978–1–56973–548–0, Call No.: IUCN-CD-003. https://portals.iucn.org/library/node/8204. Accessed 27 July 2024

Richardson SD, Reynolds JM (2000) An overview of glacial hazards in the Himalayas. Quat Int 65–66:31–47

Sattar A, Goswami A, Kulkarni AV, Emmer A (2020) Lake evolution, hydrodynamic outburst flood modeling and sensitivity analysis in the central Himalaya: a case study. Water 12(1):237. https://doi.org/10.3390/w12010237

SGIM (2010) Snow and glaciers of the Himalayas: inventory and monitoring. Space Applications Centre Indian Space Research Organization, Ahmedabad Collaboration with Ministry of Forests and Environment, New Delhi, 90 https://www.indiawaterportal.org/sites/default/files/iwp2/Snow_and_Glaciers_of_the_Himalayas_Discussion_Paper_II_ISRO_SAC_2011.pdf. Accessed 27 July 2024

Silverio W, Jaquet JM (2005) Glacial cover mapping (1987–1996) of the Cordillera Blanca (Peru) using satellite imagery. Remote Sens Environ 95(3):342–350. https://doi.org/10.1016/j.rse.2004.12.012

Wang J, Li W (2003) Comparison of methods of snow cover mapping by analyzing the solar spectrum of satellite remote sensing data in China. Int J Remote Sens 24(21):4129–4136. https://doi.org/10.1080/0143116031000070409

WGM (2018) WGM service updated and earlier reports, WMO-No. 1233, 19–22. https://wgms.ch/faqs/. Accessed 22 July 2024

Winther JG, Hall DK (1999) Satellite derived snow coverage related to hydropower production in Norway: present and future. Int J Remote Sens 20(15–16):2991–3008. https://doi.org/10.1080/014311699211570

Wouters B, Gardner AS, Moholdt G (2019) Global glacier mass loss during the GRACE satellite mission (2002–2016). Front Earth Sci 7:1–11 (Article- 96)

Xiao X, Shen Z, Qin X (2001) Assessing the potential of vegetation sensor data for mapping snow and ice cover: a normalized difference snow and ice index. Int J Remote Sens 22:2479–2487

Chapter 3
To Track Spatial and Temporal Patterns of Change of Forests with a Case Study, Western Ghat, India

Komal Rai, Gulab Singh, and S. Sreelekshmi

Abstract Monitoring forests using Synthetic Aperture Radar (SAR) imagery is crucial for tracking spatial and temporal changes, aiding in resource management and climate change analysis. SAR's ability to penetrate cloud cover and provide consistent, high-resolution data makes it an essential tool for forest monitoring. Our study outlines a methodology for detecting forest changes using SAR, with a case study on the Western Ghats. The approach integrates SAR image acquisition, pre-processing and indices to assess forest cover and health. Change detection algorithms applied to temporal SAR data map forest changes over short period of time. In the case study, Sentinel-1C SAR data from April and October 2023 was used for its high temporal resolution. Pre-processing included radiometric calibration, speckle filtering, and terrain correction. Temporal analysis employed the Radar Vegetation Index (RVI), adapted from NDVI, to detect changes. Significant deforestation events were observed, especially in correlation with increased deforestation periods in the Upper Western Ghats. Spatial analysis revealed deforestation hotspots near rivers and roads, indicating the impact of human infrastructure. Seasonal trends showed higher deforestation rates during the dry season, highlighting easier access to remote areas. The integration of SAR data with advanced analytical techniques provides a robust framework for monitoring forests, offering valuable insights for policymakers and conservationists. Hope this chapter helps in basic understanding of SAR imagery pre-processing and its application in forest change studies.

Keywords SAR · Pre-processing · Forest disturbances · Radar vegetation index (RVI) · Normalized difference vegetation index (NDVI)

K. Rai (✉) · G. Singh · S. Sreelekshmi
CSRE, Indian Institute of Technology, Bombay, Maharashtra, India
e-mail: komalresearch7@gmail.com

V. Saritha et al. (eds.), *Remote Sensing for Environmental Monitoring*,
https://doi.org/10.1007/978-981-96-5546-5_3

3.1 Introduction

Forest ecosystems are critical for maintaining ecological balance and supporting biodiversity, yet they remain highly vulnerable to disturbances such as forest fires. These fires, often exacerbated by climate change and human activities, pose significant threats to ecological stability and biodiversity-rich regions like the Western Ghats and Uttarakhand in India (Gupta et al. 2018; Prasanna Kumar et al. 2013). Accurately assessing and mapping the impacts of forest fires is essential for effective management and conservation strategies. Remote sensing technologies, particularly the integration of multispectral and Synthetic Aperture Radar (SAR) data, have proven effective in monitoring forest fire dynamics and assessing burnt area severity (Engelbrecht et al. 2017; Suresh et al. 2018). Advanced methodologies, such as fuzzy overlay analysis and Sentinel datasets, offer innovative approaches to risk zonation and post-fire assessment, enabling researchers and policymakers to mitigate fire risks and plan recovery measures (Rai et al. 2022; Imperatore et al. 2017). This study builds on these advancements to explore spatial and temporal patterns of forest fire impacts, contributing to a deeper understanding of their ecological consequences.

Tracking the spatial and temporal patterns of forest change is essential for understanding the complex dynamics of ecosystems, managing natural resources, and addressing global environmental challenges, particularly climate change. Forests are critical for maintaining ecological balance; they act as major carbon sinks, regulate water cycles, prevent soil erosion, and provide habitat for a wide range of species (FAO 2020; Bonan 2008). Despite their importance, forests worldwide face severe threats from deforestation, degradation, and land-use changes driven by human activities like agriculture, infrastructure development, and resource extraction (Hansen et al. 2013). Accurately monitoring forest changes over time and across landscapes is vital for developing effective conservation strategies and promoting sustainable land management.

The ability to monitor these changes has significantly improved with the advent of advanced remote sensing technologies, particularly Synthetic Aperture Radar (SAR) imagery. SAR has become a preferred tool for forest monitoring due to its unique capabilities. Unlike optical sensors that rely on visible light and are hindered by cloud cover and weather conditions, SAR operates in the microwave spectrum, allowing it to penetrate through clouds and collect data regardless of weather or daylight (Balzter 2001; Henderson and Lewis 1998). This makes SAR particularly useful in tropical and mountainous regions, where clouds are prevalent, and continuous monitoring is otherwise challenging.

SAR imagery provides high-resolution data that is valuable for detecting forest structure, changes in biomass, and patterns of deforestation or degradation (Petar et al. 2025). The ability to track these changes over time enables researchers and policymakers to gain insights into the health and extent of forests, identify areas of concern, and implement timely interventions (Enmanuel et al. 2024). SAR's consistency in data acquisition also ensures that temporal analyses can be conducted effectively, enabling the detection of gradual or rapid changes in forest cover.

This chapter presents a comprehensive methodology for tracking forest changes using SAR imagery, focusing on both spatial and temporal dynamics. The approach combines advanced image processing techniques and machine learning algorithms to extract meaningful insights from SAR data. Key steps include SAR image acquisition, pre-processing to reduce noise and enhance signal quality, identify forest cover, and change detection over time (Zilin et al. 2024). This methodology offers a robust framework for understanding how forested landscapes evolve and how human activities and environmental factors drive these changes.

In the case study, Sentinel-1C SAR data was used, spanning a period from April and October 2023. Sentinel-1C was chosen for its high temporal resolution, which is crucial for capturing the rapid and seasonal changes occurring in this region (ESA 2023). Pre-processing steps included radiometric calibration, speckle filtering to remove noise, and terrain correction to ensure accurate spatial representation (Wei et al. 2025). Temporal change detection was performed using a modified version of the Normalized Difference Vegetation Index (NDVI) known as the Radar Vegetation Index (RVI), which is adapted for SAR data (Chang et al. 2018). This index provided a valuable metric for assessing forest health and detecting changes in vegetation cover over time.

Through this integrated approach, the chapter highlights the importance of SAR imagery in understanding forest dynamics in forested landscapes like the Western Ghats.

3.2 Study Area

The Western Ghats, as shown in Fig. 3.1, also known as the Sahyadri Hills, is one of the most important biodiversity hotspots in the world, stretching along the western coast of India. It runs for approximately 1600 km through six Indian states—Gujarat, Maharashtra, Goa, Karnataka, Kerala, and Tamil Nadu—spanning an area of around 160,000 km^2. The mountain range plays a critical role in regulating the climate of peninsular India and significantly influences the monsoon weather patterns that are vital for the region's agriculture and water resources.

The Western Ghats experience a range of climatic conditions, from tropical and subtropical to temperate in the higher altitudes. Rainfall is high on the western slopes due to the monsoons, while the eastern slopes are drier, leading to varied forest types on either side. The region's altitudinal variation, ranging from sea level to over 2600 m at the highest peak (Anamudi), results in a diverse range of ecosystems and microclimates. These variations support a wide range of flora and fauna, contributing to the ecological richness of the region.

The Ghats are also the source of numerous rivers, including the Godavari, Krishna, and Kaveri, which provide water to millions of people in southern India. The region's forests play a crucial role in regulating the hydrological cycle, contributing to groundwater recharge, preventing soil erosion, and reducing the risk of floods.

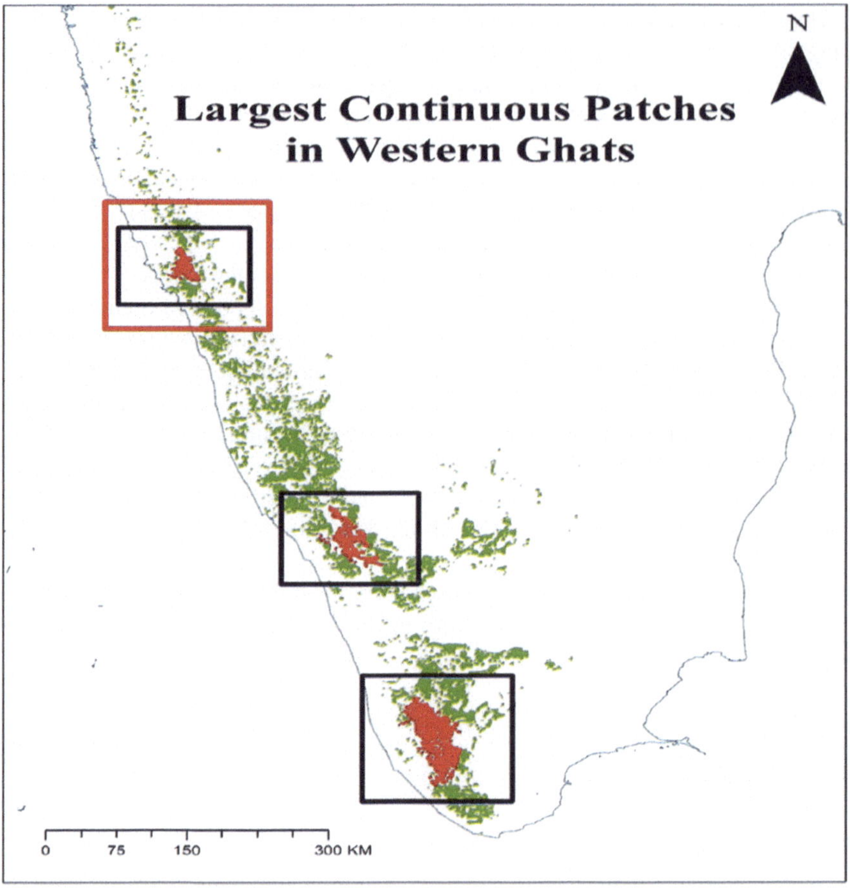

Fig. 3.1 Study area of Western Ghats

3.3 Methodology

Pre-processing of SAR images involves several crucial steps to enhance data quality and accuracy. Initially, radiometric calibration corrects for sensor-specific variations, ensuring consistent signal interpretation. This is followed by geometric correction, aligning the SAR image with ground coordinates, often using a digital elevation model (DEM) to account for terrain distortions. Speckle noise reduction is applied to minimize interference from radar signal scattering, improving image clarity. Finally, terrain correction is performed to rectify distortions caused by varying elevation. These pre-processing steps as depicted in Fig. 3.2, collectively refine the SAR data, making it suitable for accurate analysis and interpretation in applications like land monitoring and environmental studies.

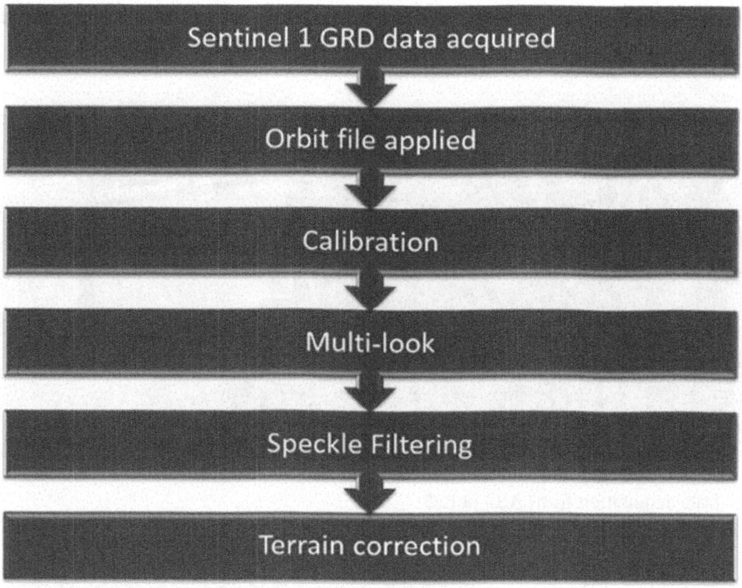

Fig. 3.2 Flowchart for pre-processing

3.3.1 Data Acquisition

In our study first step is Data acquisition; to acquire data from Sentinel-1C, as shown in Fig. 3.3 one can utilize the Alaska Satellite Facility (ASF) Data Search tool, which offers access to Synthetic Aperture Radar (SAR) data. The process begins by navigating to the ASF Data Search portal and selecting Sentinel-1 from the available missions. Users can specify parameters such as area of interest, acquisition dates, and polarization modes. After defining these criteria, the search tool generates a list of relevant data products. Each dataset can be previewed, and users can download them in formats compatible with analysis tools like SNAP or GIS software.

3.3.2 Data Visualization and Polarization Orientations

Data visualization in synthetic aperture radar (SAR) is critical for interpreting and analyzing complex datasets. SAR uses different polarization orientations (like horizontal-horizontal (HH), vertical–vertical (VV), and cross-polarizations like HV or VH) to capture distinct properties of the Earth's surface. Visualizing these polarizations helps in distinguishing between various surface types, such as vegetation, water, and urban structures. For instance, dual- and quad-polarization SAR data provide insights into forest density, soil moisture, or glacier dynamics. Effective visualization of polarization orientations enhances pattern recognition, enabling better

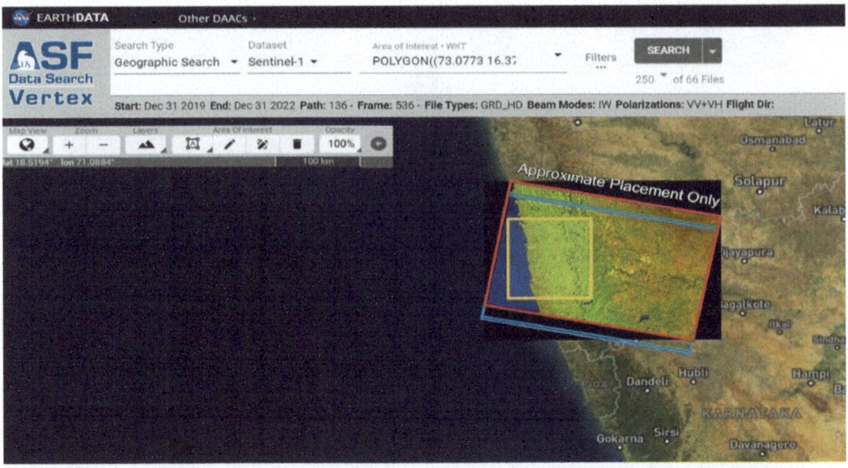

Fig. 3.3 Data acquisition from ASF portal

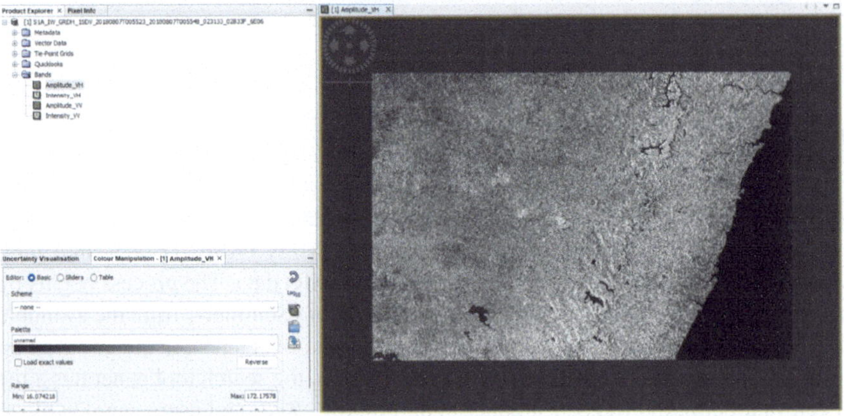

Fig. 3.4 Data visualization on SNAP

understanding of natural phenomena and improving decision-making in environmental monitoring, disaster management, and other applications. Figure 3.4 helps us visualize the upper western ghats raw image.

3.3.3 Apply Orbit File

Applying an orbit file in satellite data processing as shown in Fig. 3.5, involves integrating precise satellite position and velocity information into remote sensing data analysis. Orbit files, typically derived from tracking data, provide crucial details about

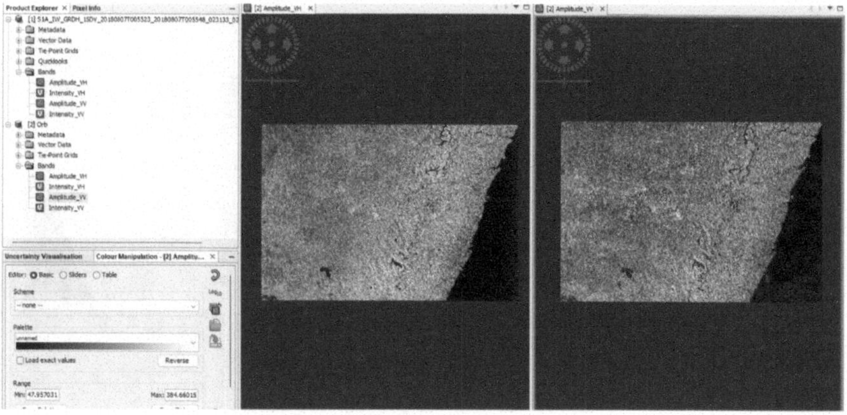

Fig. 3.5 Applying orbit file

the satellite's trajectory during data acquisition. By incorporating these files, analysts can correct for satellite motion, ensuring accurate georeferencing of the imagery. This correction is essential for applications such as change detection, mapping, and monitoring. Accurate orbit information improves the alignment of satellite data with ground truth, enhancing the reliability of analysis and ensuring that the spatial resolution and positioning of features in the images are precise.

3.3.4 Calibration

Calibration in SAR is essential for ensuring accurate and reliable measurements of the Earth's surface. It involves adjusting SAR data to account for systematic errors and variations in sensor performance. This process includes radiometric calibration as seen in Fig. 3.6, which corrects for signal intensity discrepancies caused by factors like sensor drift or environmental conditions, and geometric calibration, which aligns the data to the correct spatial coordinates. Calibration also addresses noise reduction and compensates for any distortions, improving the quality and consistency of SAR images. Proper calibration enhances the ability to interpret data accurately, crucial for applications such as environmental monitoring and disaster management.

3.3.5 Multilooking

Multilooking in SAR as shown in Fig. 3.7, is a processing technique used to improve image quality by averaging multiple radar looks or echoes to reduce speckle noise and enhance signal-to-noise ratio. This process involves combining several adjacent radar

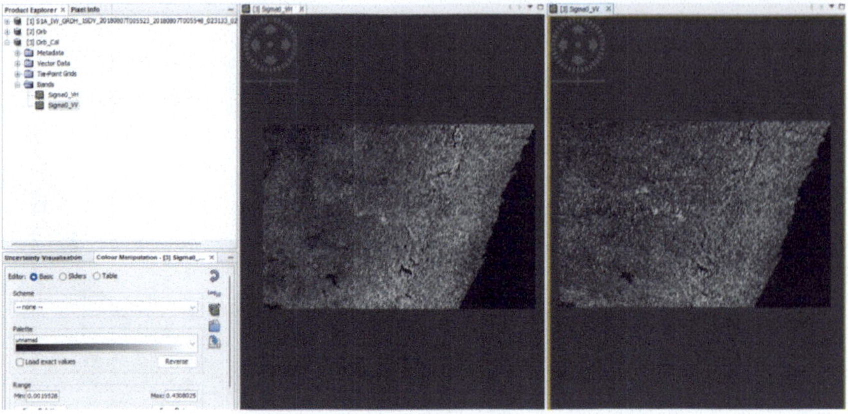

Fig. 3.6 Radiometric calibration

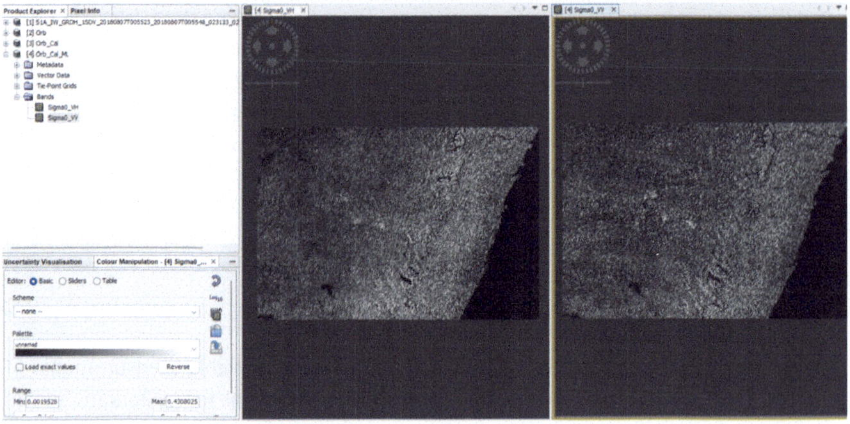

Fig. 3.7 Multi looking

range cells into a single, larger cell, effectively smoothing out noise while preserving the overall image structure. Multilooking also reduces spatial resolution but increases the clarity and interpretability of the image. By averaging different looks, multi-looking helps in better distinguishing features and improving the accuracy of land cover classification, object detection, and other SAR-based analyses.

3.3.6 Speckle Filtering

Speckle filtering in SAR is a technique used to reduce noise and improve image quality. Speckle, a granular noise caused by the interference of radar waves scattered

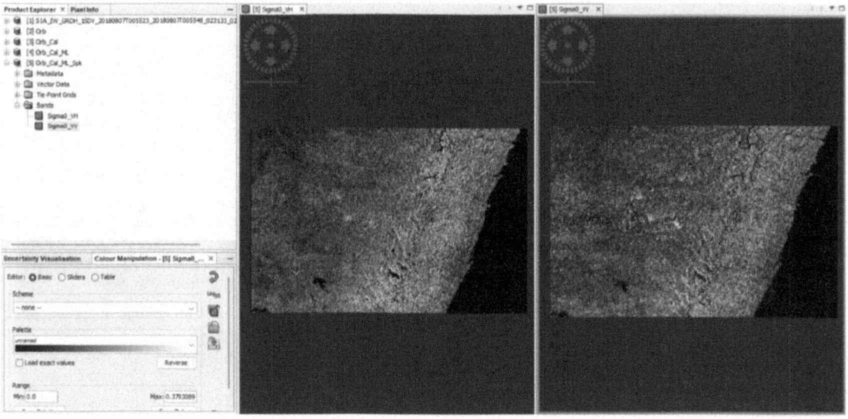

Fig. 3.8 Speckle filter

by the rough surface, can obscure meaningful details. Filtering methods, such as the Lee, Frost, or Kuan filters, work by averaging pixel values within a neighborhood while preserving edges and important features. This process smooths out the speckle noise while retaining the structural integrity of the image, enhancing the clarity of features like land cover, vegetation, and urban structures. Effective speckle filtering is crucial for accurate interpretation and analysis of SAR data as shown below in Fig. 3.8.

3.3.7 Range Doppler Terrain Correction

Range-Doppler terrain correction in SAR processing (in Fig. 3.9) adjusts for distortions caused by varying terrain elevations. SAR images can suffer from geometric inaccuracies due to the slant range geometry and Earth's topography. The range-Doppler method involves correcting these distortions by mapping the SAR image onto a reference digital elevation model (DEM). This process accounts for variations in the radar's line-of-sight and terrain relief, aligning the radar data to true ground positions. The correction ensures that features in the SAR image accurately represent their geographic locations, enhancing the image's usability for applications such as topographic mapping, land use analysis, and environmental monitoring.

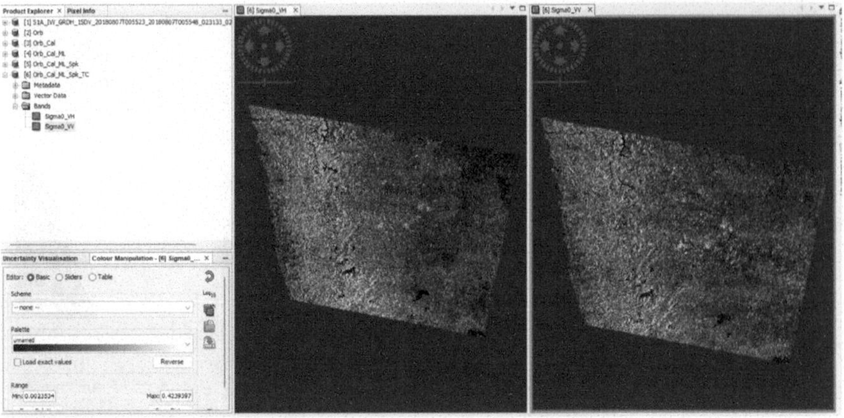

Fig. 3.9 Terrain correction

3.3.8 RVI and NDVI

The Radar Vegetation Index (RVI) and the Normalized Difference Vegetation Index (NDVI) are both used to assess vegetation health and cover, but they utilize different types of remote sensing data.

NDVI, derived from optical satellite imagery, calculates vegetation health by measuring the difference between near-infrared (NIR) and red reflectance. It is expressed as

$$\mathrm{NDVI} = \frac{\mathrm{NIR} - \mathrm{Red}}{\mathrm{NIR} + \mathrm{Red}}$$

Healthy vegetation reflects more NIR light and absorbs more red light, resulting in higher NDVI values. NDVI is widely used in agriculture, forestry, and climate studies for monitoring vegetation growth, drought effects, and land cover changes.

In contrast, the Radar Vegetation Index (RVI) uses synthetic aperture radar (SAR) data, which is sensitive to vegetation structure and moisture content.

$$RVI = \frac{4 * VH}{VV + VH}$$

SAR systems measure the backscattered radar signal, which varies with the density and structure of the vegetation. The RVI is calculated from SAR polarimetric data, reflecting the vegetation's radar backscatter characteristics. It is particularly useful in areas with frequent cloud cover, where optical sensors may be limited. RVI can offer insights into vegetation biomass, soil moisture, and structural changes, complementing optical indices like NDVI and providing valuable information for monitoring and managing vegetation in diverse environmental conditions.

3.4 Results and Discussion

3.4.1 Pre-disturbance RVI and NDVI

Before forest disturbances occur, vegetation indices like the Radar Vegetation Index (RVI) and Normalized Difference Vegetation Index (NDVI) serve as key indicators of healthy, dense forest cover. The RVI, derived from radar backscatter, reflects the structural integrity and biomass of vegetation. Dense forests with robust structures exhibit high RVI values due to strong radar returns from leaves, branches, and trunks. Similarly, NDVI, calculated from the near-infrared (NIR) and red reflectance, shows high values in healthy forests because of the strong NIR reflectance and low red-light absorption by chlorophyll.

Monitoring temporal changes in these indices enables the detection of forest disturbances and degradation. Decreasing trends in RVI often signal the loss of vegetation structure, while reduced NDVI values indicate diminished photosynthetic activity and canopy cover. The combined analysis of RVI and NDVI provides a comprehensive picture of forest health, capturing structural and physiological changes.

In April 2023, observations from the Upper Western Ghats revealed a consistent trend between RVI and NDVI values, with both indices indicating dense, healthy vegetation. This alignment demonstrates their reliability in assessing forest conditions and highlights the region's ecological stability before disturbances. The parallel trends in these indices underline their complementary role in monitoring forest ecosystems (Fig. 3.10).

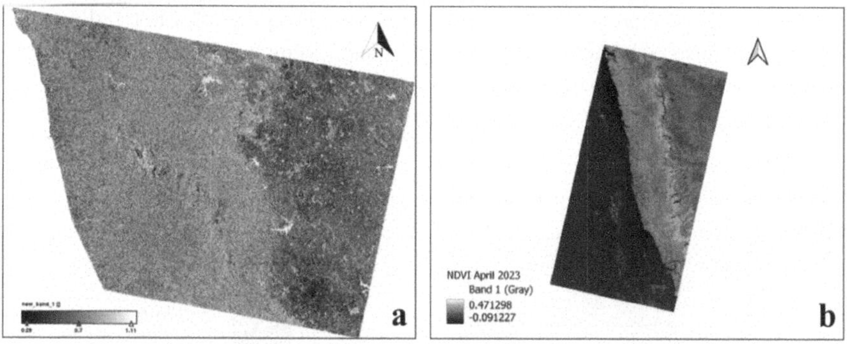

Fig. 3.10 April 2023 pre-disturbance indices **a** RVI and **b** NDVI

3.4.2 Post-disturbance RVI and NDVI

Post-forest disturbances, indices like the Radar Vegetation Index (RVI) and Normalized Difference Vegetation Index (NDVI) serve as crucial indicators for assessing changes in vegetation health and structure. These indices typically show a marked decline in response to disturbances such as deforestation, fires, or storms. The RVI, derived from radar backscatter, decreases due to the loss of vegetation structure, canopy cover, and biomass, which significantly reduce the radar signal's ability to scatter within the vegetation. Similarly, NDVI, which relies on the contrast between near-infrared (NIR) light reflected by healthy vegetation and red light absorbed during photosynthesis, diminishes as the proportion of healthy vegetation declines.

This trend is evident in the upper Western Ghats during October 2023, as illustrated in Fig. 3.11. Both RVI and NDVI values show a significant reduction, reflecting the impact of recent disturbances. The decline in RVI indicates structural degradation, while the reduced NDVI values signal a loss of photosynthetically active vegetation. Such analyses are essential for quantifying the spatial and temporal extent of disturbances, enabling targeted restoration efforts. The combined interpretation of RVI and NDVI provides a robust framework for understanding vegetation dynamics post-disturbance, supporting sustainable forest management strategies in ecologically sensitive regions.

Forest disturbances in the Upper Western Ghats, a biodiversity hotspot, have significant ecological implications. Using synthetic aperture radar (SAR) imagery, these disturbances can be detected and analyzed, even under cloudy conditions common in tropical regions. SAR provides detailed insights into changes in forest structure, biomass loss, and land cover alterations, which are key indicators of disturbances like deforestation, forest degradation, or human encroachment.

Processing SAR imagery for this analysis involves several steps. First, radiometric calibration ensures consistent radar backscatter measurements. Geometric correction aligns the imagery with geographic coordinates, often using digital elevation models (DEMs) to account for terrain variations. Speckle noise reduction enhances image

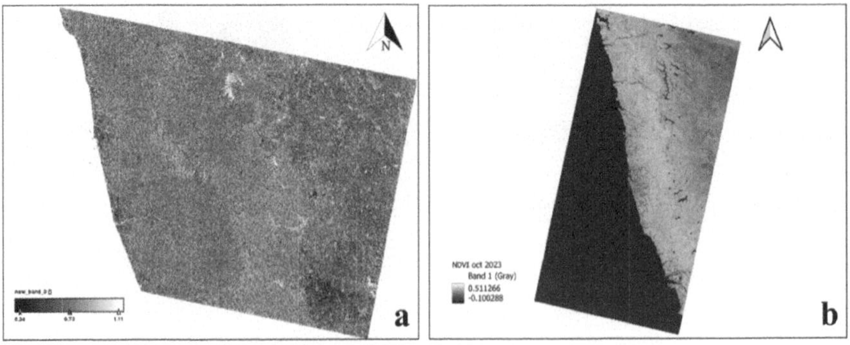

Fig. 3.11 October 2023 post disturbance indices **a** RVI and **b** NDVI

clarity by minimizing radar signal scattering. Terrain correction further improves spatial accuracy by correcting distortions caused by elevation changes.

Once pre-processed, Synthetic Aperture Radar (SAR) data becomes a powerful tool for monitoring forest ecosystems, particularly when combined with indices such as the Radar Vegetation Index (RVI). RVI is sensitive to structural changes within the forest, as it captures variations in radar backscatter caused by canopy cover, biomass, and vegetation density. Post-disturbance, RVI values typically decrease, signalling a loss of forest structure due to events like deforestation, fires, or severe storms. These structural changes are invaluable for assessing the magnitude and spatial extent of disturbances.

To further validate these findings, SAR-based results can be cross-referenced with optical data, such as the Normalized Difference Vegetation Index (NDVI). NDVI, derived from the differential reflectance of near-infrared (NIR) and red light, is an established measure of vegetation health and density. Like RVI, NDVI also shows a significant decline post-disturbance, as the loss of healthy, photosynthetically active vegetation reduces the amount of NIR light reflected. This dual analysis of RVI and NDVI enables researchers to comprehensively evaluate both the structural and physiological impacts of forest disturbances.

As depicted in Fig. 3.12, the integration of SAR and optical data reveals a marked decline in both RVI and NDVI values in the Upper Western Ghats following a disturbance event in October 2023. The RVI highlights structural degradation, while the NDVI reflects the loss of vegetative vigor. This combined approach ensures greater accuracy and reliability in monitoring, as it leverages the all-weather capability of SAR with the high sensitivity of optical indices.

Such integrated methodologies are crucial for conservation planning, as they allow for detailed assessments of disturbance impacts, identification of vulnerable areas, and informed decision-making to promote ecological restoration and sustainable forest management in biodiversity-rich regions like the Western Ghats.

The table illustrates the changes in NDVI and RVI values between April and October 2023 in the study area. NDVI values, which range from − 0.091 to 0.471 in April, show a slight increase to − 0.1 to 0.511 in October, indicating seasonal variations in vegetation health. Meanwhile, RVI values remain relatively stable, ranging

	APRIL 2023	OCTOBER 2023
NDVI	-0.091 to 0.471	-0.1 to 0.511
RVI	0.29 to 1.1	0.31 to 1.1

Fig. 3.12 Values of indices pre- and post-disturbance 2023

from 0.29 to 1.1 in April and 0.31 to 1.1 in October, reflecting minimal structural changes in forest cover. These trends suggest that while vegetation health fluctuates seasonally, the overall forest structure remains consistent, underlining the utility of combined NDVI and RVI indices for comprehensive forest monitoring.

3.5 Conclusion

In conclusion, this study underscores the critical role of Synthetic Aperture Radar (SAR) imagery in monitoring forest changes, particularly in biodiversity-rich and ecologically sensitive regions like the Western Ghats. By leveraging SAR's ability to penetrate cloud cover and provide consistent, high-resolution data acquisition regardless of weather or daylight, the study effectively captures the spatial and temporal dynamics of deforestation. The integration of advanced pre-processing techniques and indices like the Radar Vegetation Index (RVI) offers a robust framework for detecting and analyzing forest changes over time, enabling accurate assessments of structural and physiological impacts on vegetation (Mullissa et al. 2024; Le et al. 2019).

The case study utilizing Sentinel-1C SAR data from April to October 2023 demonstrates the utility of high-temporal-resolution datasets in identifying seasonal and anthropogenic patterns of forest disturbance. The findings reveal significant deforestation hotspots near rivers and roads, as well as heightened disturbance rates during the dry season, which align with increased human accessibility to forested areas. Such insights highlight the need for targeted conservation measures and sustainable land management strategies (ESA 2023).

This chapter emphasizes the potential of SAR technology, combined with advanced analytical techniques, to support sustainable forest management and climate change mitigation. By providing actionable insights for policymakers and conservationists, it lays a solid foundation for future applications of SAR in forest monitoring and environmental management.

Competing Interests The authors declare no potential conflicts of interest.

References

Adugna M, Sassan S, Ricardo D, Tyler E, Naomi P, Fiona O, Aleena A, Violet M, Daniel M (2024) LUCA: a Sentinel-1 SAR-Based global forest land use change alert. Remote Sens 16(12):2151. https://doi.org/10.3390/rs16122151

Balzter H (2001) Forest mapping and monitoring with interferometric synthetic aperture radar (InSAR). Prog Phys Geogr 25(2):159–177

Bonan GB (2008) Forests and climate change: forcings, feedbacks, and the climate benefits of forests. Science 320(5882):1444–1449

Chang JG, Maxim S, Oh Y (2018) Polarimetric radar vegetation index for biomass estimation in desert fringe ecosystems. IEEE Trans Geosci Remote Sens 56(12):7102–7108

Engelbrecht J, Teron A, Vhengani L, Kemp J (2017) A simple normalized difference approach to burnt area mapping using multi polarisation C-Band SAR. Remote Sens 9:764.

Enmanuel RP, Martin S, Achim R, Johannes S, Thomas, U (2024) Forest disturbance characterization in the era of earth observation big data: a mapping review. Int J Appl Earth Obs Geoinf 128:103755. https://doi.org/10.1016/j.jag.2024.103755

ESA (2023) Sentinel-1 SAR user guide. European Space Agency

FAO (2020) Global forest resources assessment 2020. Food and Agriculture Organization of the United Nations

Gupta S, Roy A, Bhavsar D et al (2018) Forest fire burnt area assessment in the biodiversity rich regions using geospatial technology: Uttarakhand Forest Fire Event 2016. J Indian Soc Remote Sens. 46:945–955

Hansen MC et al (2013) High-resolution global maps of 21st-century forest cover change. Science 342(6160):850–853

Henderson FM, Lewis AJ (1998) Principles and applications of imaging radar. In: Manual of remote sensing, 2

Imperatore P et al (2017) Effect of the vegetation fire on backscattering: an investigation based on Sentinel-1 observations. IEEE J Sel Top Appl Earth Obs Remote Sens 10:4478–4492

Le W, Mingming J, Dameng Y, Jinyan T (2019) A review of remote sensing for mangrove forests: 1956–2018. Remote Sensing of Environment 231:111223. https://doi.org/10.1016/j.rse.2019.111223

Prasanna Kumar SP, Hariprasad P, Brijesh Singh S, Gowtham HG, Niranjana SR (2013) Structural and functional diversity of rhizobacteria associated with Rauwolfia spp. across the Western Ghat regions of Karnataka, India. World J Microbiol Biotechnol

Petar D, Hong W, Shuhong Q, Xiuneng L, Meng Z, Sisi L, Xin W (2025) Cross-modal fusion approach with multispectral LiDAR and SAR data for forest canopy height mapping in mountainous region. Phys Chem Earth (Pt A/B/C) 137:103819. https://doi.org/10.1016/j.pce.2024.103819

Rai K, Mishra N, Mishra S (2022) Forest fire risk zonation mapping using fuzzy overlay analysis of Nainital district. In: 2022 International mobile and embedded technology conference (MECON), pp 522–526

Suresh B, Roy A, Ridhika A (2018) Mapping of forest fire burned severity using the Sentinel datasets. ISPRS XLII(5):469–474

Wei W, Yanyu Y, Guochao C, Yuming Z, Xin Y, Lei Z, Yanning Z (2025) SAR remote sensing image segmentation based on feature enhancement. Neural Networks 185:107190. https://doi.org/10.1016/j.neunet.2025.107190

Zilin Y, Jiangping L, Tingchen Z, Bingbing L, Hui L (2024) L-Band synthetic aperture radar and its application for forest parameter estimation 1972 to 2024: a review. Plants 13(17):2511. https://doi.org/10.3390/plants13172511

Chapter 4
Revolutionizing Environmental Monitoring with Cutting-Edge Imaging Technologies

Md. Khalid Hasan Milu, Nishat Tasnim Safa, Samiha Mobaswira, Jaynal Abedin Tarun, Mahfuzul Islam, Israt Jahan, Md. Ashiquzzaman, Md. Arifur Rahman, Md. Farhadur Rahman, and Hasan Muhammad Abdullah

Abstract The rapid development of remote sensing technology has dramatically advanced our ability to monitor and examine changing environments precisely. Global satellite technologies, led by space agencies like NASA, ESA, and JAXA, along with private enterprises, advances climate monitoring, disaster resilience, and oceanic carbon sequestration, promoting environmental sustainability through innovative missions and shared data. This chapter explores the nuances and uses of thermal, hyperspectral, and multispectral infrared imaging in environmental assessment. Multispectral imaging, which collects data in many specific spectral bands using sensors like Landsat, Sentinel series, and others, is essential for classifying various land cover types and assessing vegetation indices. Characterized by the ability to collect records in a mass of contiguous spectral bands, hyperspectral imaging enables the identification and classification of specified material, enabled by sensors that include Hyperion, AVIRIS, and CASI. Using sensors like drones and satellites, thermal infrared imaging, which quantifies the thermal radiation emitted, is vital for monitoring Earth's temperatures and emissivity. The chapter will begin by explaining the essential standards of these imaging technologies, accompanied by an in-depth discussion on methodologies consisting of data acquisition techniques, preprocessing steps like radiometric and atmospheric corrections, and advanced record fusion methods that integrate multiple imaging modalities for a more suitable evaluation. Analytical methods, including spectral unmixing, algorithms, and thermal anomaly detection, could be explored. The chapter will then spotlight the various imaging technology programs in environmental evaluation. Thermal imaging

---

Md. K. H. Milu (✉) · N. T. Safa · S. Mobaswira · J. A. Tarun · M. Islam · I. Jahan · Md. Ashiquzzaman · Md. A. Rahman · Md. F. Rahman · H. M. Abdullah (✉)
GIS and Remote Sensing Lab, Bangabandhu Sheikh Mujibur Rahman Agricultural University, Gazipur, Bangladesh
e-mail: khalid.official.98@gmail.com

H. M. Abdullah
e-mail: hasan.abdullah@bsmrau.edu.bd

V. Saritha et al. (eds.), *Remote Sensing for Environmental Monitoring*,
https://doi.org/10.1007/978-981-96-5546-5_4

helps detect plant stress and water content; on the other hand, multispectral and hyperspectral data are used in vegetation and forest monitoring to evaluate vegetation health, identify species, and estimate biomass precisely. These technologies also monitor indicators including algal blooms, turbidity, and chlorophyll concentration to measure the quality of the water; thermal imaging detects thermal pollution and its consequences on aquatic ecosystems widely. Hyperspectral imaging describes soil features, including its organic matter and mineral content, whereas multispectral and thermal data mainly evaluate the ground cover's erosion, desertification, and deterioration. This chapter will also discuss the use of multispectral and hyperspectral data to evaluate the effects of climate change on various ecosystems, as well as the investigation of urban heat islands and their prevention through thermal infrared imaging, which is a very concerning issue nowadays. The utility and effectiveness of these technologies will be validated through real-world case studies that include urban heat island research in major cities, water assessment in water bodies, and deforestation monitoring in forest areas. Along with discussing the capacity of future use for imaging technology, the chapter will also discuss the critical present issues in remote sensing, including the large extent of data, processing complexity, and sensor constraints. Increasing environmental monitoring capabilities involves the advent of sensors with better resolutions, more robust data analytics, and the integration of artificial intelligence. This chapter aims to provide a thorough overview of multispectral, hyperspectral, and thermal infrared imaging in a categorized manner. It will also highlight the essential function these technologies play in improving our comprehension and management of environmental resources and their ability to cope with new environmental problems.

Keywords Remote sensing · Multispectral imaging · Hyperspectral imaging · Thermal infrared imaging · Environmental monitoring · Vegetation assessment · Climate change · Urban heat islands

4.1 Introduction

Environmental monitoring has traditionally been done using conventional methods such as manual field surveys, and basic remote sensing. Although these methods have served as the foundation for the collection of environmental data, they are becoming less suited to address the evolving demands of recent times. In-situ measurements are precise, but they are time-consuming and labor-intensive (Günther et al. 1995; Mertikas et al. 2021). Human errors will likely occur in manual field surveys useful for ground truthing purposes and cover only small areas (Johnson et al. 2015). Although spatial coverage was improved to some degree by the old remote sensing methods, it was still limited by low-resolution imagery; hence, use for small-scale environmental monitoring became impossible (Cohen and Goward 2004; Wang et al. 2021). Therefore, in these rapidly changing ecosystems, the methods often fail to provide high-resolution data on time for large-scale environmental management because they are usually old. The slow data collection process of traditional approaches also contributes greatly to the limitations that affect real-time decision-making. Further, it takes time to manually collect field data, and by the time the analysis starts; environmental conditions might have changed (Lillesand et al. 2015). Indistinctly, it is hard to track inaccessible and vast lands like mountains or huge water masses because of the little spatial resolution in conventional remote sensing (Schultz and Engman 2012). Therefore, there is a need for superior technologies that can offer quick, correct, and extensive environmental information.

How we monitor our environment has changed over these years predominantly due to advancements made with multispectral (MS), hyperspectral (HS), and thermal infrared imaging (TIR) techniques. This has led to an improvement in speed, precision, and scale of environmental monitoring because with them it is possible to capture very high-resolution multi-dimensional data instantaneously (Thenkabail et al. 2014). Multispectral imaging has evolved to detect slight changes in the environment like those linked with different water quality and vegetation health through capturing a few spectral bands (Zarco-Tejada et al., 2014). Hyperspectral imaging takes this a notch higher by capturing hundreds of spectral bands which help in identifying minute differences in material properties thus allowing for the rapid detection of environmental stressors (Vane and Goetz 1988; Matese et al. 2024).

Temperature variability data provided by thermal infrared imaging supplements the understanding of processes such as surface energy exchange and evapotranspiration. Essentially, these technologies can be said to possess three great advantages over others, namely; accuracy like no other means, availability of up-to-date information, and extensibility. For example, multispectral and hyperspectral imaging have been used to develop better predictive models for assessing vegetation health or tracking deforestation. These technologies have the potential to span great distances, allowing extensive surveillance that was impossible before this way of doing things. If such data is gathered and processed in real-time, it makes it possible for environmental officers to take timely measures against new risks regardless if they are climate change-dependent or manmade (Fuller et al. 2021; Liu et al. 2023).

This chapter explores the far-reaching environmental applications of multispectral, hyperspectral, and thermal infrared imaging. This technology has revolutionized precision agriculture through improved water usage and crop health monitoring (Fakhar and Khalid 2023). It is used in forestry to detect forest degradation and monitor carbon sequestration. Additionally, water resources management, urban planning, and climate change analysis will be discussed in this chapter thus showing how these technologies make it possible for proactive, efficient, and accurate environmental planning. Besides that; the application of these technologies in disaster management systems including early warning systems before destruction happens and post-disaster damage assessments will be given. They are very significant for resilience building at the level of ecosystems as well as human societies (Jensen 2013; Pande and Moharir 2023).

4.2 Imaging Technologies: Fundamentals and Evolution

4.2.1 *Multispectral Imaging*

Multispectral imagery is widely employed in satellite and airplane-based remote sensing, which outperforms conventional broadband visible–spectrum imagery for vegetation monitoring in many applications. Such vegetation indices (VIs) which incorporate near-infrared portions of the spectrum can therefore provide more extensive information in spectral discrimination between various plant types and growth stages. These VIs can be very valuable in the estimation of certain biophysical variables, such as vegetation productivity or Leaf Area Index (Wehrhan et al. 2016; Sahu et al. 2024), as well as for other kind of vegetation classification purposes (Ahmed et al. 2017). The use of multispectral images has proven effective in monitoring water bodies, and rivers, detecting changes, as well as extracting water features (Dash et al. 2002). Eutrophication is a scientific terminology that defines algal blooms and associated results induced through the response of aquatic ecosystems to massive nutrient loads. Chlorophyll-a (Chl-a) is the maximum essential photosynthetic pigment in phytoplankton organisms, which has been considered a great indicator of nutrient enrichment (Agwanda and Iqbal 2019; Singh et al. 2025, 2024).

The Sentinel-2 MultiSpectral Instrument (S2 MSI) provides precise spatial decisions and distinct spectral bands positioned to assess chlorophyll a, an indication of water quality and trophic condition that enables the monitoring of small water bodies. The retrieval of phytoplankton chlorophyll-a concentration (chl-a) from remotely sensed facts is one of the key problems in aquatic remote sensing. The spectral signature of chl-a is characterized by robust absorption within the blue (443 nm) and red wavelengths (close to 675 nm) and excessive reflectance in green (550–555 nm) and red-edge spectrum regions (685–710 nm). These spectral capabilities had been used to expand numerous band ratios to quantify chl-a concentration in inland and near-coastal transitional waters through empirical tactics, imparting timely and accurate information (Gitelson 1992).

4.2.2 *Hyperspectral Imaging*

The advantage of hyperspectral imaging is that there is no need for any sample expertise by the operating personnel since at each place a complete spectrum is recorded and postprocessing allows mining for all information in the dataset. Hyperspectral imaging provides ample data to deal with different agricultural issues such as the detection of diseases, weeds, and stress, monitoring of crops, transportation of nutrients, soil mineralogy, estimation of yield, and sorting applications (Ran et al. 2017). The improvement of precision agriculture could be facilitated by using the short, non-unfavorable, and real-time tracking of soil nutrient modifications that hyperspectral remote sensing can offer. Hyperspectral remote sensing technology can screen soil organic matter speedy, in actual time, and without inflicting damage, it has gained importance as a device for researchers assessing soil qualities (Panigrahi and Das 2018). The ratio of the reflected flux to the incoming flux at a particular wavelength is known as the spectral reflectance. The soil spectral characteristic curve is created by way of the way exceptional molecules within the soil take in and mirror mild at numerous wavelengths. By the usage of numerous spectrum transformation strategies to reduce noise in the unique spectral statistics and arrange the correlation among the spectral information and the chemical measurement values, hyperspectral records can be expressed more efficiently. In this way, soil organic matter, mineral content (N, P, K) as well as soil properties change can be monitored (Hong et al. 2018).

4.2.3 Thermal Infrared Imaging

Thermal imaging provides a discreet, scalable approach to monitoring soil surface temperatures based on the idea of soil moisture analysis. Thermal cameras are beneficial in figuring out field portions that can be below moisture stress given that they could hit upon temperature changes between wet and dry soil. By accurately applying water where it is needed, these facts can improve irrigation schedules, resulting in more green water usage and less waste. Thermal cameras are beneficial in figuring out field portions that can be below moisture stress given that they could hit upon temperature changes between wet and dry soil. By accurately applying water where it is needed, these facts can improve irrigation schedules, resulting in more green water usage and less waste. Infrared imaging is a kind of remote sensing where thermal cameras, especially with uncooled microbolometer detectors are used to measure the quantity of heat radiation radiated from the earth's surface. The geographic distribution of heat radiation is represented via thermograms, which might be thermal pictures taken by using the sensors. A map of radiant temperature is created while the statistics are corrected based on exclusive variables like object emissivity and surrounding situations (Frodella et al. 2020).

Thermal imaging can determine the effect of soil components, such as biochar, on soil temperature changes. This presents precious insights into how these methods can beautify soil best and water retention (Blanco et al. 2023).

4.3 Data Acquisition, Processing, and Integration

Accurate collection of environmental data historically has been a cornerstone to understanding ecological systems and is cheaper than other methods. It is still bound by localized measurements, slow processes, and human error. However, this poses a significant limitation as ground-based surveys can gather environmental data only for localized areas and largely depend on narrow spatial coverage and substantial time lags in the processing of data. Those traits make a big splash by capturing ecosystem changes over time (Fascista 2022).

Currently, we live in a technology-driven world, and only with the use of modern gadgets to gather accurate data. Satellites, on the other hand, offer a bird's-eye view—literally—with large-scale and continuous surveillance across Earth's geology; however, this technology also has trouble providing significant resolution due to both technical limitations and clouds. UAVs have low-altitude, short-duration, high-load advantages to obtain detailed and flexible data acquisition within a relatively limited range and time of flight; ground-based instruments can provide accurate local measurements but always occupy space. These all combined to develop one big environmental surveillance mechanism, thus improving overall data collection (Fagan and DeFries 2009). Nowadays, new and emerging technologies such as Landsat, Sentinel satellites, UAVs, and other ground-based systems fundamentally change the ways how data can be captured by facilitating automatic, as well as timely, data harvesting. Continuous monitoring at a high level of detail is possible using satellites, while quite narrow-area operations conducted using UAVs allow for conducting detailed surveys. On-site sensors provide immediate accurate data that is essential for the validation of simulations. Overall, these technologies increase the effectiveness and accuracy in a broad range of activities such as environmental monitoring, agriculture, and disaster relief (Zhang and Zhu 2023).

In such a context, however, it is of primary importance to address the quality and preparation of satellite, UAV, and ground-based data for analysis by using advanced pulp-processing techniques. Data cleaning, georeferencing, normalization, resampling, dimensionality reduction, image segmentation, and interpolation are a few of the key techniques that contribute to increasing data quality by increasing accuracy and reliability with these processes (Illarionova et al. 2022). Atmospheric And Radiometric Corrections That Improve the Reliability and Accuracy of Remote Sensing Data These corrections tackle long-standing issues to improve the quality and accuracy of data and allow for greater comparability between instruments and across different conditions. This is a fundamental requirement in sound environmental monitoring and analysis (Ahern Ran et al. 1987). Data fusion combines multispectral/hyperspectral and thermal data for increased accuracy, insights, and decision-making in environmental analysis. A Comprehensive model is more able to extract features and understand complex phenomena as a whole which makes it of great importance for agriculture, resource management, and environmental monitoring (Zhou et al. 2020).

Multispectral, Hyperspectral, and Thermal data integration enables a better understanding of environmental systems that reflect diversified properties and dynamic features. The fusion of both not only provides a sophisticated environmental assessment but also supports informed decision-making to advance deeper comprehension in understanding complex ecological relationships (Halog and Manik 2011).

Finally, there have been great advances made in data analytics techniques such as those involving AI and spectral unmixing which can greatly enhance the analyses of complex raster-based remote sensing datasets. The characteristic extraction, scalability, and accuracy of machine learning along with the systematic nature of

GEP were successfully merged to bring about data-driven decision-making and real-time monitoring in environmental management and research. Along the same line, AI-based spectral unmixing shortens delays in decision-making and enhances precision, efficiency, and adaptability to eventually support the sustainable management of natural resources (Janga et al. 2023).

So, the traditional way of collecting data has gone forward to conserve and document the historical record well, while current technologies provide real-time snapshot monitoring on a larger scale much more efficiently to help inform our management and conservation efforts.

4.4 Applications in Agriculture and Precision Farming

Traditional agricultural monitoring techniques, such as manual field surveys, have frequently been criticized for their labor-intensive nature and restricted scalability. These methods fall short of addressing the increased demand for precision in farm management, which necessitates quick and accurate reactions to crop stress, nutrient shortages, and pest outbreaks. Food security and environmental conservation are two of the most important concerns confronting the world today. Food production is predicted to increase by at least 70% by 2050 to support continued population growth, even though global agricultural acreage remains mostly unaltered (Sona et al. 2016). Precision agriculture is a technique for boosting productivity to meet rising food demand while lowering the economic and environmental expenses of food production (Sethy et al. 2022). Emerging technologies such as multispectral, hyperspectral, and thermal infrared imaging have revolutionized agricultural surveillance by providing precise, real-time data, significantly enhancing the effectiveness of precision agriculture. Traditional crop monitoring systems rely mainly on physical labor and visual inspections, which are sometimes inefficient for large-scale production. These systems are constrained by the time required to respond to stress indicators, which can affect crop health and output. Multispectral and hyperspectral imaging technologies are significant advances in remote sensing for agriculture. Multispectral imaging collects data from a small number of spectral bands, generally in the visible and near-infrared (NIR) ranges. Vegetation indices derived from multispectral data, such as the Normalized Difference Vegetation Index (NDVI), are widely used to assess crop vigor, identify diseases, and follow vegetation growth cycles. These techniques allow for early diagnosis of crop stress, leading to timely actions. Hyperspectral imaging, on the other hand, offers a more comprehensive view by collecting data from hundreds of small spectral bands. This higher spectral resolution allows for more advanced investigations of plant physiology, such as early identification of nutrient deficits and insect epidemics. For example, Vegetation Characterization is a crucial indication of ecosystem resilience to environmental change (Singh 2022).

More accurate fertilization is possible by utilizing certain spectral bands to detect early-stage shortages in potassium, phosphorus, and nitrogen. In addition, hyperspectral imaging improves disease identification by identifying subtle physiological changes in crops, which enables more targeted and early treatment than conventional techniques. More proactive and accurate agricultural management is made possible by the capacity to forecast pest and disease outbreaks using hyperspectral imaging. Because it can identify soil temperature and plant water stress, thermal infrared imaging has become more and more common in agricultural applications. When plants experience a shortage of water because of a soil water deficit or an excessive amount of atmospheric evaporative demand, it is referred to as drought stress or water-deficit stress (Gerhards et al. 2019). Planning irrigation schedules and identifying droughts depend heavily on surface temperature readings provided by thermal sensors. Because it makes accurate water management possible in real-time, thermal imaging is particularly helpful in drought-prone areas as it lowers waste and boosts crop yield. In addition to identifying water stress, thermal imaging is utilized to gauge soil temperature. Crop output, root growth, and seed germination are all significantly influenced by soil temperature. Thermal infrared (TIR) data can reveal additional information on the temperature and spectral emissivity of several environmental elements, including important minerals that create rocks and soil, particular gaseous constituents, and flora (Schlerf et al. 2012). By giving information regarding soil conditions, thermal infrared imaging can assist farmers in improving planting schedules and germination rates. Combining thermal imaging data with other precision agriculture technologies can help farmers manage their crops and soil more sustainably and efficiently, hence increasing resource utilization. A move away from labor-intensive procedures and toward data-driven decision-making has been made possible by the integration of multispectral, hyperspectral, and thermal imaging technologies into automated monitoring systems. Using drones, satellites, or stationary ground-based equipment, these remote sensing technologies are frequently used to gather data from vast agricultural areas. With the help of automated methods, broad territories can be efficiently administered with little to no human interaction thanks to real-time monitoring that eliminates the need for manual field surveys. Because they empower farmers to make data-driven, well-informed decisions that maximize crop health, productivity, and resource efficiency, these technologies are essential to the future of precision farming. Advanced imaging technology will become more crucial as agriculture develops in supporting sustainability and tackling global concerns like environmental preservation and food security (Zhang et al. 2020).

Case Study I: Effect of various nitrogen levels on maize yield performance using UAS technology

This study investigates the effect of different nitrogen levels on maize yield performance using Unmanned Aerial System (UAS) technology. By capturing multispectral imagery at various growth stages, UAS technology enabled precise monitoring of nitrogen's impact on crop health and productivity. Yield estimates, ranging from 0.07 to 9.1 tons/ha, were derived using advanced tools like R and QGIS (Fig. 4.1). The yield map revealed how nitrogen variability influenced maize growth, highlighting areas with both high and low productivity. This approach demonstrates the effectiveness of UAS technology in optimizing nitrogen use, improving yield performance, and promoting resource-efficient farming practices.

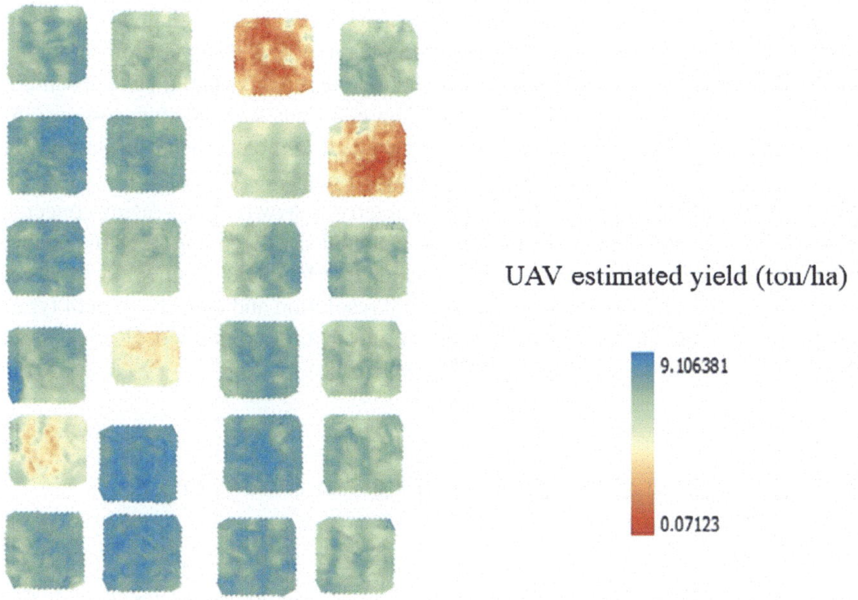

Fig. 4.1 UAV-based maize yield estimation using multispectral imagery for precision farming

4.5 Forest Monitoring and Conservation

4.5.1 Conventional Methods of Monitoring

For many years, deforestation rate and biodiversity changes have relied on surveys both forest ground and satellite-based. Ground-based surveys are quite efficient in such measurements as the diameter of the tree, identifying species, or even calculating forest density; however, they involve a lot of time, require many resources, and cover only a limited area. Remote sensing is cheap and effective through a wide range of optical, infrared, and radar pictures aimed at monitoring large areas, however, such data can be poor resolution-wise and clogged by clouds. The combination of ground-based and satellite-based approaches provides all-rounded forest management (Leckie 1990).

4.5.2 Limitations to Traditional Monitoring

Coverage and Reach Problems. In the case of field surveys, their geographical extent can be a limiting factor in tracking and monitoring wide stretches of land such as forests, whereas remote sensing provides general information but does not provide information of interest regarding small changes. Ground truthing provides accurate information in small regions, but because of its limited scope, it tends to make blanket assumptions over larger regions. At times remote sensing equipment yields images that are of a very low quality, thus making the sensitive parameters obscured. Furthermore, reaching the field to collect data contributes to the slow response to threats such as illegal logging and bushfires, while remote sensing is limited by the presence of clouds; thus, there are delays in analysis (Mitchell et al. 2017). Multispectral and hyperspectral imaging are relatively new remote sensing concepts that involve collecting data from multiple wavelengths, where imaging in the visible near-infrared and short-wave infrared for vegetation health is cheap while imaging in all wavelengths allows for very fine-spectral resolution and identification of types of species. Both techniques contribute to the physical interpretation of forests and their condition over time (Fischer and Kakoulli 2006).

4.5.3 Changing the Perspective of Forests Resilience Monitoring

The imaging epiphany that has taken over biomass estimation, assessment of vegetation state, and species composition includes multispectral and hyperspectral imaging. They promote management by allowing ancillary information on the forests to be captured much earlier than the deterioration of the forest becomes evident (Seidl et al. 2016).

4.5.4 Monitoring Carbon Stocks and How Deforestation Can Be Controlled

A combination of the two imaging techniques is necessary for forest carbon stock assessment as well as deforestation assessment by evaluating landscape biomass and vegetation status for facilitating efficient forestry management and reserve programs (Ling et al. 2017).

4.5.5 Thermal Infrared Imaging (TIR)

An Imaging technique using thermal radiation to evaluate forest vitality, find evidence of water stress, and map out temperature hotspots that may indicate areas of potential fire outbreak. It also measures the rate of evapotranspiration adding to the comprehension of forest dynamics. Though TIR offers useful information, it has a depth of spectral range limitation and is weather-dependent (Qi and Diakides 2003).

1. **Wildfire Management**: TIR in the management of wildfires aids in the prevention and suppression activities of wildfires by providing means of spotting the active fire in progress in real-time (Riggan 2003).
2. **Heat Stress Detection**: Heat Stress detection in forests aimed at alleviating risks of fire and drought is where TIR plays an important role. It aids in averting damages and appropriate management of resources in readiness for periods of prolonged dry weather thereby enhancing the health of the ecosystem and practices of management (Ghezzi et al. 2024).

So, the use of these modern monitoring methods facilitates better management of forests, combats problems related to the environment, and advances their conservation.

Case Study II: Structural difference of Sal-forest in different land cover based on GEDI-lidar

This particular scenario incorporates the identification of changes in land use land cover (LULC) over the Madhupur Sal Forest region between the years 2013 and 2023 using multi-spectral sensors, Landsat imagery, and random forest machine learning classification techniques (Table 4.1, Fig. 4.2). It was established that a considerable amount of deforestation took place in the Sal Forest area that measured 4962.78 eh in the year 2013 decreasing to 2937.24 eh in the year 2023, whereas several other lands were brought for agricultural purposes. The Random Forest algorithm provided a good classification and showed encroachment into farmlands and developments of homesteads. Also, the GEDI data added the dimension of the landscape's vertical profile showing lower canopy heights in farming areas as opposed to the rest of the forest. The study therefore demonstrates the effective role of remote sensing and machine learning in assessing forest loss and land use changes.

Table 4.1 Temporal changes in land cover: Madhupur Sal Forest (2013–2023)

Categories	2013 Area (ha)	2018 Area (ha)	2023 Area (ha)
Sal forest	4962.78	3300.03	2937.24
Rubber plantation	2686.77	2391.48	2243.61
Agriculture	6359.76	9228.24	8184.33
Homestead	22,618.89	21,646.26	23,209.02
Waterbodies	2153.79	1387.8	1525.41
Others	718.74	1546.92	1402.92
Total	39,500.73	39,500.73	39,502.53

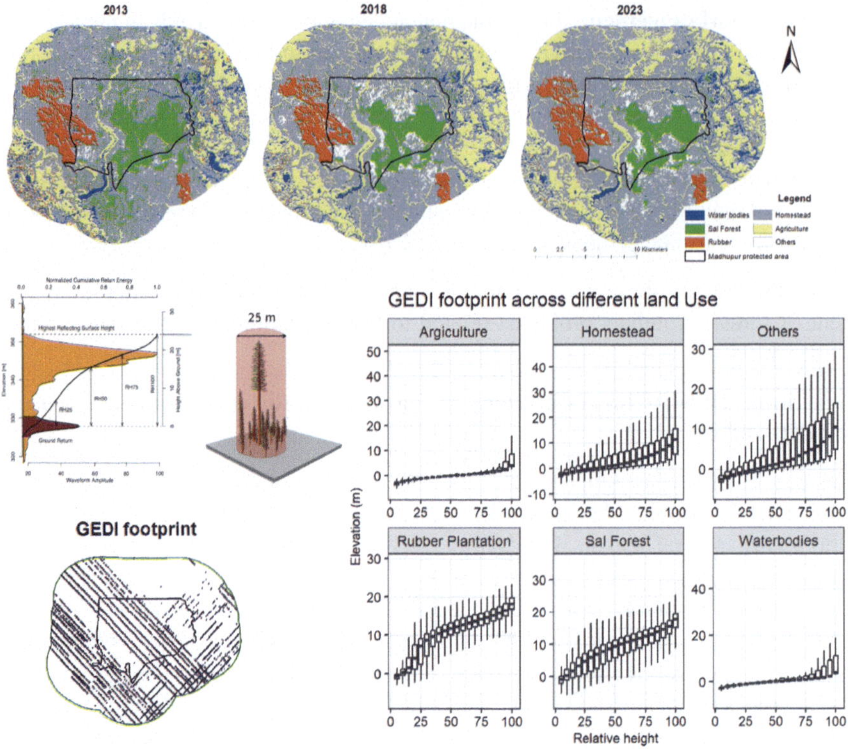

Fig. 4.2 Evaluating structural variations of Sal forests about different land cover types using GEDI lidar

4.6 Water Resource and Marine Science Applications

Oceans are the most important of all because they hold and store about 25 percent of the annual worldwide carbon emissions. Technologies that improve the effectiveness of monitoring this primary process are frontier developments. Applications under NASA's PACE (Plankton, Aerosol, Cloud, and ocean Ecosystem) satellite launched in 2024 will be enabling detailed monitoring of ocean color and phytoplankton populations-the critical indicators of what is happening in the ocean and possible carbon sinking (NASA 2024a). Photosynthesizing phytoplankton and seaweeds play a very significant role in the marine carbon cycle by absorbing CO_2. ESA's Sentinel-2 and Sentinel-3 missions take a step further in measuring seaweed biomass and natural kelp forests at a global scale, thereby giving critical information in terms of assessment of their role in climate change mitigation (ESA 2024a). Added to that is the GOSAT-2 mission by the Japan Aerospace Exploration Agency which monitors the methane and CO_2 concentrations over oceans; thus, providing emission events from volcanic activities, ocean-based industries, and bed sediments of the seabed. These missions collectively facilitate the assessment of oceanic sequestration of carbon and advancing the knowledge about the role of marine ecosystems in the fight against climate change (JAXA 2024a).

Conventional techniques are fundamentally established by sampling and observing water samples, and the analysis is conducted in a laboratory setting. Mistakes may transpire during the process of specimen preparation within the laboratory environment (Amrita et al. 2018). Traditional laboratory techniques require a longer time, incur higher costs, involve chemical materials, and do not provide real-time results (Zulkifli et al. 2018). Modern approaches offer greater advantages compared to traditional methods because they are capable of generating output results and conducting real-time analysis of water quality parameters. UAV-based remote sensing technology is mobile, flexible, and cost-effective. Its high spatial and temporal resolution is gaining popularity, making it ideal for small-scale water quality monitoring and quantitative analysis of water quality parameters (Hu and Zhou 2020).

UAV-based multispectral images are responsive to changes in dissolved oxygen and turbidity, making them useful for monitoring water quality parameters (Sun 2019). Hyperspectral imaging is useful for monitoring water quality parameters like Chlorophyll-a, Blooms, total phosphorus, and lake bathymetry (Premkumar et al. 2021; Wolny et al. 2020; Niu et al. 2021; Zhang et al. 2021). Hyperspectral imaging can effectively capture spectral signatures of marine flora, such as macroalgae and seagrasses, which serve as bioindicators of ecosystem health (Varini et al. 2024). Combining hyperspectral and multispectral data enhances spatial and spectral resolution, allowing for more accurate biodiversity assessments and the calculation of alpha and beta diversity. Low-cost hyperspectral imaging has been successfully used to assess coral reefs, demonstrating its potential for identifying habitat types and monitoring environmental changes (Teague et al. 2023).

The technique of thermal infrared remote sensing offers a compelling option for measuring water temperatures and analyzing spatial distributions of water temperature across various spatial scales (Handcock et al. 2006). Thermal imaging technology can accurately identify marine mammals, including killer whales, in coastal areas, which is essential for enforcing conservation measures like speed limits and area closures (Richter et al. 2023). Infrared thermography (IRT) plays a crucial role in observing how intertidal organisms react to heat stress, aiding in the identification of thermal refuges that are vital for species to adapt to climate change. In addition to aerial surveys, this technology can be utilized to generate comprehensive thermal maps of coastal areas. These maps aid in evaluating the appropriateness of habitats (Lathlean et al. 2017).

4.7 Urban and Regional Planning: Urban Heat Island Mitigation

4.7.1 Traditional Urban Environmental Monitoring Techniques

Conventional methods of monitoring the environment comprise the fixed station (Manfreda et al. 2018) and sampling which gives regional information in the town but fails to record changes in the environmental gradient (Nabiollahi et al. 2018; Reba and Seto 2020). These methods are time-consuming are confined to selective areas only and provide only point estimates. But, as cities expand and environmental issues evolve, there are other more extensive and real-time applicative methods, called remote sensing (Li et al. 2020), that have developed to augment these two classical ones, allowing for broader survey, monitoring, and ability for more accurately develop observational scales (Cao and Lam 2023) and quantification of environmental conditions (de Araujo Barbosa et al. 2015).

4.7.2 TIR Imaging and UHI Monitoring

TIR has become a powerful tool for UHI monitoring (Coutts et al. 2016; Shi et al. 2021) through the detection of the emitted infrared radiation for measurement of the surface temperature in real time (Wan et al. 2021). This allows proper distribution mapping of heat in the urban area, and hence, defines areas of high heat density and prevents heat through modifying building roofs to 'cool roofs and planting more trees. Communities such as LA have been able to use TIR (Fu and Weng 2015) and measure the effectiveness of reflective surface measures by reducing the surface temperature.

- **Real-Time Data Analytics and AI**: With AI and machine learning able to parse vast thermal datasets in real-time (Sadiqbatcha et al. 2021), the system can detect parts or regions with atypical high-temperature activity (Ahmed et al. 2024) and even project future heat development, giving opportunities for early caloric interference.
- **Community Thermal Sensors**: Small thermal sensors installed by the community extend the scope of satellite and drone data with ground-level information (Zhang et al. 2018) that can benefit residents of policymakers in combating UHI impact and pursuing energy conservation.

4.7.3 Multispectral and Hyperspectral Imaging in Urban Planning

Multispectral and hyperspectral imaging technologies are accurately mapped for the environment with spectral layers across multiple bands (Lu et al. 2020) to Track Air Quality, Urban Infrastructure, and land usage in detail (Nisha and Anitha 2022). These sensors can measure the presence of particulate matter, NO_2, and SO_2 (He et al. 2019; Kheiralipour et al. 2024) important for understanding urban air pollution. Hyperspectral imaging also aids in infrastructure monitoring (Shimoni et al. 2019) because signs of abrasion of roads, bridges, and buildings are early detected. These sensors help in measuring the impacts of some extent of urbanization on natural environments (Lynch et al. 2020). These can help city planners bring sustainable management to urban growth through the detecting of changes in these areas as well as help ensure clean water supplies through improvements of multispectral and hyperspectral imaging (Ghezzi et al. 2024).

Case Study III: Urban Heat Island

UHI is combated by utilizing Thermal Infrared (TIR) imagery or using LA's Urban Heat Island critics such as the "Cool Roof" program annuals or financing vegetation in the urban heat island. Information gathered from TIR was used to plot heat density and evaluate reflective rooftops that caused cooling of between 3 degrees within the local region (Santamouris et al. 2012). Additional parks and open spaces, as well as the planting of trees in the specific locations impacting heat island effects, the reduction was recorded at 1.9 °C according to the Los Angeles Urban Cooler Collaborative (2020). Supplemental entrenchment utilizing TIR sensors fixed to satellites and drones, including information from the Landsat 8 satellite, enabled real-time monitoring of these intercessions. Through combined efforts, consumption of energy for cooling has been reduced by 15 percent across the entire city (Sailor et al. 2021).

All in all, TIR, multispectral, and hyperspectral sensors are among the most recent imaging technologies that hold the potential for substantial improvement in environmental monitoring with real-time high-resolution data necessary for tackling urban issues for sustainable development.

4.8 Climate Change Monitoring and Earth Observation

The old ways of watching climate change like tracking surface temps or basic remote sensing aren't always good enough anymore—they miss too much nuance! Things like desertification or changing vegetation patterns often slip through the cracks using just these methods. But now things have changed—thanks to high-tech tools like thermal infrared imaging as well as multispectral and hyperspectral imaging that offer extensive real-time data needed for meaningful insights into climate changes. Traditional techniques mostly deal with surface temperature readings paired with limited satellite data; they lack the overall resolution needed at times for accuracy when spotting crucial climate shifts—like soil degradation or vegetation health deterioration. As climate change speeds up—finer instruments are desperately required! They provide precise updates on how ecosystems cope. By gathering various wavelengths through multispectral and hyperspectral imaging—we can see ecosystem changes incredibly well! Researchers have found that we might be nearing dangerous climate tipping points right here on Earth (Ustin and Middleton 2021). Hyperspectral insights see even tiny shifts happening within soil makeup while also seeing signs of water quality drops or vegetation stress early before it's too late compared to just a handful of indicators seen through multispectral strategies (Rast and Painter 2019).

These breakthroughs help monitor plant health changes linked directly back to global climate patterns as we map out erosion events or desert expansion there also spot carbon emissions from struggling veggie life often linked closely together by respiratory rates. Especially handy during 'heat-wave moments' that happen around polar spots where thawing permafrost patterns lead directly concerning melting glaciers affected heavily by rising atmospheric temps combined assessments–this utilizes heat emissions observed across broad surfaces providing insight into warming impacts felt globally. Merging these three amazing techniques allows us to further examine environments thoroughly observing how they react dynamically to stream-energy keeping track all along adjusting strategies toward resiliency moving forward towards sustainable initiatives.

Environmental monitoring is in a position to make unprecedented advancements in the coming century because of collaborations between space agencies and private enterprises, creating all new approaches to satellite technologies in the field to tackle climate change and encourage sustainable development. For example, the NASA GeoXO mission on Geostationary Extended Observations is scheduled to kick off in about 2032 and continue into the 2050s, to improve atmospheric and oceanic measurements, weather forecasting, and analysis of climate systems. Agreed incidental document under NEON (Near-Earth Orbit Network), a jointly funded NASA and NOAA project will have QuickSounder spacecraft launched in 2026, using its small and medium-sized satellites to revolutionize disaster management and weather forecasting (NASA 2024b). "Copernicus Sentinel-2C," ESA's program of the continuation of the largest environmental monitoring program in the world, launched in 2024, equipped exclusively for deforestation and urban growth monitoring (ESA 2024b). Another advanced land observing satellite, launched by JAXA in 2024, simultaneously augments Japanese satellite missions other than the nationally launched satellites for disaster response and environmental monitoring ability (JAXA 2024b).

Private sector intervention in environmental monitoring is quite significant. SpaceX was one of the private firms that pioneered launching a satellite into orbit for environment monitoring in 2024, MethaneSAT, which involves the initiative of shoring up data on the world's methane emissions as part of activities critical to addressing climate change. Accompanying the launch of MethaneSAT is that of the Weather System Follow-on Microwave (WSF-M) satellite, which serves to boost environmental observations from Sun-synchronous orbit (Environmental Defense

Fund 2024; AmericaSpace 2024). The €10.6 billion project IRIS[2], which combines secure satellite communication and environmental monitoring, is proof of Europe awaiting more holistic space-based solutions and is expected to be completed within the year 2030 (ESA 2024c).

4.9 Ecosystem and Biodiversity Conservation

4.9.1 Traditional Ecosystem Monitoring: Challenges

Consequently, conservationists have been dependent for some time upon basic field methods for ecosystem health assessments, requiring extensive surveys and collections of specimens (Hughes and Peck 2008). Although useful as they provide high confidence in having ground truth image matching of the local data, these methods are time consuming, labor intensive and suffer from human error (Faheem et al. 2024). Geological biodiversity assessments have substantial geographical data gaps with entire ecosystems such as rainforests and wetlands having inadequate monitoring (Brown et al. 2024; Junk et al. 2024).

4.9.2 Role of Multispectral and Hyperspectral Imaging:

For remote sensing, multispectral or hyperspectral imaging can provide data points beyond what is seen in our sun's output. Multispectral images access a few separated bands (i.e. band, green, near-infrared) (De Petris et al. 2024), whereas hyperspectral images collect hundreds of contiguous bands offering comprehensive reflectance data (Tejasree and Agilandeeswari 2024). These tools make it possible to monitor slight changes in vegetation health, soil moisture, or water quality on an ecosystem scale (Adam et al. 2010).

1. **Mapping Species and Habitat Types**: One major application of sensors tailored for hyperspectral imaging is species mapping (Matese et al. 2024) and habitat type mapping (McCraine et al. 2024) at landscape scales, as each spectral signature carries information on the particular species (Hesketh and Sánchez-Azofeifa 2012). It is especially useful in communities like rainforests where traditional surveys may simply miss rare species (Pang et al. 2024).
2. **Vegetation Health and Land Use Monitoring**: Multispectral and hyperspectral imaging can detect signs of stress from drought, disease, or pollution in plants by measuring near-infrared reflectance (Zhang et al. 2024) and allowing for preemptive risk assessments in the environment. Moreover, these techniques enable the monitoring of land-use modification (de-as well as urbanization), therefore greatly enhancing conservation efforts (Yang et al. 2024).

4.9.3 The Use of Thermal Infrared Imaging in Wildlife and Habitat Groups Conservation

Thermal infrared imaging (TIR) involves mapping the infrared emitted by an object (Upadhyay et al. 2024) to determine thermal signatures of the ecosystem (Smigaj et al. 2023), or monitor animal activities (Barrios et al. 2024). Peculiar to TIR, it is possible to determine temperature differences within landscapes, which are important for relating species distribution and environmental conditions (Smigaj et al. 2023). TIR is best used in monitoring animals that are active at night as it does not disturb the flow of movement. It is important to monitor the big predator species as well as the endangered species to understand their movements and the requisite environments (Dawlings 2024).

Applications

1. **Illegal Logging Detection in Brazil**: In the case of thermal and multispectral imagery, they have succeeded in halting whatalué logging in the Amazon (Souza et al. 2005), thus avoiding large tracts of forest being cut down.
2. **Coral Reef Health Monitoring**: Focal plane array hyperspectral imagery that is taken of the Great Barrier Reef tells the conservationist common occurrences such as coral bleaching and algal blooms (Hafizt et al. 2023).
3. **Deforestation in Tropical Rainforests**: Remotely sensed data by multispectral cameras addresses the effects of deforestation on the biome and contributes to reforestation strategies (Haq et al. 2024).
4. **Wetland Restoration**: TIR aids in the evaluation of temperatures and water levels concerning wetlands and drives restoration especially in sensitive areas like the Everglades (Reid Nichols et al. 2025).

4.10 Disaster Management and Environmental Risk Assessment

Traditional knowledge is usually restricted to particular communities and areas, with little to no wide-ranging, networked supervision. Since the risks and consequences of disasters often cross local boundaries, traditional approaches struggle to provide an accurate large-scale viewpoint (Macnight Ngwese et al. 2018). The ability of sensors and ground-based surveys to deliver timely information in a quickly changing crisis may be constrained. Disasters are frequently detected and monitored more quickly across a larger area when remote sensing techniques employing satellite or aerial imagery are used (Ye 2022).

Remote sensing monitoring is widely utilized in various disaster management fields and offers clear advantages over traditional methods regarding timeliness, space, and affordability (Abdullah et al. 2019). The use of multispectral images has proven effective in monitoring water bodies, and rivers, detecting changes, as well as extracting water features (Buma et al. 2018). Optical remote sensing can also be utilized to map flooded areas enabling quick and accurate identification of hazardous locations and facilitating the implementation of flood coping methods and response operations even in the presence of cloud cover. Thorough information on flood inundation styles may be acquired using combining multispectral and hyperspectral facts with other remote sensing technologies, along with Cyclone Global Navigation Satellite System (CYGNSS) (Gerlein-Safdi and Ruf 2019; Zhang et al. 2020). Hyperspectral imaging has been used to identify drought stress in plants by examining spectral signatures and indicators such as the red-edge slope. This enables the early detection of the effects of drought on vegetation. In the visible and near-infrared spectral regions, low-cost image hyperspectral cameras can now collect high-resolution spectral data, creating new opportunities for high-throughput plant phenotyping. This makes it possible to fast and effectively determine how agriculture is responding to environmental stressors like drought. The use of less expensive, portable hyperspectral cameras is one capability improvement that can raise the technology accessibility for programs including environmental tracking and disaster preparedness (Genangeli et al. 2023). Ground-based thermal infrared (TIR) remote sensing is widely adopted for volcanological research and surveillance. They are used to investigate volcanic plumes and gases, lava flows, lava lakes, and fumarole fields. Increasing interest has been in developing better monitoring instruments and processing long TIR time series of infrared (IR) images on volcanic areas. That makes it possible to monitor temporal variations in thermal anomalies on the surface, and potentially more eruptive activity (Sansivero and Vilardo 2019). Infrared (IR) cameras can measure heat radiation emitted from individuals or groups of objects in the immediate area. Being able to do so will identify these or better thermal sources as a very legitimate target, and at the end of the day, that gives them equal value and utility in both daytime and nighttime scenarios. Multiple fire detection scenarios could be covered by this, allowing us to capture data at various resolutions. Thermal Cameras allow hotspot detection and shorter fire response times it is of even greater importance in places where there is no satellite surveillance, such as remote areas (Carta et al. 2023).

By measuring the temperature distribution and detecting radiation energy on the exterior walls of buildings, we can find cracks or fractures using infrared thermal imaging technology After a tragedy, this type of detection is important to doing away with capability threats and safeguarding people's property. This technology is useful for nighttime emergency rescue operations following earthquakes and different calamities because it can oppose the temperature facts of a ground item's floor by detecting its emission energy, and is unaffected by harsh weather. For emergency monitoring following failures, accumulating wall fracture facts and crack records from broken homes in regions of intense screw-ups offers a particular advantage (Zhang et al. 2020).

4.11 Future of Remote Sensing: Technological Advancements and Challenges

Remote sensing data, with its heterogeneous and sometimes incomplete nature, presents a complex challenge for analysis (Murugan and Haldorai 2020). Generating high-quality pixel-level labels for deep learning models remains a significant challenge due to the resource-intensive nature of the process (Li et al. 2024). The remarkable increase in remote sensing big data underscores the need for advanced computational techniques, such as machine learning, to effectively process and analyze the data (Zhang et al. 2020).

Compared to conventional techniques, machine learning algorithms are better at handling the volume and complexity of remote sensing data, resulting in accurate and reliable predictions. Integration of Machine learning with remote sensing data enables us to predict climatic and environmental conditions, advanced resources, and disaster management. Thus, scientific understanding and practical solutions for global challenges are achieved, enhancing the necessity of further research (Shaik et al. 2024). Artificial intelligence (AI)–based models and techniques are frequently employed to improve the efficiency of remote sensing technology. Due to their great performance and effectiveness, deep learning (DL) models are the most studied AI-based models (Chen et al. 2023). As deep learning (DL) approaches perform exceptionally well in feature extraction, data categorization, and interpretation, they have been used in many studies on remote sensing technology. Building a high-performing deep learning model for certain use cases necessitates a large amount of pertinent labeled data. DL approaches are very helpful in remote sensing since massive data processing is a major component of most applications of these technologies (LeCun et al. 2015). The implementation of low-cost sensors for the environment has received a lot of interest because of its low cost and potential to expand environmental monitoring networks. Calibration is difficult for low-cost sensors because of the variety of sensing materials, transducer designs, and environmental factors (Salehabadi et al. 2023). The Fraunhofer Lighthouse Project 'eHarsh' is a collaboration of eight Fraunhofer Institutes to address the challenges of extreme hostile environments (Kappert et al. 2022). Environmental monitoring continues to gain popularity due to its major impact on natural resource management, the economy, and human life and health. While these systems play a vital role in our society, their implementation may generate various types of security and privacy concerns, thereby hindering the establishment of potential applications related to the environment (Vimercati et al. 2013). The implementation of sensors for workplace exposure assessment highlights issues regarding data exploitation and the necessity for the involvement of stakeholders in the process of making decisions. The possibility of privacy violations caused by ongoing monitoring emphasizes the significance of transparency and ethical concerns in sensor applications (Goede et al. 2021).

4.12 Conclusion

Multispectral, hyperspectral, and thermal infrared imaging technologies have revolutionized environmental monitoring through their unmatched levels of precision, real-time information delivery, and scalability. Together with traditional methods, these innovations provide better options for monitoring a wider range of changing ecosystems. By using multispectral imaging, it is possible to monitor the health of plants and the quality of water through different spectral bands that are available (Zarco-Tejada et al. 2014). Hyperspectral imaging allows to capture of hundreds of spectral bands which makes it possible to detect small environmental stressors, e.g. soil degradation and early signs of crop disease (Vane and Goetz 1988). The understanding of surface energy exchanges and processes such as evapotranspiration is enhanced, which are important in water resource management and climate studies, through thermal infrared imaging that adds another vital dimension by capturing temperature variations (Dash et al. 2002). All these technologies together are a multidimensional way to deal with the environment, helping to take action in different areas such as farming, timber production, water supply and city planning (Lambers et al. 2008).

The prospects of these imaging technologies seem more hopeful. Due to rapid sensor technology advancements, extensive and continuous global ecology monitoring can be done in real time and based on ecosystem changes (Thenkabail et al. 2014). In addition, the inclusion of artificial intelligence (AI) and machine learning within these systems is going to strengthen their analytic capability, allowing them to automatically find and anticipate the changes that occur in the environment (Mutanga and Kumar 2007). Over time, integrating artificial intelligence along with hyperspectral and thermal infrared information can lead to heightened precision in early warning systems for natural calamities, enhanced resource management, and sustainable city planning. These technologies will be deployed increasingly often for data-centric insights given the expanding scope of environmental issues and their complexity (Anderson et al. 1997).

Competing Interests The authors declare no potential conflicts of interest.

References

Abdullah HM, Islam I, Miah MdG, Ahmed Z (2019) Quantifying the spatiotemporal patterns of forest degradation in a fragmented, rapidly urbanizing landscape: a case study of Gazipur, Bangladesh. Remote Sens Appl Soc Environ 13:457–465

Adam E, Mutanga O, Rugege D (2010) Multispectral and hyperspectral remote sensing for identification and mapping of wetland vegetation: a review. Wetl Ecol Manag 18(3):281–296

Agwanda PO, Iqbal MM (2019) Engineering control of eutrophication: potential impact assessment of wastewater treatment plants around Winam Gulf of Lake Victoria in Kenya. J Coast Res 91(sp1):221

Ahern Ran FJ, Brown RJ, Cihlar J, Gauthier R, Murphy J, Teillet PM (1987) Review article radiometric correction of visible and infrared remote sensing data at the Canada Centre for Remote Sensing. Int J Remote Sens 8(9):1349–1376

Ahmed OS, Adam Shemrock RW, Chabot D, Dillon C, Williams G, Franklin SE (2017) Hierarchical land cover and vegetation classification using multispectral data acquired from an unmanned aerial vehicle. Int J Remote Sens 38(8–10):2037–2052

Ahmed I, Asif M, Alhelou HH, Khalid M (2024) A review on enhancing energy efficiency and adaptability through system integration for smart buildings. J Build Eng 89:109354. https://doi.org/10.1016/j.jobe.2024.109354

AmericaSpace (2024) SpaceX launches next-generation environmental monitoring satellite to orbit—AmericaSpace. https://www.americaspace.com/2024/04/11/spacex-launches-next-generation-environmental-monitoring-satellite-to-orbit/. Accessed 21 Dec 2024

Amrita CM, Babiyola D, Nadu T (2018) Analysing the water quality parameters from traditional to modern methods in aquaculture. Int J Sci Environ Technol 7(6):1954–1961

Anderson MC, Norman JM, Diak GR, Kustas WP, Mecikalski JR (1997) A two-source time-integrated model for estimating surface fluxes using thermal infrared remote sensing. Remote Sens Environ 60(2):195–216

Barrios DB, Valente J, van Langevelde F (2024) Monitoring mammalian herbivores via convolutional neural networks implemented on thermal UAV imagery. Comput Electron Agric 218:108713

Blanco V, Willsea N, Campbell T, Howe O, Kalcsits L (2023) Combining thermal imaging and soil water content sensors to assess tree water status in pear trees. Front Plant Sci 14:1197437

Brown C, Sjögersten S, Ledger MJ, Parish F, Boyd D (2024) Remote sensing for restoration change monitoring in tropical peat swamp forests in Malaysia. Remote Sens 16(15):2690

Buma WG, Lee S-I, Seo JY (2018) Recent surface water extent of lake Chad from multispectral sensors and GRACE. Sensors 18(7):2082

Cao C, Lam NS-N (2023) Understanding the scale and resolution effects in remote sensing and GIS. In: Scale in remote sensing and GIS. Routledge, pp 57–72

Carta F, Zidda C, Putzu M, Loru D, Anedda M, Giusto D (2023) Advancements in forest fire prevention: a comprehensive survey. Sensors 23(14):6635

Chen YN, Fan KC, Chang YL, Moriyama T (2023) Special issue review: artificial intelligence and machine learning applications in remote sensing. Remote Sens 15(3):1–10

Cohen WB, Goward SN (2004) Landsat's role in ecological applications of remote sensing. Bioscience 54(6):535

Coutts AM, Harris RJ, Phan T, Livesley SJ, Williams NS, Tapper NJ (2016) Thermal infrared remote sensing of urban heat: hotspots, vegetation, and an assessment of techniques for use in urban planning. Remote Sens Environ 186:637–651

Dash P, Göttsche F-M, Olesen F-S, Fischer H (2002) Land surface temperature and emissivity estimation from passive sensor data: theory and practice-current trends. Int J Remote Sens 23(13):2563–2594

Dawlings F (2024) Novel use of remote sensors to improve population monitoring and ecological understanding of threatened vertebrates for the purpose of conservation management. Ph.D. Thesis, Monash University

de Araujo Barbosa CC, Atkinson PM, Dearing JA (2015) Remote sensing of ecosystem services: a systematic review. Ecol Indic 52:430–443

De Petris S, Ruffinatto F, Cremonini C, Negro F, Zanuttini R, Borgogno-Mondino E (2024) Exploring the potential of multispectral imaging for wood species discrimination. Eur J Wood Wood Prod 82:1541–1550

Environmental Defense Fund (2024) MethaneSAT now in orbit after SpaceX launches groundbreaking mission to protect the climate. https://www.edf.org/media/methanesat-now-orbit-after-spacex-launches-groundbreaking-mission-protect-climate. Accessed 22 Dec 2024

ESA (2024a) The Sentinel missions. https://www.esa.int/Applications/Observing_the_Earth/Copernicus/The_Sentinel_missions. Accessed 21 Dec 2024

ESA (2024b) ESA releases new strategy for Earth observation. https://www.esa.int/Applications/Observing_the_Earth/ESA_releases_new_strategy_for_Earth_observation. Accessed 21 Dec 2024

ESA (2024c) ESA confirms kick-start of IRIS^2 with European Commission and SpaceRISE | ESA CSC. https://connectivity.esa.int/news/esa-confirms-kickstart-iris%C2%B2-european-commission-and-spacerise. Accessed 22 Dec 2024

Fagan M, DeFries R (2009) Measurement and monitoring of the world's forests: a review and summary of remote sensing technical capability, 2009–2015. https://media.rff.org/documents/RFF-Rpt-Measurement20and20Monitoring_Final.pdf

Faheem MA, Zafar N, Kumar P, Melon MMH, Prince NU, Al Mamun MA (2024) AI and robotic: about the transformation of construction industry automation as well as labor productivity. Remit Rev 9(S3):871–888

Fakhar M, Khalid M (2023) Satellites to agricultural fields: the role of remote sensing in precision agriculture. Biol Agric Sci Res J 2023(1):14

Fascista A (2022) Toward integrated large-scale environmental monitoring using WSN/UAV/crowdsensing: a review of applications, signal processing, and future perspectives. Sensors 22(5):1824

Fischer C, Kakoulli I (2006) Multispectral and hyperspectral imaging technologies in conservation: current research and potential applications. Stud Conserv 51(sup1):3–16

Frodella W, Lazzeri G, Moretti S, Keizer J, Verheijen FGA (2020) Applying infrared thermography to soil surface temperature monitoring: case study of a high-resolution 48 h survey in a Vineyard (Anadia, Portugal). Sensors 20(9):2444

Fu P, Weng Q (2015) Temporal dynamics of land surface temperature from Landsat TIR time series images. IEEE Geosci Remote Sens Lett 12(10):2175–2179

Fuller MR, Ebersole JL, Detenbeck NE, Labiosa R, Leinenbach P, Torgersen CE (2021) Integrating thermal infrared stream temperature imagery and spatial stream network models to understand natural spatial thermal variability in streams. J Therm Biol 100:103028

Genangeli A, Avola G, Bindi M, Cantini C, Cellini F, Grillo S, Petrozza A, Riggi E, Ruggiero A, Summerer S, Tedeschi A, Gioli B (2023) Low-cost hyperspectral imaging to detect drought stress in high-throughput phenotyping. Plants 12(8):1730

Gerhards M, Schlerf M, Mallick K, Udelhoven T (2019) Challenges and future perspectives of multi-/hyperspectral thermal infrared remote sensing for crop water-stress detection: a review. Remote Sens 11(10):1240

Gerlein-Safdi C, Ruf CS (2019) A CYGNSS-based algorithm for the detection of inland waterbodies. Geophys Res Lett 46(21):12065–12072

Ghezzi MD, Napolitano F, Casas-Alvarado A, Hernández-Ávalos I, Domínguez-Oliva A, Olmos-Hernández A, Pereira AMF (2024) Utilization of infrared thermography in assessing thermal responses of farm animals under heat stress. Animals 14(4):616

Gitelson A (1992) The peak near 700 nm on radiance spectra of algae and water: relationships of its magnitude and position with chlorophyll concentration. Int J Remote Sens 13(17):3367–3373

Goede H, Kuijpers E, Krone T, Le Feber M, Franken R, Fransman W, Duyzer J, Pronk A (2021) Future prospects of occupational exposure modelling of substances in the context of time-resolved sensor data. Ann Work Expo Health 65(3):246–254

Günther O, Radermacher FJ, Riekert W-F (1995) Environmental monitoring: models, methods, and systems. In: Avouris NM, Page B (eds) Environmental informatics. Springer, Netherlands, Dordrecht, pp 13–38

Hafizt M, Adi NS, Munawaroh M, Wouthuyzen S, Adji AS (2023) Coral reef health index calculation from remote sensing data: A review. Int J Conserv Sci 14(1):247–264

Halog A, Manik Y (2011) Advancing integrated systems modelling framework for life cycle sustainability assessment. Sustainability 3(2):469–499

Handcock RN, Gillespie AR, Cherkauer KA, Kay JE, Burges SJ, Kampf SK (2006) Accuracy and uncertainty of thermal-infrared remote sensing of stream temperatures at multiple spatial scales. Remote Sens Environ 100(4):427–440

Haq B, Jamshed MA, Ali K, Kasi B, Arshad S, Kasi MK, Ali I, Shabbir A, Abbasi QH, Ur-Rehman M (2024) Tech-driven forest conservation: combating deforestation with Internet of Things, artificial intelligence, and remote sensing. IEEE Internet Things J 11(14):24551–24568

He X, Xu X, Zheng Z (2019) Optimal band analysis of a space-based multispectral sensor for urban air pollutant detection. Atmosphere 10(10):631

Hesketh M, Sánchez-Azofeifa GA (2012) The effect of seasonal spectral variation on species classification in the Panamanian tropical forest. Remote Sens Environ 118:73–82

Hong Y, Chen Y, Yu L, Liu Y, Liu Y, Zhang Y, Liu Y, Cheng H (2018) Combining fractional order derivative and spectral variable selection for organic matter estimation of homogeneous soil samples by VIS–NIR spectroscopy. Remote Sens 10(3):479

Hu ZT, Zhou Y (2020) Research on urban water quality monitoring method based on low altitude multispectral remote sensing. Geospat Inf 18:4–8

Hughes RM, Peck DV (2008) Acquiring data for large aquatic resource surveys: the art of compromise among science, logistics, and reality. J North Am Benthol Soc 27(4):837–859

Illarionova S, Shadrin D, Ignatiev V, Shayakhmetov S, Trekin A, Oseledets I (2022) Estimation of the canopy height model from multispectral satellite imagery with convolutional neural networks. IEEE Access 10:34116–34132

Janga B, Asamani G, Sun Z, Cristea N (2023) A review of practical AI for remote sensing in earth sciences. Remote Sens 15(16):4112

JAXA (2024a) GOSAT-2 (Greenhouse gases Observing Satellite-2)/Ibuki-2—eoPortal. https://www.eoportal.org/satellite-missions/gosat-2#gosat-2-greenhouse-gases-observing-satellite-2--ibuki-2. Accessed 21 Dec 2024

JAXA (2024b) Status of advanced land observing satellite-4 "DAICHI-4" (ALOS-4). In: JAXA Jpn. Aerosp. Explor. Agency. https://global.jaxa.jp/press/2024/07/20240701-2_e.html. Accessed 22 Dec 2024

Jensen JR (2013) Remote Sensing of the Environment: an earth resource perspective. Pearson New International Edition. Pearson Education

Johnson JB, Reynolds HT, Mycoff JD (2015) Political science research methods. SAGE Publications

Junk WJ, Piedade MTF, Schöngart J, Da Cunha CN, Goncalves SRA, Wantzen KM, Wittmann F (2024) Riparian wetlands of low-order streams in Brazil: extent, hydrology, vegetation cover, interactions with streams and uplands, and threats. Hydrobiologia 851(7):1657–1678

Kappert H, Schopferer S, Saeidi N, Döring R, Ziesche S, Olowinsky A, Naumann F, Jägle M, Spanier M, Grabmaier A (2022) Sensor systems for extremely harsh environments. IMAPSource Proceedings 2022 (HiTEN): 1–13. https://doi.org/10.4071/001c.89680

Kheiralipour K, Al-Ansari N, Dibs H (2024) Monitoring air quality using ground and remote sensing based imaging techniques. In: The future of imaging technology. Nova Science Publishers, pp 2140

Lambers H, Chapin FS, Pons TL (2008) Plant physiological ecology. Springer, New York, New York, NY

Lathlean JA, Seuront L, Ng TPT (2017) On the edge: the use of infrared thermography in monitoring responses of intertidal organisms to heat stress. Ecol Indic 81:567–577

Leckie DG (1990) Advances in remote sensing technologies for forest surveys and management. Can J for Res 20(4):464–483

LeCun Y, Bengio Y, Hinton G (2015) Deep learning. Nature 521(7553):436–444

Li W, Batty M, Goodchild MF (2020) Real-time GIS for smart cities. Int J Geogr Inf Sci 34(2):311–324

Li B, Gong A, Zhang J, Fu Z (2024) From image-level to pixel-level labeling: A weakly-supervised learning method for identifying aquaculture ponds using iterative anti-adversarial attacks guided by aquaculture features. Int J Appl Earth Obs Geoinformation 132:104023

Lillesand T, Kiefer RW, Chipman J (2015) Remote sensing and image interpretation. Wiley

Ling F, Foody G, Du H, Ban X, Li X, Zhang Y, Du Y (2017) Monitoring thermal pollution in rivers downstream of dams with Landsat ETM+ thermal infrared images. Remote Sens 9(11):1175

Liu Y, Sun L, Liu B, Wu Y, Ma J, Zhang W, Wang B, Chen Z (2023) Estimation of winter wheat yield using multiple temporal vegetation indices derived from UAV-based multispectral and hyperspectral imagery. Remote Sens 15(19):4800

Lu B, Dao PD, Liu J, He Y, Shang J (2020) Recent advances of hyperspectral imaging technology and applications in agriculture. Remote Sens 12(16):2659. https://doi.org/10.3390/rs12162659

Lynch P, Blesius L, Hines E (2020) Classification of urban area using multispectral indices for urban planning. Remote Sens 12(15):2503. https://doi.org/10.3390/rs12152503

Macnight Ngwese N, Saito O, Sato A, Agyeman Boafo Y, Jasaw G (2018) Traditional and local knowledge practices for disaster risk reduction in Northern Ghana. Sustainability 10(3):825

Manfreda S, McCabe MF, Miller PE, Lucas R, Pajuelo Madrigal V, Mallinis G, Ben Dor E, Helman D, Estes L, Ciraolo G (2018) On the use of unmanned aerial systems for environmental monitoring. Remote Sens 10(4):641. https://doi.org/10.20944/preprints201803.0097.v1

Matese A, Czarnecki JMP, Samiappan S, Moorhead R (2024) Are unmanned aerial vehicle-based hyperspectral imaging and machine learning advancing crop science? Trends Plant Sci 29(2):196–209

McCraine D, Samiappan S, Kohler L, Sullivan T, Will DJ (2024) Automated hyperspectral feature selection and classification of wildlife using uncrewed aerial vehicles. Remote Sens 16(2):406. https://doi.org/10.3390/rs16020406

Mertikas SP, Partsinevelos P, Mavrocordatos C, Maximenko NA (2021) Chapter 3—Environmental applications of remote sensing. In: Mohamed A-MO, Paleologos EK, Howari FM (eds) Pollution assessment for sustainable practices in applied sciences and engineering. Butterworth-Heinemann, pp 107–163. https://doi.org/10.1016/B978-0-12-809582-9.00003-7

Mitchell AL, Rosenqvist A, Mora B (2017) Current remote sensing approaches to monitoring forest degradation in support of countries measurement, reporting and verification (MRV) systems for REDD+. Carbon Balance Manag 12(1):9

Murugan S, Haldorai A (2020) Role of machine intelligence and Big Data in remote sensing. In: Big Data analytics for sustainable computing. IGI Global, pp 118–130. https://doi.org/10.4018/978-1-5225-9750-6.ch007

Mutanga O, Kumar L (2007) Estimating and mapping grass phosphorus concentration in an African savanna using hyperspectral image data. Int J Remote Sens 28(21):4897–4911

Nabiollahi K, Golmohamadi F, Taghizadeh-Mehrjardi R, Kerry R, Davari M (2018) Assessing the effects of slope gradient and land use change on soil quality degradation through digital mapping of soil quality indices and soil loss rate. Geoderma 318:16–28

NASA (2024a) PACE—NASA Science. https://science.nasa.gov/mission/pace/. Accessed 21 Dec 2024

NASA (2024b) GeoXO. https://science.nasa.gov/mission/geoxo/. Accessed 22 Dec 2024

Nisha A, Anitha A (2022) Current advances in hyperspectral remote sensing in urban planning. In: 2022 Third international conference on intelligent computing instrumentation and control technologies (ICICICT). IEEE, pp 94–98

Niu C, Tan K, Jia X, Wang X (2021) Deep learning based regression for optically inactive inland water quality parameter estimation using airborne hyperspectral imagery. Environ Pollut 286:117534

Pande CB, Moharir KN (2023) Application of hyperspectral remote sensing role in precision farming and sustainable agriculture under climate change: a review. Clim Change Impacts Nat Resour Ecosyst Agric Syst:503–520

Pang Y, Räsänen A, Wolff F, Tahvanainen T, Männikkö M, Aurela M, Korpelainen P, Kumpula T, Virtanen T (2024) Comparing multispectral and hyperspectral UAV data for detecting peatland vegetation patterns. Int J Appl Earth Obs Geoinformation 132:104043

Panigrahi N, Das BS (2018) Canopy spectral reflectance as a predictor of soil water potential in rice. Water Resour Res 54(4):2544–2560

Premkumar R, Venkatachalapathy R, Visweswaran S (2021) Bio-optical studies on chlorophyll-a concentration in Hooghly River, India. Mater Today Proc 47:488–492

Qi H, Diakides NA (2003) Thermal infrared imaging in early breast cancer detection-a survey of recent research. In: Proceedings of the 25th annual international conference of the IEEE engineering in medicine and biology society (IEEE Cat. No.03CH37439), vol 2, pp 1109–1112

Ran L, Zhang Y, Wei W, Zhang Q (2017) A hyperspectral image classification framework with spatial pixel pair features. Sensors 17(10):2421

Rast M, Painter TH (2019) Earth observation imaging spectroscopy for terrestrial systems: an overview of its history, techniques, and applications of its missions. Surv Geophys 40(3):303–331

Reba M, Seto KC (2020) A systematic review and assessment of algorithms to detect, characterize, and monitor urban land change. Remote Sens Environ 242:111739

Reid Nichols C, Donelson Wright L, Zarillo G (2025) Integrated coastal resilience. Springer Nature Switzerland, Cham

Richter BD, Ao Y, Lamsal G, Wei D, Amaya M, Marston L, Davis KF (2023) Alleviating water scarcity by optimizing crop mixes. Nat Water 1(12):1035–1047

Riggan PJ (2003) FireMapper: a thermal-imaging radiometer for wildfire research and operations. IEEE Aerospace Conference Proceedings (Cat. No.03TH8652), pp 4_1843–4_1854

Sadiqbatcha S, Zhang J, Amrouch H, Tan SX-D (2021) Real-time full-chip thermal tracking: a post-silicon, machine learning perspective. IEEE Trans Comput 71(6):1411–1424

Sailor DJ, Anand J, Kalkstein L (2021) Potential overall heat exposure reduction associated with implementation of heat mitigation strategies in Los Angeles. Int J Biometeorol 65(3):407–418

Salehabadi A, Enhessari M, Ahmad MI, Ismail N, Gupta BD (2023) Environmental sensors. In: Metal Chalcogenide Biosensors. Woodhead Publishing, pp. 99–120

Sansivero F, Vilardo G (2019) Processing thermal infrared imagery time-series from volcano permanent ground-based monitoring network. Latest methodological improvements to characterize surface temperatures behavior of thermal anomaly areas. Remote Sens 11(5):553

Santamouris M, Gaitani N, Spanou A, Saliari M, Giannopoulou K, Vasilakopoulou K, Kardomateas T (2012) Using cool paving materials to improve microclimate of urban areas—design realization and results of the flisvos project. Build Environ 53:128–136

Schlerf M, Rock G, Lagueux P, Ronellenfitsch F, Gerhards M, Hoffmann L, Udelhoven T (2012) A hyperspectral thermal infrared imaging instrument for natural resources applications. Remote Sens 4(12):3995–4009

Schultz GA, Engman ET (2012) Remote sensing in hydrology and water management. Springer, Berlin Heidelberg

Seidl R, Spies TA, Peterson DL, Stephens SL, Hicke JA (2016) REVIEW: Searching for resilience: addressing the impacts of changing disturbance regimes on forest ecosystem services. J Appl Ecol 53(1):120–129

Sethy PK, Pandey C, Sahu YK, Behera SK (2022) Hyperspectral imagery applications for precision agriculture—a systemic survey. Multimed Tools Appl 81(2):3005–3038

Shaik AS, Shaik N, Priya Dr CK (2024) Predictive modeling in remote sensing using machine learning algorithms. Int J Curr Sci Res Rev 07(06):4116–4123

Sahu H, Purohit P, Srivastava A, Singh R, Mishra AP, Arunachalam K, Kumar U (2024) Comparative assessment of soil parameters and ecological dynamics in the Western Himalayan wetland and its surrounding periphery. Environ Qual Manage 34(1):e22283

Shi H, Xian G, Auch R, Gallo K, Zhou Q (2021) Urban heat island and its regional impacts using remotely sensed thermal data—a review of recent developments and methodology. Land 10(8):867

Shimoni M, Haelterman R, Perneel C (2019) Hypersectral imaging for military and security applications: combining myriad processing and sensing techniques. IEEE Geosci Remote Sens Mag 7(2):101–117

Singh S (2022) Role of hyperspectral imaging for precision agriculture monitoring. ADBU J Eng Technol 11(1): 011010008

Singh R, Saritha V, Pande CB (2024) Dynamics of LULC changes, LST, vegetation health and climate interactions in wetland buffer zone: a remote sensing perspective. Phys Chem Earth, Parts A/B/C:103660. https://doi.org/10.1016/j.pce.2024.103660

Singh R, Saritha V, Mishra AP, Pande CB, Sahu H (2025) A comprehensive analysis of water quality index in a wetland ecosystem supporting drinking water to major cities in Rajasthan, India. J Clean Prod:144593. https://doi.org/10.1016/j.jclepro.2024.144593

Smigaj M, Agarwal A, Bartholomeus H, Decuyper M, Elsherif A, De Jonge A, Kooistra L (2023) Thermal infrared remote sensing of stress responses in forest environments: a review of developments, challenges, and opportunities. Curr for Rep 10(1):56–76

Sona G, Passoni D, Pinto L, Pagliari D, Masseroni D, Ortuani B, Facchi A (2016) UAV multispectral survey to map soil and crop for precision farming applications. ISPRS—Int Arch Photogramm Remote Sens Spat Inf Sci XLI-B1:1023–1029

Souza CM, Roberts DA, Monteiro A (2005) Multitemporal analysis of degraded forests in the southern Brazilian Amazon. Earth Interact 9(19):1–25

Sun SY (2019) Research on inversion of water quality parameters in Miyun reservoir based on multisource remote sensing and machine learning. Ph.D. Thesis, Beijing Forestry University

Teague J, Day JCC, Allen MJ, Scott TB, Hochberg EJ, Megson-Smith D (2023) A demonstration of the capability of low-cost hyperspectral imaging for the characterisation of coral reefs. Oceans 4(3):286–300

Tejasree G, Agilandeeswari L (2024) An extensive review of hyperspectral image classification and prediction: techniques and challenges. Multimed Tools Appl 83:80941–81038

Thenkabail PS, Gumma MK, Teluguntla P, Mohammed IA (2014) Hyperspectral remote sensing of vegetation and agricultural crops. Photogramm Eng Remote Sens: 97–709

Upadhyay A, Sharma M, Mukherjee P, Singhal A, Lall B (2025) A comprehensive survey on synthetic infrared image synthesis. Infrared Phys Technol 147:105745

Ustin SL, Middleton EM (2021) Current and near-term advances in Earth observation for ecological applications. Ecol Process 10(1):1

Vane G, Goetz AFH (1988) Terrestrial imaging spectroscopy. Remote Sens Environ 24(1):1–29

Varini F, Madonia A, Scanu S, Piermattei V, Marcelli M (2024) Spectral signatures of submerged vegetation for remote sensing mapping and benthic coastal marine ecosystem quality assessment. In: EGU general assembly conference abstracts, p 18582

Vimercati SDCD, Genovese A, Livraga G, Piuri V, Scotti F (2013) Privacy and security in environmental monitoring systems: issues and solutions. In: Computer and information security handbook. Elsevier, pp 959–977

Wan Q, Brede B, Smigaj M, Kooistra L (2021) Factors influencing temperature measurements from miniaturized thermal infrared (Tir) cameras: a laboratory-based approach. Sensors 21(24):8466

Wang Z, Ma P, Zhang L, Chen H, Zhao S, Zhou W, Chen C, Zhang Y, Zhou C, Mao H, Wang Y, Wang Y, Zhang L, Zhao A, Weng G, Hu K (2021) Systematics of atmospheric environment monitoring in China via satellite remote sensing. Air Qual Atmosphere Health 14(2):157–169

Wehrhan M, Rauneker P, Sommer M (2016) UAV-based estimation of carbon exports from heterogeneous soil landscapes—a case study from the CarboZALF experimental area. Sensors 16(2):255

Wolny JL, Tomlinson MC, Schollaert Uz S, Egerton TA, McKay JR, Meredith A, Reece KS, Scott GP, Stumpf RP (2020) Current and future remote sensing of harmful algal blooms in the Chesapeake Bay to support the shellfish industry. Front Mar Sci 7:337

Yang Y, Meng Z, Zu J, Cai W, Wang J, Su H, Yang J (2024) Fine-scale mangrove species classification based on UAV multispectral and hyperspectral remote sensing using machine learning. Remote Sens 16(16):3093

Ye P (2022) Remote Sensing Approaches for Meteorological Disaster Monitoring: Recent Achievements and New Challenges. Int J Environ Res Public Health 19(6):3701

Zarco-Tejada PJ, Hubbard N, Loudjani P (2014) Precision agriculture: an opportunity for EU-farmers—potential support with the CAP 2014–2020. European Parliamentary Research Service

Zhang Z, Zhu L (2023) A review on unmanned aerial vehicle remote sensing: platforms, sensors, data processing methods, and applications. Drones 7(6):398

Zhang Y, Chen D, Wang S, Tian L (2018) A promising trend for field information collection: an air-ground multi-sensor monitoring system. Inf Process Agric 5(2):224–233

Zhang R, Li H, Duan K, You S, Liu K, Wang F, Hu Y (2020) Automatic detection of earthquake-damaged buildings by integrating UAV oblique photography and infrared thermal imaging. Remote Sens 12(16):2621

Zhang D, Guo Q, Cao L, Zhou G, Zhang G, Zhan J (2021) A multiband model with successive projections algorithm for bathymetry estimation based on remotely sensed hyperspectral data in Qinghai Lake. IEEE J Sel Top Appl Earth Obs Remote Sens 14:6871–6881

Zhang K, Yan F, Liu P (2024) The application of hyperspectral imaging for wheat biotic and abiotic stress analysis: a review. Comput Electron Agric 221:109008

Zhou Y, Rangarajan A, Gader PD (2020) An integrated approach to registration and fusion of hyperspectral and multispectral images. IEEE Trans Geosci Remote Sens 58(5):3020–3033

Zulkifli SN, Rahim HA, Lau W-J (2018) Detection of contaminants in water supply: a review on state-of-the-art monitoring technologies and their applications. Sens Actuators B Chem 255:2657–2689

Chapter 5
Transforming Geriatric Care: The Role of Remote Sensing Technologies in Nursing for Older Adults

Tiago Horta Reis Da Silva

Abstract The integration of remote sensing technologies in nursing care for older adults holds significant promise for improving health monitoring, enhancing patient outcomes, and addressing the unique challenges faced by this growing population. As the global demographic shift towards an ageing population continues, innovative approaches are needed to manage the complex health needs of older adults. Remote sensing, involving the acquisition and analysis of data from a distance, offers transformative potential through continuous monitoring of vital signs, chronic conditions, and environmental factors. This chapter explores the application of remote sensing in geriatric nursing, highlighting current uses such as vital sign monitoring, fall detection, cognitive health assessment, and environmental monitoring. It addresses the technical, ethical, and operational challenges associated with these technologies, including data accuracy, privacy concerns, and usability for older adults. Additionally, the chapter examines future trends and advancements in sensor technologies, the integration of artificial intelligence, and the expansion of telehealth services. As healthcare systems evolve to meet the demands of an aging population, remote sensing technologies are poised to play a crucial role in enhancing the quality of care for older adults, ensuring their health and well-being through innovative, data-driven solutions.

Keywords Remote sensing · Geriatric nursing · Older adults · Health monitoring · Fall detection · Chronic conditions · Sensor technology · Telehealth · Artificial intelligence · Patient outcomes · Privacy · Usability · Cognitive health

---

T. H. R. Da Silva (✉)
King's College London, London, UK
e-mail: tiago.horta_reis_da_silva@kcl.ac.uk

V. Saritha et al. (eds.), *Remote Sensing for Environmental Monitoring*,
https://doi.org/10.1007/978-981-96-5546-5_5

5.1 Introduction

The global ageing phenomenon constitutes one of the most critical demographic trends observed in the twenty-first century, marked by a rising percentage of older individuals within the population (Reis da Silva 2024a, b). As reported by the World Health Organization (WHO), it is anticipated that by the year 2050, the number of individuals aged 60 years and above will ascend to 2 billion, an increase from 1 billion in 2020 (Smith et al. 2021). The demographic transition brings many challenges to the healthcare system, particularly with regard to older patient care. Ageing patients often face complex health issues, including chronic diseases, decline in mental capability, and loss of physical strength, and thus need specialised healthcare services (Reis da Silva 2024c, 2024d, e). The increasing prevalence of diseases, such as dementia and frailty, complicates care and requires not only medical interventions but also ancillary services focused on addressing the psychosocial needs of older adults (Teófilo et al. 2018).

The healthcare system faces significant challenges in adapting to the needs of an ageing population. Currently, there is a great demand for health professionals with training in geriatrics; however, most nursing curricula lack the comprehensive curricula to equip the student with the necessary preparation for complexity when caring for older adults (Gibney et al. 2015; Reis da Silva 2024b, c). Research conducted by Bahçecioğlu et al. (2022), Huh and Shin (2021) emphasises the necessity of incorporating gerontological education within nursing curricula to promote favorable perspectives regarding geriatric care and to stimulate a greater number of nurses to pursue careers in this domain.

Furthermore, the insufficiency of healthcare professionals, especially in rural and underserved regions, intensifies the challenges associated with delivering sufficient care to the older population (Alsufyani et al. 2019). This will continue to rise, and it becomes imperative that the healthcare system thinks of novel strategies and adjusts to meet the challenges posed efficiently.

Application of various remote-sensing technologies has tremendous potential for addressing several healthcare problems affecting the older population. The use of remote sensing—a procedure for gathering information about any object or phenomenon without being in direct physical contact with the object—has gained considerable momentum over the past years (Bahçecioğlu et al. 2022). The purposes for which these technologies serve vary within many fields; in health, for instance, they monitor health status, assess the influence of the environment on health, and enhance care (Bahçecioğlu et al. 2022). For instance, remote sensing technology could facilitate better monitoring of the specific environmental conditions that affect the well-being of older adults, such as air quality and temperature fluctuations, important for the self-management of chronic conditions like asthma and cardiovascular disease (Stewart et al. 2020). Additionally, a combination of remote sensing with various other technologies such as the Internet of Things might also enable the development of smart care systems that could provide instant data about the conditions of older patients, thus enabling timely intervention (Sarabia-Cobo et al. 2021).

This chapter provides an overall review of the global trend in ageing and health challenges among older people and introduce different remote-sensing technologies and their possible applications within the realm of geriatric care, hence highlighting the importance of such technologies in the practice of geriatric nursing. The chapter, therefore, scopes the current literature on issues related to ageing, health needs, and the contribution of remote sensing toward improvement in the care availed for older people. By integrating these elements, this chapter aims to outline an urgent need for innovation in geriatric care and the potential benefits arising from remote-sensing technologies in improving health outcomes among older adults.

The role of remote sensing in geriatric nursing cannot be underestimated. Indeed, there is an ever-growing need for nurses to be capable in their practice and cognisant of emerging technologies. Remote sensing will facilitate the nursing practice in terms of providing valuable data that informs care decisions, improves patient monitoring, and allows for better communication amongst healthcare providers (MacLeod et al. 2021). For instance, by leveraging remote sensing technologies, nursing practitioners would be able to identify potential environmental hazards posing potential threat to the health of the older population and therefore establish preventive measures that reduce risks (Lundberg et al. 2020). In addition, remote sensing in telehealth programs will facilitate access to health professionals by the older adults, especially those who live in far distances or hard-to-reach areas and ensure they get the services and care needed (Lundberg et al. 2020).

5.2 The Role of Remote Sensing in Geriatric Health Monitoring

5.2.1 Definition and Fundamentals of Remote Sensing in Healthcare

In modern times, remote sensing technology has emerged as a transformative tool in many sectors including healthcare in general, and more so in the monitoring and management of geriatric health (Ajaz et al. 2022). In healthcare, remote sensing refers to the different high-tech collection and analytical methods for gathering data on the health status of patients and/or environmental conditions over distances. It encompasses a host of technologies-satellite imaging, aerial photography, and sensor networks-that may be applicable to the monitoring of vital signs, the detection of falls, the assessment of cognitive status, and the exploration of the environment surrounding the health of the older adults (Hajela 2023). It helps integrate remote sensing into geriatric care for improving the quality not only of health care itself but even health management is made more proactive, thus ensuring betterment in the quality of life during later life.

5.2.2 Key Applications in Geriatric Care

One of the significant applications of remote sensing in the context of geriatric health surveillance is the continuous monitoring of vital physiological parameters, such as heart rate and blood pressure. Wearable technology within remote monitoring systems can measure vital signs of health unceasingly and allow health professionals to detect anomalies that may indicate a deterioration in health status. For instance, studies have shown that the use of mobile health (mHealth) solutions significantly improves the management of chronic diseases among the older individuals through the timely medical intervention and reduction in hospitalization rates (Anbazhagan et al. 2020).

Furthermore, the adoption of telehealth services has been proven to increase access to health services among older patients, particularly within those areas with scarce health resources in rural settings (Ajaz et al. 2022). The ability to monitor vital signs from a distance would give patients the opportunity for increased self-management of their health, while at the same time allowing for prompt intervention by caregivers for any disturbing changes in the course of their illness.

Another critical application of remote sensing in geriatric health is fall detection and prevention, a major concern for older adults (Horta Reis da Silva 2022; Reis da Silva 2023a). Such falls result in a severe kind of injury, which includes but is not limited to fractures and head traumas. It is also considered one of the top class morbidity and mortality factors in older adults (Reis da Silva 2024f). Remote monitoring of older people's activities within their homes or assisted living facilities may also use other motion detection devices and other video analytical tools to monitor older people (Reis da Silva 2023b). By analysing motion, these systems can detect events signifying a fall and alert caregivers to take action. Research indicates that the implementation of fall detection systems can reduce the incidences of falls and improve the lives of independent older individuals significantly (Hajela 2023). Furthermore, environmental sensors, through observation of lighting and surface conditions elements, may apply to further reduce fall incidents by giving real-time feedback to both the patients and the caregivers.

Another place where the remote sensing technology is very intrinsic to the care given is in cognitive health assessments (Fortuna et al. 2019). Cognitive decline is a factor among older adults, and as such, it requires early detection to curtail the disorder effectively. Remote sensing instruments include cognitive assessment applications and virtual reality systems that allow the performance of cognitive evaluations by non-invasive means. This could be done by offering the patient various forms of entertainment while assessing cognitive functions such as memory, attention, and problem-solving, therefore providing substantial information about their condition (Fortuna et al. 2019). Additionally, the integration of artificial intelligence in these assessments can enhance the accuracy of cognitive evaluations by analyzing patterns and trends over time, allowing for personalized care plans tailored to the individual needs of older patients (Javadi-Pashaki et al. 2021).

Indeed, environmental monitoring is considered a very important part of geriatric health, which remote-sensing technologies can considerably enhance. Environmental factors, including air quality, temperature, and humidity, are quite easily affecting factors of older's health and wellbeing, especially in cases of prior disability (Chamat et al. 2021). Remote sensing technologies allow observation of environmental parameters continuously to obtain valid current data related to healthcare decision support. The findings indicated that poor air quality has been related to the exacerbation of respiratory diseases among the older individuals, meaning the increasing rates of hospitalisation (Chamat et al. 2021). It is the beginning of remote-sensing technologies observing environmental condition applications that allow healthcare practitioners to implement preventive strategies, such as modifying indoor air quality or temperature regulation, to make home environments for the safety of the geriatric population.

Besides, the introduction of remote sensing technologies into the control of geriatric health corresponds to the general tendency of making medicine personalised and patient-oriented. As health is becoming an ever-evolving field, much emphasis has been laid on tailoring interventions to meet the unique needs of individual patients. In this regard, remote sensing will help health professionals with massive data that will improve decision-making and development of care plans. For instance, data from EHRs integrated with data collected by remote monitoring will offer a much finer resolution of the patient's health trajectory. Secondly, remote-sensing technology has increasingly become a strategic tool in many fields, including healthcare, particularly in aged care health management and monitoring (Chamat et al. 2021).

In general, remote sensing is understood as a health care service that utilises advanced technologies for the acquisition and analysis of information about the state of the patients' health and the ambient environment from a distance (Anbazhagan et al. 2020). The technology area consists of various methods-satellite images, aerial photography, and sensor networks-that alone might be used for patient monitoring of vital signs, detection of falls incidents, cognitive health assessments, or the study of ambient variables affecting the welfare of the older adults. On-site sensing technology in geriatric care boosts the quality of health care services and allows taking up a more active attitude to health management that can surely raise the quality of life among the older people (Anbazhagan et al. 2020).

Another very important application of remote sensing in geriatric health monitoring involves the continuous monitoring of life parameters like heart rate and blood pressure. Wearable devices can enable remote monitoring systems to continuously track vital signs of patients and thus facilitate early detection of abnormal readings, which would imply deterioration of health status. For instance, studies have shown that mHealth technologies can significantly improve the management of chronic diseases in older patients by facilitating timely interventions that reduce admission to hospitals (Anbazhagan et al. 2020).

Furthermore, the inclusion of telehealth services has increased access to care for older patients, particularly in rural communities with more limited resources within the healthcare system (Ajaz et al. 2022). The ability to monitor vital signs remotely not only empowers patients to take an active role in their health management but also enables caregivers to respond promptly to any alarming changes in the patient's condition. Another critical use of remote sensing in geriatric health is in the detection and prevention of falls, which is among the major concerns amongst older population (Reis da Silva 2024f). Injury from falls among older adults often involves a severe fracture of bones and head injuries apart from being a leading cause of morbidity and mortality in older adults (Reis da Silva 2023a, 2024f). This can be achieved through remote sensing technologies, such as motion sensors and video analytics, that monitor the movements of older individuals within residence or care environments (Hajela 2023). These systems monitor these different activities, enabling them to identify fall risk activities and alert the caregiving staff of a situation where a patient may have fallen. Researchers believe that the implementation of such fall detection systems can reduce the occurrence of falls and increase the safety of older patients living independently (Hajela 2023). Besides, the number of fall incidents could be minimised much more by using environment sensors, like monitoring factors such as light and floor conditions in providing direct feedback to patients and attendants (Reis da Silva 2023c).

In addition, the area of geriatric care utilises remote sensing technology in the evaluation of cognitive health. The older population has considerable instances of cognitive loss; hence, early detection offers an intervention key. Remote sensing technologies, such as cognitive assessment applications and virtual reality platforms, have the potential to enable cognitive evaluations without invasive procedures. These innovative tools can involve patients in interactive activities that measure memory, attention, and problem-solving abilities, thereby offering significant insights into their cognitive well-being (Fortuna et al. 2019).

Moreover, the integration of AI in such assessments can ensure greater accuracy regarding cognitive diagnostics through pattern and trend analysis over time, thus helping in focused care planning according to the specific requirements of each older individual (Javadi-Pashaki et al. 2021; Reis da Silva 2025a, b).

Environmental monitoring constitutes a crucial element of geriatric health that can be considerably improved via remote sensing technologies. Such factors include air quality, temperature, and humidity that may drastically affect the overall health of the older residents, especially those with predisposing illnesses. These ambient factors may be continually monitored through a remote-sensing system installed in a specific area, which delivers current data that are helpful in decision-making in healthcare. The literature shows that air quality can impact the exacerbation of respiratory diseases among older adults, which has been associated with higher hospital admissions (Chamat et al. 2021; Reis da Silva 2025c).

By applying remote sensing to observe environmental factors, health professionals can mobilise prevention strategies that will adjust the indoor air quality or adjust the indoor temperature configuration to make homes a safer place for the older people. Besides, the inclusion of remote-sensing technologies into geriatric health monitoring also reflects a larger trend toward personalised and patient-centred treatment (Chen et al. 2019). As the transformation of the healthcare environment continues, increasing attention is being placed on tailoring interventions to meet patients' unique needs. In this transformation, remote sensing plays a key role in feeding extensive data to practitioners to support decision-making processes and care strategies. For instance, integrating EHRs with remote monitoring information can enhance the understanding of the patient's health course and therefore support better-informed clinical decisions (Chen et al. 2019; Reis da Silva 2025d).

This holistic approach to caring for geriatric patients provides improvement not only in health outcomes but also allows the older patients to feel independent and more empowered.

5.3 Current Technologies and Implementations

5.3.1 Review of Existing Remote Sensing Technologies in Use

The integration of remote sensing technology into old-age health care would mean a raised standard of health, with improved awareness of healthy living to create a better quality of life for the geriatric population.

5.3.1.1 Wearable Devices

Wearable technology, such as smartwatches and health monitoring systems, has become prominent in the oversight of geriatric health (Suwanthanma et al. 2019). These contain sensors that can continuously monitor vital signs, levels of physical activity, and a variety of other health-related parameters in real time. For instance, smartwatches can monitor heart rate and blood pressure and even detect abnormal heart rhythms. It thus provides vital information that may be forwarded to doctors and other clinicians (Suwanthanma et al. 2019). These authors present the ease with which wearable technology allows the older people to maintain their independence but still ensure that their health is constantly monitored (Reis da Silva 2025e).

Furthermore, studies have shown that wearable technology can even facilitate adherence to healthy behaviour among older adults. For example, one mobile application implemented in a care home to monitor daily exercise levels was able to demonstrate that such systems go a long way toward improving activity levels among older residents in the care home setting (Joosen et al. 2018). Such innovation helps promote not only physical but also mental health, as social interaction and engagement in physical activities become easier. The advancement of wearable technology has facilitated the emergence of novel innovations, including fall detection systems embedded within these devices (Reis da Silva 2024f). These systems employ accelerometers and gyroscopes to identify falls and promptly notify caregivers or emergency services (Hu 2023). This functionality is essential, given that falls represent a predominant cause of injury in the geriatric population, and prompt intervention can markedly diminish the likelihood of adverse outcomes (Reis da Silva 2025f).

Artificial intelligence integrated into wearable technology enhances the wearables' functionalities considerably by allowing predictive analytics, through which potential health problems can be identified before they worsen (Lin et al. 2023).

5.3.1.2 In-Home Monitoring Systems and Smart Sensors

In-home monitoring systems with intelligent sensors aim to develop a living space suitable for older adults (Reis da Silva 2023b). Such systems can monitor various parameters of movement, temperature, air quality, among others, to make that space healthy enough for living. For example, smart home systems can detect changes in the pattern of daily activities by residents, such as periods of no activity that may indicate possible health issues or accidents (Maswadi et al. 2020). Empirical evidence demonstrates that smart home technologies have actually helped the independent living of older adults. A study in this regard focused on a smart home model that could detect safety and risk parameters with much efficiency among older residents and thus always offered real-time warnings to caregivers (Chiridza et al. 2019). In this regard, the predictive mode of monitoring makes sure about better safety as well as addressing the concern of family members about their older relatives staying alone.

Furthermore, the use of IoT technology in home monitoring systems will enable communication between devices without any interruptions. For instance, a home environment monitoring system might use a cloud platform to both collect and analyse data from various sensors, hence providing detailed information about the resident's health and safety to caregivers (Cai 2023). This level of integration is vital to establishing a clear understanding of an older person's health status, hence enabling timely interventions when necessary.

5.3.1.3 Telehealth and Remote Monitoring Platforms

Telehealth has fundamentally transformed the delivery of healthcare services to senior patients, especially regarding remote monitoring practices (Gong et al. 2022). Telehealth platforms facilitate virtual consultations for healthcare providers, allowing for the remote assessment of patients' health and the provision of timely interventions, all without necessitating in-person visits (Gong et al. 2022). This approach proves particularly advantageous for older individuals who may experience mobility constraints or reside in isolated regions with restricted access to healthcare resources (Ceraulo et al. 2022).

Application of telehealth systems has improved the health outcomes of the older population. Various studies have shown that telehealth services manage chronic diseases of older patients with reduced readmission to the hospital and improved overall health status (Gong et al. 2022). On the other hand, telehealth platforms have the capability to improve communication between patients and caregivers that would ensure timely attention to changes in their health status and possibly reduce loneliness and social isolation (Reis da Silva 2024g). Increased integration among telehealth and wearable devices has also improved the efficiency of health monitoring remotely. For instance, the integration of wearable health monitoring technology and telehealth services provides the immediate vital signs and activity metrics of the patients to the healthcare practitioner for making informed decisions for the care of the patients (Tang et al. 2019). The integration between telehealth and wearable devices reflects an important advancement of the care of health among geriatric patients (Reis da Silva 2025g).

5.3.2 Empirical Evidence and Illustrative Examples from Current Nursing Practice

Many case studies show the appropriate use of remote sensing technologies in the nursing domain. The use of a smart ball-driven serious game to measure grip strength for independent living older people is a notable example. This innovative approach not only encourages patients to engage in physical activities but also provides essential data on their strength and movement, which will be useful in tailoring rehabilitation programs accordingly (Lunardini et al. 2020).

Another case study implemented an integrated remote health monitoring system at a nursing home; the system utilised a combination of wearable devices, in-home sensors, and telehealth platforms to enable the sustained monitoring of resident health. These findings showed significant reductions in the number of emergency department transfers along with hospitalisation, which will reveal the effectiveness of integrated remote monitoring in yielding better health outcomes for seniors living within nursing facilities (Zhiling, and Fuchun 2022).

The idea also sounds very encouraging in developing a smart home environment for independent living older adults. Research has indicated that such environments are capable of effectively assisting in risk surveillance and improving the overall quality of life for older residents by delivering prompt notifications and enabling interaction with caregivers (Arar et al. 2021). This method not only enhances safety but also promotes a sense of autonomy in older individuals, permitting them to uphold their independence while accessing essential support.

5.4 Benefits and Challenges in Integrating Remote Sensing Technologies

Integration of remote sensing technologies into healthcare, in particular, geriatric care, has both advantages and challenges that face it in equal measure. Accomplishment of the remote sensing technologies' implementation in the monitoring of geriatric health calls for an understanding of attendant benefits and challenges.

5.4.1 Advantages of the Use of Remote Sensing Technologies

5.4.1.1 Improved Patient Outcomes and Preventive Care

Some of the major benefits of integrating remote sensing technologies into the field of geriatric care are improved patient outcomes enabled by early interventions (Al-Naher et al. 2022). Remote monitoring systems facilitate the ongoing observation of vital signs and health indicators, thus allowing healthcare professionals to identify potential health concerns prior to their development into severe conditions. For example, wearable technology that tracks heart rate, blood pressure, and physical activity can notify caregivers of irregularities that might necessitate prompt intervention (Hardin et al. 2018). This cannot only enhance healthcare quality but also cut down hospital readmission and emergency department admission as it would allow for timely intervention using the latest available data (Al-Naher et al. 2022).

Moreover, studies also documented that older patients who have used remote health monitoring systems reported to have better satisfaction and also improved health outcomes (Al-Naher et al. 2022). For example, it was highlighted in a systematic review that the integration of remote health monitoring significantly reduces the development of complications in patients with chronic diseases, such as heart failure, chronic kidney failure and diabetes (Esquivel et al. 2018; Reis da Silva 2024h). In enhancing the management of chronic diseases through early detection and timely intervention, remote sensing technologies contribute to an increase in the quality of life of older adults.

5.4.1.2 More Data-Driven Clinical Decision Making

Remote sensing technologies grant healthcare practitioners access to extensive datasets that can be scrutinised to guide clinical decision-making processes. The incorporation of artificial intelligence (AI) and machine learning algorithms within remote monitoring systems facilitates the examination of intricate datasets, thereby empowering healthcare professionals to discern trends and patterns that might not be readily visible (Hardin et al. 2018).

It has indeed helped in making clinical decisions evidence-based by embedding data-driven insights into informed treatment plans and interventions. For instance, predictive analytics can be used to determine the fall likelihood of older patients based on their mobility trends and ambient environmental conditions (Martínez-Martín and Costa 2021). Data derived from wearable devices and intelligent sensors provides healthcare professionals with the ability to create personalised care plans that address a patient's specific needs and risks. This level of personalisation allows for improved patient outcomes and better implementation of a patient-centered approach in healthcare service delivery.

5.4.1.3 Increased Patient Safety and Independence:

The use of remote sensing technologies significantly enhances patient safety, particularly among the older who live independently. For example, fall detection systems use motion sensors combined with wearable devices to monitor the movement of older patients (Reis da Silva 2023a). In the event of a fall, such systems are able to trigger an alarm automatically via caregivers or an emergency response system, ensuring that assistance is deployed in a timely manner (Shahbazi et al. 2021). This aspect is quite critical, considering that timely interventions can prevent serious injuries and other complications that result from falls.

In addition, remote sensing technologies improve the independence and self-sufficiency of the older individuals. Smart home sensor-enabled systems can monitor environmental conditions, such as temperature and quality of air. This can allow real-time responses for the patients themselves and their caregivers too (Ahmadi 2023). With this information, seniors can stay comfortably and safely at their homes while receiving the needed support from a caregiver (Reis da Silva 2023a, b).

These remote sensing technologies promote autonomy and make a great contribution to improved mental health and quality of life for aged patients.

5.4.2 Challenges in the Integration of Remote Sensing Technologies

5.4.2.1 Problems with Data Accuracy and Reliability

Notwithstanding those tremendous benefits that come with remote sensing technologies, concerns about accuracy and reliability remain paramount. The effectiveness of remote monitoring systems depends mainly on the quality of the data collected. Incorrect or inconsistent data could lead to misinterpretation and inappropriate clinical decisions that could put a patient's safety at risk (Ahmadi 2023). For instance, wearables can provide incorrect measurements because of technical faults or user error; this can lead to false alarms or neglected health issues.

Moreover, the integration of different data sources brings with it challenges that may contribute to inaccuracies in health assessments. Variations in data formats and standards across devices and platforms could impede the flow of information and further complicate the process by which health professionals aim to gain comprehensive insights into a patient's condition (Wang and Wang 2015). Addressing these issues of data accuracy and reliability is instrumental in the development of remote sensing technologies in geriatric care.

5.4.2.2 Ethical Implications and Privacy Concerns About the Use of Data and Patient Consent

The application of different remote sensing technologies raises several ethical and privacy concerns on the use of data and consent from patients. The acquisition and exchange of such vital health information should be handled with strict data protection techniques that will ensure the privacy of the patient data. Most older adults however have little knowledge about the interpretation of their data hence they may provide consents for health information sharing unaware of the risk involved (Esquivel et al. 2018).

There is also a significant risk associated with data breaches and unauthorised intrusion into personal health information. As remote-sensing technologies continue to permeate everyday life, healthcare professionals should raise awareness on the implementation of efficient security measures while protecting patients from cyber threats (Biswas 2021). For all patients, it should be emphasized that proper information about rights related to their data use and consent is necessary for the continuity of trust in remote health monitoring systems.

5.4.2.3 Usability Issues of Older People

The difficulties associated with usability represent a considerable obstacle to the effective incorporation of remote sensing technologies within the realm of geriatric care. Numerous older individuals may encounter difficulties stemming from the intricate nature of contemporary technologies, resulting in feelings of frustration and a withdrawal from remote monitoring systems (Bi 2020). Various elements, including inadequate technological familiarity, cognitive deterioration, and physical limitations, can impede older adults' capacity to proficiently operate remote sensing devices (Schmidt et al. 2018).

In addressing the usability challenges identified, there is a need for the development of remote sensing technologies that are sensitive to the needs of older people. This includes the development of intuitive interfaces, clear instructions, and even training and support so that older people are able to use new technologies proficiently. According to Jin and Shi (2022), this may involve the involvement of older people in the design process to ensure that the remote sensing systems are tailored to their needs and capabilities, enhancing their usability and effectiveness.

In trying to integrate these remote-sensing technologies into healthcare, particularly geriatric care, a number of technical, ethical, and regulatory challenges first need to be overcome for proper implantation and patient safety. This paper discusses some of the technical barriers, including interoperability and data integration; ethical concerns, including consent and protection of privacy; and how to maintain a balance between technological monitoring while trying to preserve patient dignity. The present work takes into consideration the remote sensing of health care from the point of view of regulatory and compliance issues.

5.5 Technical, Ethical, and Regulatory Considerations

5.5.1 Technical Barriers

5.5.1.1 Interoperability Issues

One major barrier-technical-challenges in the integration of such technologies into health care remains the ability of interoperability (Kheirkhahan et al. 2019). Fundamentally, interoperability implies that various systems, equipment, and applications can communicate with each other and transit information smoothly. In the case of remote patient monitoring, this issue becomes very critical since many devices and platforms are used within health care settings.

For instance, different manufacturers' wearables may store data in proprietorial data formats, which are difficult to integrate into one system. Daley et al. (2019) illustrated that incomplete data will affect clinical decisions since inability to standardise data will impede the acquisition of complete patient data by health providers. Efforts like the Integrating the Healthcare Enterprise framework have been realised in promoting interoperability through setting standards for the interchange of health information (Daley et al. 2019). However, pursuing full adoption of such standards remains challenging as many health organisations may lack the proper resources or technical expertise for successful integration.

Moreover, the swift progression of technological innovations in remote sensing instruments frequently surpasses the formulation of corresponding standards, resulting in persistent interoperability challenges (Kheirkhahan et al. 2019).

5.5.1.2 Data Integration Issues

Data integration is one of the main technical challenges associated with the adoption of the remote sensing technologies in the health industry. According to Fu (2023), information from wearable devices, EHR, and telehealth has to be integrated in order to attain a whole understanding of the condition of the patient. However, the extent of the volume and complexity of data emanating from such systems could well be beyond the capacity of the existing healthcare infrastructure; therefore, it is an issue in managing and analysing the data (Fu 2023).

The lack of uniformity in data formats, not to mention even protocols, can be an obstacle to integration. For instance, remote monitoring systems generate data into different formats, some of which pose certain complications in their effective integration and analysis. Daley et al. (2019) add that addressing such issues of data integration requires the development of robust data management systems that are capable of managing large volumes of varied data without compromising data accuracy and reliability.

5.6 Ethical Considerations

5.6.1 Consent and Safeguarding Privacy

The use of remote sensing technologies in the healthcare industry introduces significant ethical concerns; the two most prevalent issues involve patient consent and personal privacy. Patients should be adequately informed about the collection of their health information, the intended use of that information, and the intended disclosures; they should have given their consent prior to participating in telemonitoring programs. However, many older patients are poorly educated about the implications related to the disclosure of their information; this could lead to potential problems with regard to consent (Baratta et al. 2022).

This could be considered an important issue because collecting and sharing sensitive health-related information can put the patients at risk of data breach and unauthorised access. Hence, the healthcare organisations should implement stringent security measures to ensure data protection for the patients and assure the compliance with the relevant regulatory frameworks related to patient data protection, such as HIPAA in the United States, as pointed out by Nicolau et al. (2023). It is also very important to provide information on the state of patients' data usage rights and the consideration for maintaining privacy.

5.6.2 Reconciling Technological Surveillance with Patient Respect

Another ethical issue in the integration of remote sensing technologies pertains to weighing the balance between technological monitoring and patient dignity. As much as remote monitoring enhances patient safety and results in better health outcomes, it may also be associated with negative feelings of surveillance and erosion of autonomy among the elderly (Baratta et al. 2022). This, however, should make healthcare providers sensitive to the very idea of remote monitoring, being respectful and valuing of patients in their opinions and thoughts throughout the whole process.

In response to this issue, it is imperative for healthcare organisations to emphasise patient-centred methodologies that engage patients in the decision-making processes related to their care. This entails a thorough discussion of both the advantages and drawbacks of remote monitoring technologies, as well as providing patients with the opportunity to articulate their preferences and concerns (Baratta et al. 2022). By cultivating transparent communication and collaborative practices, healthcare providers can alleviate perceptions of surveillance and enhance patients' sense of autonomy.

5.7 Regulatory and Compliance Considerations

5.7.1 Regulatory Structures Relating to Remote Sensing Technologies

The integration of remote sensing technologies into the care industry is subject to various regulatory mechanisms that govern the use of medical devices and health information. In the United States, the Food and Drug Administration plays a critical role in regulating medical devices, including wearable health monitors and remote patient monitoring systems (Nicolau et al. 2023). Such devices must undergo rigorous testing and evaluation before they are released for use by consumers to ensure their safety and effectiveness in application (Mallinis et al. 2020). On top of that, healthcare establishments must follow the policies related to data privacy and security, such as HIPAA, which established standards for the protection of patient data.

Compliance with these policies is important to maintain patients' trust and ensure that remote-sensing applications are implemented in an ethical manner in healthcare (Nicolau et al. 2023). The regulatory landscape is, however, quite complex to navigate, and this poses serious difficulties for healthcare providers, particularly the smaller entities with limited means.

5.7.2 Compliance with Data Protection Legislation

Considering that remote sensing technologies generate significant volumes of health-related data, strict adherence to data protection regulations is increasingly important. It has become necessary to create appropriate data governance frameworks by healthcare organisations in describing modes of patient data collection, storage, and dissemination. It will involve anonymisation of data where possible, establishing appropriate controls in access to information, and frequent auditing to ensure compliance with legislation on data protection (Nicolau et al. 2023).

Moreover, health professionals have to be aware of the ever-evolving legislative framework regarding data protection and privacy. For instance, the General Data Protection Regulation within the European Union sets very strict criteria on how personal information is collected and processed, which may impact the adoption of remote sensing technologies into health settings (Mallinis et al. 2020). Due to this fact, the management of an organisation should not wait but proactively seek to achieve change in practice in order not to face any potential legal and economic repercussions (Reis da Silva 2024i).

5.8 New Trends and Innovations in Geriatric Care

Geriatric care is a fast-evolving domain for several reasons, primarily because of the rapid changes mandated by technology, particularly in fields related to remote sensing and wearable technology (Mirjalali et al. 2021; Reis da Silva 2024b; Reis da Silva and Mitchell 2024).

5.8.1 Emerging Sensor Technologies and Wearable Advances

New sensor technologies are at the forefront of transforming elders' care. Wearable technologies, such as smartwatches and devices used for health monitoring, have significantly evolved, making continuous health monitoring and real-time data collection possible (L'Hommedieu et al. 2019; Reis da Silva 2025f). Such devices are embedded with advanced sensors that can monitor various physiological parameters of a patient, like heart rate, blood pressure, oxygen saturation, and levels of physical activity (L'Hommedieu et al. 2019). Integration of wireless communication technologies makes them capable of sharing such wearables data to facilitate timely interventions and personalised care with the health care provider.

Recent progress within sensor technology has allowed for the creation of multifunctional wearables that can track multiple health parameters simultaneously. Some were developed with state-of-the-art biosensors capable of even detecting biochemical markers by analysing sweat, thus providing important details related to hydration status, electrolyte balance, and metabolic function among others (Mirjalali et al. 2021). This feature is quite useful in an older population, as such health conditions can be monitored in a non-invasive fashion, which otherwise may require more invasive methods of investigation (Reis da Silva 2025a).

Furthermore, the emergence of adaptable and fabric-based sensors has created novel opportunities for the integration of wearable technology within the realm of geriatric care. Such sensors can be seamlessly incorporated into garments, thereby enhancing comfort and minimising intrusiveness for senior users (Pérez and Zeadally 2021). As these wearable devices evolve to become increasingly user-friendly and visually appealing, it is anticipated that their acceptance among older adults will rise, consequently improving their overall efficacy in health monitoring and management (Reis da Silva 2025a, b).

5.8.2 Role of Artificial Intelligence and Machine Learning

The fields of artificial intelligence and machine learning are set to transform geriatric healthcare by augmenting predictive abilities and refining clinical decision-making processes. Algorithms driven by AI possess the capacity to scrutinise extensive datasets produced by wearable technology and various health monitoring systems, thereby detecting patterns and trends that could signify potential health concerns (Javadi-Pashaki et al. 2021). For instance, machine learning models can predict the probability of falls among older patients by analysing data related to their locomotion, environmental factors, and any previous history of falls (Javadi-Pashaki et al. 2021; Reis da Silva 2023a, 2024j, 2025h).

The integration of AI in geriatric care also facilitates personalised treatment plans tailored to the unique needs of each patient. By leveraging predictive analytics, healthcare providers can anticipate health deteriorations and implement preventive measures before complications arise (Reis da Silva 2024k, l). This proactive approach not only improves patient outcomes but also reduces healthcare costs associated with emergency interventions and hospitalisations (Javadi-Pashaki et al. 2021). Besides, AI can facilitate telehealth services by carrying out simple tasks, including setting appointments or sending follow-up messages. In this way, healthcare professionals will be free to dedicate their time to providing quality healthcare services and ensure that the patients are actively involved in managing their health condition. Javadi-Pashaki et al. (2021) argue that once AI technologies are developed further, more active use of AI in geriatric care is likely to result in more effective and efficient healthcare provision.

5.8.3 Growth and Development of the Telehealth Service

The COVID-19 pandemic further accelerated this adoption of telehealth services, pointing to the potential for telehealth to enhance access to care among older adults. Telehealth provides health professionals with an avenue through which they can offer virtual consultations, manage patients remotely, and deliver clinical care without necessarily relying on physical appointments (Chen et al. 2022). For older adults with potential mobility complications or those living in remote areas with limited healthcare facilities, this is a welcome development.

With the development of telehealth services, increasing attention has been paid to the integration of remote monitoring technologies into telehealth platforms. Such integration enables healthcare professionals to access, in real time, wearable device data during virtual consultations, which further improves the quality of clinical decisions (Chen et al. 2022). For instance, a healthcare professional can review a patient's vital signs and level of physical activity during a telehealth appointment, thus facilitating a more comprehensive assessment of his or her general condition (Reis da Silva 2025c, d). In the future, more specialist telehealth services, such as mental health, rehabilitation, and chronic disease management, will be added. It would contribute to the comprehensive care of elderly patients with several health needs and hence improve their quality of care (Chen et al. 2022). With continued advancements in technology, telehealth in the future shall be an integral part of improving accessibility and outcomes for the elderly (Reis da Silva 2025e).

5.8.4 *Predictive Analytics and Personalised Care Models*

Predictive analytics are considered a strong weapon in the armamentarium of geriatric care, truly enabling practitioners to anticipate complications that may arise in older adults and tailor interventions to prevent such occurrences (Reis da Silva 2024k, l). By analysing historical health records with real-time data from wearable devices, predictive models identify individuals who are at risk for experiences such as falls, readmission to the hospital, and exacerbation of chronic diseases (Javadi-Pashaki et al. 2021). In this way, timely interventions may prevent complications and improve outcomes.

Personalised care models, as underpinned by predictive analytics, continue to find applications in geriatric care settings. These care models emphasise what differentiates the individual needs and preferences of each patient, taking into account the particular aspects of their medical history, lifestyle behaviour, and social determinants of health (Reis da Silva 2024k; l). For example, a personalised care plan for an aged patient with several comorbidities may comprise certain interventions relating to medication management, physical activity promotion, and nutrition counselling that take into account predictive analytics (Javadi-Pashaki et al. 2021). Predictive analytics embedded in geriatric care ensure better outcomes for the patient while also promoting a more patient-centred approach to healthcare. Also, by involving the patients in the planning and decision-making about their care, healthcare professionals are able to achieve higher involvement and adherence to prescribed treatment plans, thereby yielding better health outcomes (Javadi-Pashaki et al. 2021).

5.9 Strategic Recommendations for Nursing Practice

With further developments in remote-sensing technologies and telehealth services, nursing practice must position itself to effectively integrate such innovation into care.

5.9.1 *Approaches for Effective Execution*

Interdisciplinary Collaboration: Seamless integration of remote sensing technologies into nursing practice will require interdisciplinary collaboration by health care professions. Developing and implementing processes for the integration of wearables and telehealth platforms in providing care to geriatric patients will involve the collaborative efforts of all nurses, physicians, and technology specialists (Reis da Silva 2024b, j).

Patient Education and Engagement: There should be a system for educating the patient about the benefits and functionality of remote monitoring technologies to increase acceptance and engagement. Give clear instructions on how to use wearable devices and a telehealth platform so the patients are confident in managing their health (Reis da Silva 2024b; j, l).

Personalisation of Care Plans: Nursing should be oriented to personalise the care plan according to the needs and preferences of the patient. Using the data provided by remote monitoring technologies, nurses can offer appropriate interventions related to specific health concerns and improve self-management capability in patients (Reis da Silva 2024k; l).

5.9.2 Training for Health Professionals

- Technology Training: The health professional should be trained on the use of remote-sensing technologies and telehealth systems. A training curriculum for such technology could include wearable device functionality, how to interpret data, and best practices for remote monitoring of patients (Reis da Silva 2025a, b).
- Data Management Skills: With increased dependence on wearable devices by healthcare providers for their data, the need to undertake training in data management and analytics is compelling. Nurses should be prepared with the necessary skills to analyse and interpret the data that will support clinical decisions (Reis da Silva 2025f).
- Communication Skills: This is paramount because a telehealth consultation should, therefore, be complemented with effective communication. Nurses, therefore, need training on how to make virtual visits and discuss health issues with the patients and handle concerns about the use of technology (Reis da Silva 2025h).

5.9.3 Principles for Ethical and Patient-Centred Application

- Informed Consent: Such programs of remote monitoring demand that the health professional first obtains informed consent from the patients. About the use, sharing, and protection of data pertaining to the patient, due information must be provided (Reis da Silva 2024b).
- Privacy and Security: These two factors are particularly paramount in protecting patient privacy and ensuring data security. Healthcare organisations should take strong measures to secure sensitive health information captured through remote monitoring technologies (Reis da Silva 2024b).
- Patient-Centred Care: Application of remote sensing technologies should hold the dignity and autonomy of a patient paramount. Healthcare providers should engage a patient in their care planning and decision-making, hence respecting the patient's preferences and values (Reis da Silva 2024b, k).

5.10 Conclusion

The global ageing trend presents significant healthcare challenges that require innovative solutions. Remote sensing technologies hold great promise in transforming geriatric care by providing valuable data and insights that can enhance the delivery of care and improve health outcomes for older adults. As the nursing profession continues to adapt to these changes, it is essential to integrate remote sensing into nursing education and practice to prepare future nurses for the complexities of caring for an ageing population.

It can be used to revolutionise the monitoring of the health of the aged: vital signs, fall prevention, assessment of cognitive health, and environmental monitoring. It is believed that the use of such technology in health facilities will enhance not only the quality of care for the aged but also their proactive health management and intervention strategies, as treatment measures would be more personalised. With further growth in the population base of geriatric patients, the application of remote sensing technologies will be cardinal to ensuring the complicated health requirements of the aged patient are met and his safety secured in the digital health landscape, ultimately allowing for better clinical decisions to be afforded possible (Chen et al. 2019). A multidimensional approach to the care provided has the twofold impact of improving health outcomes while providing a sense of independence to the older patient.

Remote-sensing technology may change the face of geriatric health monitoring with regard to enhancing the observability of a patient's vital signs, prevention of falls, assessment of cognitive ability, and monitoring of environmental parameters. The integration of such technologies into health management methodologies will not only raise the quality of care provided for older adults but also facilitate active management of health and focused interventions. In fact, against the backdrop of an ever-increasing older population, remote-sensing technologies will continue to be imperative to meet the increasingly complex health needs of the older adults and ensure their well-being in a digitally evolving health system.

The cumulative application of different remote-sensing technologies in the care of older people holds immense potential for bringing dramatic change to their lives. Wearable technology, residential monitoring systems, and telehealth all aid in having an integrated approach toward health management by assuring timely intervention and promoting independent living. This will be very important, as this trend keeps on rising with the increase in the ageing population, to be able to provide for the complex health needs of older persons and protection in this increasingly digital world.

The integration of remote sensing technologies in the care of the older adults seems a promising pathway toward improved patient outcomes, improved clinical decision-making, and enhanced patient safety and autonomy. Yet, several challenges related to data precision, ethics, and friendliness to the user have to be overcome in order to maximise the benefit of such technologies. Health care providers should focus on the development of reliable, secure, and easy-to-operate remote monitoring systems to enable remote sensing technologies to bring benefits in the quality of health care for older people.

The integration of these remote-sensing technologies into health systems will bring in significant dividends in improved patient outcomes and quality of care for the older population. Full implementation will require resolution of technical issues relating to interoperability and integration, and a number of substantial ethical challenges relating to consent and privacy. More importantly, it is also important to adhere to the regulatory framework in order to ensure that the use of such technology is secure and ethical. By putting these considerations first, healthcare organisations can use the benefits of remote sensing technologies while preserving patient dignity and privacy.

Competing Interests The authors declare no potential conflicts of interest.

References

Ahmadi A (2023) Unveiling the complexity of earth's dynamic ecosystems: harnessing the power of remote sensing for environmental analysis. Indonesian J Earth Sci 3(2):A827. https://doi.org/10.52562/injoes.2023.827

Ajaz M, Bouchama M, Khalid K, Usman M, Uddin M, Alocha H, Hassanain S et al (2022) Telehealth use and models for geriatric patients since the covid-19 pandemic: a systematic review. J Adv Med Med Res:122–130. https://doi.org/10.9734/jammr/2022/v34i2231585

Al-Naher A, Downing J, Scott K, Pirmohamed M (2022) Factors affecting patient and physician engagement in remote health care for heart failure: systematic review. JMIR Cardio 6(1):e33366. https://doi.org/10.2196/33366

Alsufyani A, Aboshaiqah A, Moussa M, Baker O, Almalki K (2019) Competence of nurses relating self-directed learning in Saudi Arabia: a meta-analysis. Global J Health Sci 11(9):145. https://doi.org/10.5539/gjhs.v11n9p145

Anbazhagan S, Anbazhagan S, Jaysingh S (2020) Mhealth in comprehensive geriatric service—the way forward. Indian J Community Health 32(2):341–343. https://doi.org/10.47203/ijch.2020.v32i02.007

Arar M, Jung C, Awad J, Chohan A (2021) Analysis of smart home technology acceptance and preference for elderly in Dubai, UAE. Designs 5(4):70. https://doi.org/10.3390/designs5040070

Bahçecioğlu T, Türkben P, Bahar Ç (2022) The effect of the "geriatric nursing" course on nursing students' attitudes toward old age and aging: a comparative study. Int Arch Nurs Health Care 8(2). https://doi.org/10.23937/2469-5823/1510172

Baratta J, Brown-Johnson C, Safaeinili N, Rosas L, Palaniappan L, Winget M, Mahoney M et al (2022) Patient and health professional perceptions of telemonitoring for hypertension management: qualitative study. JMIR Form Res 6(6):e32874. https://doi.org/10.2196/32874

Bi T (2020) Optimal allocation algorithm of geological and ecological high-resolution remote sensing monitoring sampling points. Earth Sci Res J 24(1):105–110. https://doi.org/10.15446/esrj.v24n1.85531

Biswas A (2021) Can remote sensing and artificial intelligence based technologies benefit modern agriculture and food production systems? Biotechnology Kiosk 3(12):3–9. https://doi.org/10.37756/bk.21.3.12.1

Cai X (2023) Design of home environment monitoring system based on a cloud platform for the elderly living alone. J Phys: Conf Ser 2632(1):012023. https://doi.org/10.1088/1742-6596/2632/1/012023

Ceraulo S, Caccianiga P, Casto C, Ceraulo I, Caccianiga G (2022) Dental prosthetic rehabilitation interventions in elderly patients hospitalized in the nursing homes of the Lombardy region: a retrospective study. Healthcare 10(11):2328. https://doi.org/10.3390/healthcare10112328

Chamat A, Nagrale N, Bankar N (2021) Covid 19 in geriatric patients in Vidarbha region. J Pharm Res Int:245–249. https://doi.org/10.9734/jpri/2021/v33i38a32081

Chen T, Dredze M, Weiner J, Hernandez L, Kimura J, Kharrazi H (2019) Extraction of geriatric syndromes from electronic health record clinical notes: assessment of statistical natural language processing methods. JMIR Med Inform 7(1):e13039. https://doi.org/10.2196/13039

Chen W, Flanagan A, Nippak P, Nicin M, Sinha S (2022) Understanding the experience of geriatric care professionals in using telemedicine to care for older patients in response to the covid-19 pandemic: mixed methods study. JMIR Aging 5(3):e34952. https://doi.org/10.2196/34952

Chiridza T, Wesson J, Vogts D (2019) A smart home environment to support risk monitoring for the elderly living independently. South Afr Comp J 31(1). https://doi.org/10.18489/sacj.v31i1.534

Daley C, Toscos T, Mirro M (2019) Data integration and interoperability for patient-centered remote monitoring of cardiovascular implantable electronic devices. Bioengineering 6(1):25. https://doi.org/10.3390/bioengineering6010025

Esquivel K, Nevala E, Alamäki A, Condell J, Kelly D, Davies R, Tedesco S et al (2018) Remote rehabilitation: a solution to overloaded & scarce health care systems. Trends Telemed E-Health 1(1). https://doi.org/10.31031/tteh.2018.01.000503

Fortuna K, Torous J, Depp C, Jimenez D, Areán P, Walker R, Bartels S et al (2019) A future research agenda for digital geriatric mental healthcare. Am J Geriatric Psychiatry 27(11):1277–1285. https://doi.org/10.1016/j.jagp.2019.05.013

Fu Z (2023) Reducing clinical trial monitoring resources and costs with remote monitoring: retrospective study comparing on-site versus hybrid monitoring. J Med Internet Res 25:e42175. https://doi.org/10.2196/42175

Gibney J, Wright C, Sharma A, Naganathan V (2015) Nurses' knowledge, attitudes, and current practice of daily oral hygiene care to patients on acute aged care wards in two Australian hospitals. Spec Care Dentist 35(6):285–293. https://doi.org/10.1111/scd.12131

Gong Y, Zhou J, Fang D (2022) Investigating the demands for mobile internet-based home nursing services for the elderly. J Investig Med 70(3):844–852. https://doi.org/10.1136/jim-2021-002118

Hajela N (2023) Telehealth implementation and teaching strategies during covid-19 and beyond in gait, balance, and mobility clinic for community-dwelling older adults. Top Geriatric Rehab 39(4):240–252. https://doi.org/10.1097/tgr.0000000000000408

Hardin P, Lulla V, Jensen R, Jensen J (2018) Small unmanned aerial systems (SUAS) for environmental remote sensing: challenges and opportunities revisited. Giscience Remote Sensing 56(2):309–322. https://doi.org/10.1080/15481603.2018.1510088

Horta Reis Da Silva T (2022) Falls—prevention, assessment and management. In: Curr S, Fordham-Clarke C (eds) Clinical skills at glance. Wiley, pp 130–131

Hu K (2023) Application for detecting falls for elderly persons through internet of things combined with pulse sensor. Sensors Mater 35(11):3655. https://doi.org/10.18494/sam4641

Huh A, Shin J (2021) Person-centered care practice, patient safety competence, and patient safety nursing activities of nurses working in geriatric hospitals. Int J Environ Res Public Health 18(10):5169. https://doi.org/10.3390/ijerph18105169

Javadi-Pashaki N, Ghazanfari M, Karkhah S (2021) Machine learning for geriatric clinical care: opportunities and challenges. Ann Geriatric Med Res 25(2):137–138. https://doi.org/10.4235/agmr.21.0054

Jin W, Shi Y (2022) Dynamic monitoring of ecological environment quality in Gansu province supported by google earth engine. In: The international archives of the photogrammetry remote sensing and spatial information sciences, XLIII-B3–2022, pp 887–892. https://doi.org/10.5194/isprs-archives-xliii-b3-2022-887-2022

Joosen P, Piette D, Buekers J, Taelman J, Berckmans D, Boever P (2018) A smartphone-based solution to monitor daily physical activity in a care home. J Telemed Telecare 25(10):611–622. https://doi.org/10.1177/1357633x18790170

Kheirkhahan M, Nair S, Davoudi A, Rashidi P, Wanigatunga A, Corbett D, Ranka S et al (2019) A smartwatch-based framework for real-time and online assessment and mobility monitoring. J Biomed Inform 89:29–40. https://doi.org/10.1016/j.jbi.2018.11.003

L'Hommedieu M, L'Hommedieu J, Begay C, Schenone A, Dimitropoulou L, Margolin G, Narayanan S et al (2019) Lessons learned: recommendations for implementing a longitudinal study using wearable and environmental sensors in a health care organization. JMIR Mhealth Uhealth 7(12):e13305. https://doi.org/10.2196/13305

Lin H, Chen M, Lee C, Kung L, Huang J (2023) Fall recognition based on an IMU wearable device and fall verification through a smart speaker and the IoT. Sensors 23(12):5472. https://doi.org/10.3390/s23125472

Lunardini F, Borghese N, Piccini L, Bernardelli G, Cesari M, Ferrante S (2020) Validity and usability of a smart ball–driven serious game to monitor grip strength in independent elderlies. Health Informatics J 26(3):1952–1968. https://doi.org/10.1177/1460458219895381

Lundberg A, Saveman B, Boman E (2020) Characteristics of nursing encounters in primary health-care in remote areas: a survey of nurses' patient record documentation and self-report. Nordic J Nurs Res 41(2):84–91. https://doi.org/10.1177/2057158520973165

MacLeod M, Zimmer L, Kosteniuk J, Penz K, Stewart N (2021) The meaning of nursing practice for nurses who are retired yet continue to work in a rural or remote community. BMC Nurs 20(1). https://doi.org/10.1186/s12912-021-00721-0

Mallinis G, Chrysafis I, Korakis G, Pana E, Kyriazopoulos A (2020) A random forest modelling procedure for a multi-sensor assessment of tree species diversity. Remote Sensing 12(7):1210. https://doi.org/10.3390/rs12071210

Martínez-Martín E, Costa Â (2021) Assistive technology for elderly care: an overview. IEEE Access 9:92420–92430. https://doi.org/10.1109/access.2021.3092407

Maswadi K, Ghani N, Hamid S (2020) Systematic literature review of smart home monitoring technologies based on IoT for the elderly. IEEE Access 8:92244–92261. https://doi.org/10.1109/access.2020.2992727

Mirjalali S, Peng S, Fang Z, Wang C, Wu S (2021) Wearable sensors for remote health monitoring: potential applications for early diagnosis of covid-19. Adv Mater Technol 7(1). https://doi.org/10.1002/admt.202100545

Nicolau D, Bica O, Băjenaru L (2023) Data security approach in remote healthcare monitoring. Romanian Cyber Security J 5(1):45–55. https://doi.org/10.54851/v5i1y202304

Pérez A, Zeadally S (2021) Recent advances in wearable sensing technologies. Sensors 21(20):6828. https://doi.org/10.3390/s21206828

Reis da Silva TH (2023a) Falls assessment and prevention in the nursing home and community. British J Community Nurs 28:2, 68–72. https://doi.org/10.12968/bjcn.2023.28.2.68

Reis da Silva TH (2023b) Ageing in place: ageing at home and in the community. British J Community Nurs 28(5):213–214. https://doi.org/10.12968/bjcn.2023.28.5.213

Reis da Silva TH (2023c) Moving and handling in the community. British J Community Nurs 28(8):369. https://doi.org/10.12968/bjcn.2023.28.8.369

Reis da Silva T (2024a) The evolution of nursing for older adult: a historical perspective. Associative J Health Sci 3(3):000561. https://doi.org/10.31031/AJHS.2024.03.000561. https://crimsonpublishers.com/ajhs/pdf/AJHS.000561.pdf

Reis da Silva TH (2024b) Chapter 11—Integrating business essentials into gerontological nursing: enhancing care for older adults in diverse settings. In: Sedky A (ed) Resiliency strategies for long-term business success. IGI Global, pp 283–316. https://doi.org/10.4018/979-8-3693-9168-6.ch011

Reis da Silva TH (2024c) Chapter 8—Emotional intelligence in teaching geriatric nursing: humanising technology for compassionate care in higher education. In: Tikadar S, Liu H, Bhattacharya

P, Bhattacharya S (eds) Humanizing technology with emotional intelligence. IGI Global. https://doi.org/10.4018/979-8-3693-7011-7.ch008
Reis da Silva TH (2024d) Oncology and cancer medicine: understanding the complexities in older patients. Biomed J Sci Tech Res 55(3). https://doi.org/10.26717/BJSTR.2024.55.008720
Reis da Silva TH (2024e) Death and its significance in nursing practice. Palliat Med Care Int J 4(3):555640. https://juniperpublishers.com/pmcij/pdf/PMCIJ.MS.ID.555640.pdf
Reis da Silva TH (2024f) Falls prevention in older people and the role of nursing. British J Community Nurs 29(7):335–339. https://doi.org/10.12968/bjcn.2024.0005
Reis da Silva TH (2024g) Loneliness in older adults. British J Community Nurs 29(2):60–66. https://doi.org/10.12968/bjcn.2024.29.2.60
Reis da Silva TH (2024h) Chronic kidney disease in older adults: nursing implications for community nurses. J Kidney Care 9(4):174–179. https://doi.org/10.12968/jokc.2024.9.4.174
Reis da Silva TH (2024i) Chapter 6—Navigating healthcare complexity: integrating business fundamentals into nursing leadership. In: Sedky A (ed) Resiliency strategies for long-term business success. IGI Global, pp 145–168. https://doi.org/10.4018/979-8-3693-9168-6.ch006
Reis da Silva TH (2024j) The nursing process in geriatric care: a comprehensive approach. Int J Biomed Res 3(5). https://doi.org/10.31579/2834-5029/064
Reis da Silva TH (2024k) Enhancing nursing care planning through the Levett-Jones clinical reasoning cycle: a comprehensive analysis. Int J Biomed Res 3(5). https://doi.org/10.31579/2834-5029/065
Reis da Silva TH (2024l) Integrating effective strategies for teaching nursing care for older adults into the curriculum. Int J Biomed Res 3(5). https://doi.org/10.31579/2834-5029/063
Reis da Silva TM (2025a) Integrative approaches to geriatric care: enhancing physical and mental health through AI-driven insights. In: Jermsittiparsert K, Marzo R (eds) Physical health, mental health, and human well-being in the age of AI. IGI Global Scientific Publishing, pp 113–136. https://doi.org/10.4018/979-8-3693-6190-0.ch007
Reis da Silva TH (2025b) Chapter 9: AI-powered emotional intelligence in nursing: bridging the gap in older people care. In: Jermsittiparsert K, Marzo RR (eds) Physical health, mental health, and human well-being in the age of AI. IGI Global Scientific Publishing. https://doi.org/10.4018/979-8-3693-6190-0.ch9
Reis da Silva TH (2025c) Chapter 11—Integrating artificial intelligence in geriatric care: enhancing physical and mental health outcomes. In: Jermsittiparsert K, Marzo RR (eds) Physical health, mental health, and human well-being in the age of AI. IGI Global Scientific Publishing. https://doi.org/10.4018/979-8-3693-6190-0.ch11
Reis da Silva TM (2025d) Gamification as a pedagogical tool in geriatric nursing: engaging and empowering future nurses. In: Marcão R, Ribeiro Santos V (eds) Enhancing engagement with gamification: education, business, and healthcare perspectives. IGI Global Scientific Publishing, pp 103–136. https://doi.org/10.4018/979-8-3693-8322-3.ch005
Reis da Silva TH (2025e) Chapter 10: Emotional intelligence and AI in geriatric nursing: bridging technology and compassionate care. In: Jermsittiparsert K, Marzo RR (eds) Physical health, mental health, and human well-being in the age of AI. IGI Global Scientific Publishing. https://doi.org/10.4018/979-8-3693-6190-0.ch10
Reis da Silva TMH (2025f) The role of emotional intelligence in promoting mental peace, healthy work environments, and emotional well-being among entrepreneurs. In: Tunio MN (ed) Supporting psychological and emotional wellbeing among entrepreneurs. IGI Global Scientific Publishing. https://doi.org/10.4018/979-8-3693-3673-1.ch21
Reis da Silva TH (2025g) Chapter 8—Promoting emotional well-being in older adults: the role of AI in supporting emotional intelligence and mental health. In: Jermsittiparsert K, Marzo RR (eds) Physical health, mental health, and human well-being in the age of AI. IGI Global Scientific Publishing, pp 137–167. https://doi.org/10.4018/979-8-3693-6190-0.ch8

Reis da Silva TH (2025h) Chapter 8—Emotional intelligence in teaching geriatric nursing: humanising technology for compassionate care in higher education. In: Tikadar S, Liu H, Bhattacharya P, Bhattacharya S (eds) Humanizing technology with emotional intelligence. IGI Global Scientific Publishing. https://doi.org/10.4018/979-8-3693-7011-7.ch008

Reis da Silva TH, Mitchell A (2024) Chapter 4—Integrating digital transformation in nursing education: best practices and challenges in curriculum development. In: Lytras M, Serban AC, Alkhaldi A, Malik S, Aldosemani T (eds) Digital transformation in higher education, Part B cases, examples and good practices. Emerald Publishing Limited, pp 57–101. https://doi.org/10.1108/978-1-83608-424-220241004

Sarabia-Cobo C, Pérez V, Lorena P, Fernández-Rodríguez Á, González-López J, González-Vaca J (2021) Burnout, compassion fatigue and psychological flexibility among geriatric nurses: a multicenter study in Spain. Int J Environ Res Public Health 18(14):7560. https://doi.org/10.3390/ijerph18147560

Schmidt M, Locks M, Hammerschmidt K, Fernandez D, Tristão F, Girondi J (2018) Challenges and technologies of care developed by caregivers of patients with Alzheimer's disease. Revista Brasileira De Geriatria e Gerontologia 21(5):579–587. https://doi.org/10.1590/1981-22562018021.180039

Shahbazi M, Bagherian H, Sattari M, Saghaeiannejad-Isfahani S (2021) The opportunities and challenges of using mobile health in elderly self-care. J Educ Health Promotion 10(1):80. https://doi.org/10.4103/jehp.jehp_871_20

Smith J, Sawhney M, Duhn L, Woo K (2021) The association between new nurses' gerontological education, personal attitudes toward older adults, and intentions to work in gerontological care settings in Ontario, Canada. Canadian J Nurs Res 54(2):190–198. https://doi.org/10.1177/08445621211063702

Stewart N, MacLeod M, Kosteniuk J, Olynick J, Penz K, Karunanayake C, Morgan D et al (2020) The importance of organizational commitment in rural nurses' intent to leave. J Adv Nurs 76(12):3398–3417. https://doi.org/10.1111/jan.14536

Suwanthanma Y, Chutikorntaweesin O, Yodthong T (2019) Management of the residence prototype for the happiness of elderly people. Int J Bus Manage 7(1). https://doi.org/10.24940/theijbm/2019/v7/i1/bm1901-012

Tang V, Choy K, Ho G, Lam H, Tsang Y (2019) An IOMT-based geriatric care management system for achieving smart health in nursing homes. Ind Manag Data Syst 119(8):1819–1840. https://doi.org/10.1108/imds-01-2019-0024

Teófilo T, Veras R, Silva V, Cunha N, Oliveira J, Vasconcelos S (2018) Empathy in the nurse–patient relationship in geriatric care: an integrative review. Nurs Ethics 26(6):1585–1600. https://doi.org/10.1177/0969733018787228

Wang S, Wang C (2015) Research on dimension reduction method for hyperspectral remote sensing image based on global mixture coordination factor analysis. In: The international archives of the photogrammetry remote sensing and spatial information sciences, XL-7/W4, pp 159–167. https://doi.org/10.5194/isprsarchives-xl-7-w4-159-2015

Zhiling L, Fuchun H (2022) Application of the monitoring system for health of the elderly in the nursing home. Front Med Sci Res 4(3). https://doi.org/10.25236/fmsr.2022.040307

Chapter 6
Optimizing Groundwater Replenishment: A Geospatial Approach to Site Selection for Artificial Recharge in the Narmada River Basin, Madhya Pradesh

Deepak Patle, Manoj Kumar Awasthi, and Shailesh Kumar Sharma

Abstract Effective groundwater management is crucial for food security, water availability, and economic prosperity, given the global water shortage and the effects of a changing climate. Groundwater is a natural resource that is hidden underground and cannot be seen directly. As a result, mapping of this resource can be difficult. In light of the significant geological heterogeneity present in the Narmada River Basin, groundwater potential mapping is a highly complex and difficult task, and there is still much to explore. Therefore, this study aims to identify areas with groundwater potential and suitable locations for artificial groundwater replenishment in the Narmada River Basin, India, by geospatial approach. The eight thematic factors geology, geomorphology, lineament density, land use/land cover, slope, soil texture, rainfall, and drainage density were taken for groundwater potential mapping. The Analytical Hierarchy Process (AHP) is a multi-criteria decision analysis method which used to delineate groundwater potential zones. Based on index values, the integrated map identified five classes of groundwater potential: very good, good, moderate, poor, and very poor. The classification was determined by index values, ranging from high to low. The groundwater potential zones were validated through well yield data procured from Bhujal-Bhuvan Portal and found very good accuracy. Next, a critical area for artificial recharge was identified within the study region. Suitable sites for artificial groundwater recharge structures as their locations were identified in critical areas of Narmada Basin with good accuracy, including 8431 percolation tanks, 20,639 staggered contour trenches, 3582 nala bunds, and 54,803 check dams. The strategic use of geoinformatics, coupled with a robust multi-criteria decision-making framework, proved instrumental in pinpointing ideal locations for augmenting groundwater recharge.

D. Patle (✉) · M. K. Awasthi · S. K. Sharma
Department of Soil and Water Engineering, College of Agricultural Engineering, Jawaharlal Nehru Krishi Vishwa Vidyalaya, Jabalpur, Madhya Pradesh, India
e-mail: deepak.patle12@gmail.com

V. Saritha et al. (eds.), *Remote Sensing for Environmental Monitoring*,
https://doi.org/10.1007/978-981-96-5546-5_6

Keywords Groundwater recharge · Narmada river basin · Geoinformatics

6.1 Introduction

One of the vital natural assets is groundwater, which is deposited in underground geological formations in the critical region of the earth's foundations. It supplies water for residential, commercial, industrial, agricultural, and other expansion initiatives. Over the years, climate change and excessive groundwater extraction have exerted enormous pressure on the global groundwater supply. Hussein et al. (2017) states, as the global demand for potable water for human consumption, agriculture, and industry rises, so does the need to assess the potential and productivity of aquifers. A multitude of human and environmental factors influence the presence and supply of groundwater. The majority of tropical and temperate regions with concentrated populations and developed economies are plagued by significant groundwater issues (Singh et al. 2024). In a semi-arid nation like India, surface water sources may not always be readily available for various purposes, so residents of semi-arid regions must rely more heavily on subterranean supplies to survive (Awasthi and Patle 2019, 2020; Singh et al. 2025).

The Narmada River Basin has experienced a significant decrease in groundwater levels over the last ten years (2010–2019) (DGWR 2020). This decline is attributed to inconsistent rainfall and varied geological factors. To mitigate this problem, it is crucial to identify the most vulnerable areas within the basin and implement effective management strategies. Furthermore, potential sites for artificial groundwater recharge structures should be determined to improve the overall groundwater situation.

Recently, GIS-based studies have become increasingly popular in groundwater exploration (Sahu et al. 2024). This is because they are quick and provide direct insights into groundwater resources, aiding future planning and development (Patle and Awasthi 2019; Patle et al. 2020, 2022). There are so many RS-GIS based decision making approaches like Multiple techniques such as Multi Influencing Factor (MIF), Analytical Hierarchical Process (AHP), and machine learning models for assessing groundwater potential areas. In this study, we employed a combined approach of remote sensing, GIS, and AHP to delineate the groundwater potential zones. Total of 8 thematic maps including Geology, Geomorphology, Land Use/Land Cover, Lineament density, Drainage density, Rainfall, Soil, and Slope have been prepared and analysed for assessment of ground water potential zones.

Groundwater potential in the Narmada River Basin has not been thoroughly investigated. This study aims to fill this knowledge gap by identifying areas with high groundwater potential, pinpointing critical zones for artificial recharge, and selecting optimal sites for various groundwater recharge structures using geospatial techniques.

6.2 Materials and Methods

6.2.1 Study Area

The Narmada River is one of India's longest rivers and travels through multiple states, including Madhya Pradesh. The Narmada River is approximately 1312 kms long, with approximately 1077 kms flowing through Madhya Pradesh (Fig. 6.1). The river begins in the Amarkantak highlands of Madhya Pradesh, travels through the state, then continues through Maharashtra and Gujarat before emptying into the Arabian Sea. Narmada River Basin lies between 21° 28′ 41.83″ N–23° 38′ 4.81″ N latitudes and 73° 58′ 15.89″ E–81° 45′ 54.51″ E longitudes. The total topographical area of the Narmada Basin in Madhya Pradesh is approximately 85,083 km^2. It is one of the utmost substantial river basins in India due to its ecological, cultural, and economic significance. The Narmada basin population is predominately agrarian, with agriculture being the primary occupation of the local population. The basin is also home to several important wildlife reserves and protected areas, including the Satpura Tiger Reserve and the Pachmarhi Biosphere Reserve.

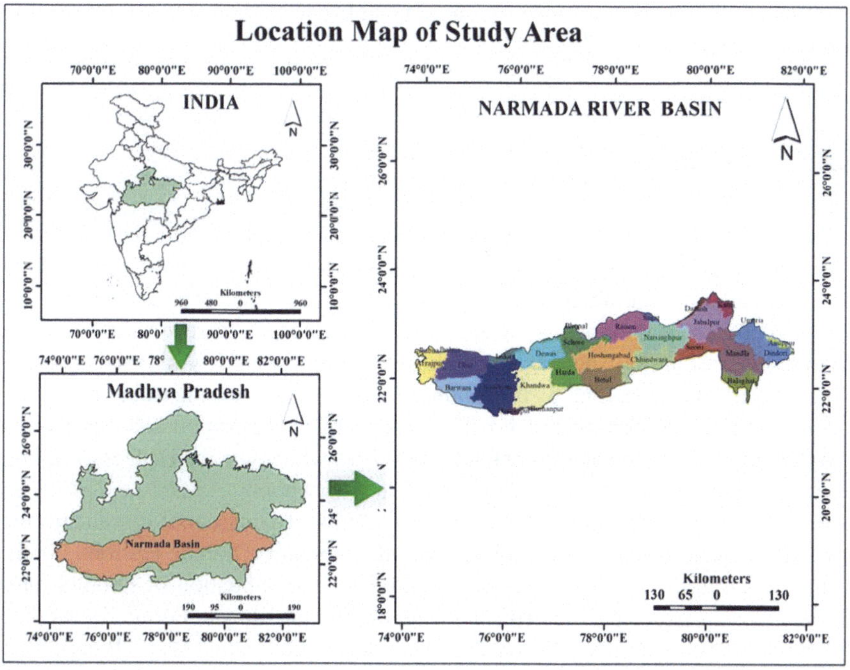

Fig. 6.1 The location map of Narmada River Basin

6.2.2 Data Collection

The Narmada River Basin map has been developed based on the scanned map collected from MPWRD website (http://mpwrd.gov.in/) and digitized on a scale of 1: 50,000 scale. Elevation data with a 30-m resolution was downloaded from the USGS EarthExplorer (https://earthexplorer.usgs.gov/). This data (SRTM DEM) was used to generate maps of slope, drainage patterns, stream order, and drainage density. Geological (1:50,000 scale) and Geomorphological (1:250,000 scale) maps were obtained from the Bhukosh portal (https://www.gsi.gov.in/). Lineament data (1:50,000 scale) was manually extracted from a WMS link of the ISRO Bhuvan geoportal (https://bhuvan.nrsc.gov.in/) using QGIS 3.16 software. Soil texture information was derived by digitizing soil map sheets (1:250,000 scale) collected from the National Bureau of Soil Survey and Land Use Planning (NBSS&LUP) in Nagpur. Normal monsoon rainfall data (0.25 $\times$ 0.25 degree grid) was acquired from the Climate Data Service Portal (CSDP) of the India Meteorological Department (IMD) in Pune (https://www.imdpune.gov.in/). To create a spatial map of land use and land cover, raw land cover tiles were acquired from the ESRI Sentinel Land Cover Explorer for the year 2021. The Sentinel Land Cover Explorer is a satellite imagery-based portal that provides land cover data year wise with a spatial resolution of 10 m. The Bhujal-Bhuvan geo-platform was used to gather well yield data for observation wells.

6.2.3 Procedure

A comprehensive literature review identified eight key factors influencing groundwater availability: geology, geomorphology, slope, lineament density, land use/land cover, soil texture, rainfall, and drainage density. Relevant data was collected, rectified, and processed into thematic maps in GIS. These maps were standardized to a 10-m spatial resolution. The Analytical Hierarchy Process (AHP) was used to prioritize and weight these factors. By integrating GIS and remote sensing techniques, each thematic map was ranked and reclassified. The weighted sum approach was applied to calculate the Groundwater Potential Index (GPI), which was then categorized into five zones: very good, good, moderate, poor, and very poor.

To evaluate the reliability of the forecast results, it is essential to validate the resulted data after making any model or classified project (Das and Pardeshi 2018). In this research, Well Yield data may be used for the validation of Groundwater Potential Zones which can be extracted from the Bhujal-Bhuvan Portal. To validate our analysis, we randomly selected 100 locations on the Ground Water Prospecting Zone (GPZ) map using ArcGIS. For each point, we extracted the existing class or zone information from the platform. Additionally, we obtained the well yield range data for these points from the Bhujal Bhuvan Portal. Both the well yield and GPZ classes need to be categorized into five levels, ranging from very good to very poor. Based

Table 6.1 Well yield ranges with category (NRSA 2011)

Classes	Bore wells yield	Open wells yield	Category
1	< 50 LPM	< 25 m^3/day	Very poor
2	50–100 LPM	25–50 m^3/day	Poor
3	100–200 LPM	50–100 m^3/day	Moderate
4	200–400 LPM	100–200 m^3/day	Good
5	> 400 LPM	> 200 m^3/day	Very good

on the Ground Water Prospect map developed by NRSA (2011), the groundwater categories are classified and presented in Table 6.1.

The accuracy of Groundwater Potential Zones (GPZs) was verified by comparing them with actual well yield data. A table was constructed to list point number, geographic coordinates, GPZ classification, well yield, and agreement status. The overall accuracy was determined by analyzing the agreement between well yield classes and GPZ classes, considering categories such as "agree condition (agree, agree-less, and agree-excess)" and "disagree" statement between well yield class and GPZ classes (Patle et al. 2024a, b).

$$\text{Accuracy}(\%) = \text{No. of points in agree condition}/\text{total no. of points} \times 100$$

According to Hosmer and Lemeshow (2000), the validation accuracy can be classified into the following categories: 0.5–0.6 (poor); 0.6–0.7 (average); 0.7–0.8 (good); 0.8–0.9 (very good); and 0.9–1.0 (excellent).

According to Patle et al. (2024a, b), a map was produced to show areas of groundwater levels where a decline in water levels is more than 0.1 m/year in post-monsoon season which help to identifying critical area. Area with depletion more than 0.1 m/year is considered as the critical areas. For identification of suitable sites for the development of Artificial Groundwater Recharge Structures, a Decision Support System (DSS) was prepared based on the different guidelines of IMSD (1995), CGWB (2007), NRSA (2011), and SAKSHAM (2017) which is given in Table 6.2.

The selection of suitable pinpointing sites for artificial groundwater recharge structures was guided by four main factors: slope, stream order, soil permeability,

Table 6.2 Criteria for site selection of groundwater recharge structures (Patle et al. 2024a, b)

S. N.	Groundwater recharge structures	Slope	Stream order	Soil permeability	Lineament
1	Check dam	3–5%	1–2[c]	Low[d]	–
2	Percolation tank	Less than 2% *	–	High[b]	Fractures[c]
3	Nala bund	Less than 1% *	3–4[b]	High[a]	–
4	Staggered contour trenches	8–25%	–	Low[d]	–

[a]IMSD (1995), [b]CGWB (2007), [c]NRSA (2011), [d]SAKSHAM (2017)

and lineament. Groundwater recharge structures are versatile and can be implemented in diverse geological settings, including hard rock and alluvial terrains. Check dams, nala bunds, and percolation tanks should typically be built in areas where the stream or nala drains between 40 to 100 hectares of land. The distance between two check dam or percolation tank sites should be approximately 250 m. The distance between two nala bund sites should be approximately 1 km. The distance between two staggered contour trench sites should be approximately 500 m. The proposed sites for artificial groundwater recharge structures were selected based on specific criteria. Critical areas were identified for constructing check dams, nala bunds, percolation tanks, and staggered contour trenches. The suitability of these sites was confirmed using Google Earth Pro software and by inclusion of 100 random samples for each type of structure.

6.3 Results and Discussions

6.3.1 Groundwater Potential Mapping

In Narmada Basin, based on the analysis of different thematic maps; a total of 25 major geological groups were identified in geology map. The total geographical area of the Narmada Basin is 85,083 km^2. It has been found that the geology of the Narmada River Basin consists of Malwa group with 56.29% followed by the Alluvium (14.20%) and Gneissic Complex with supracrystal (4.83%) respectively. The Pediment Pediplain Complex is the most extensive geomorphological unit, covering 42.36% of the total area, followed by the Dissected Plateau (30.82%) and alluvial plain (15.19%). Lineament Density was ranging from 0 to 1.41 km/sq. km over the study area. The majority of the area has a low lineament density of 0–0.30 km/sq. km. The Drainage Density obtained from 0.22 to 4.78 km/sq. km over the study area. About 68.38% area falling with the drainage density of 2.5–3.5 km/sq. km followed by 3.5–4.5 km/sq. km (16.60%) and 2.5–3.5 km/sq. km (11.60%). The dominant land use/land cover class is agricultural lands covering 48.96% of the total area, followed by Forest (46.47%) and waste lands (1.74%). Land use and land cover classes were assigned according to the 2019 NRSC classification standards (Dubey et al. 2020; Rao et al. 2020). Four different soil textural classes were identified namely: loamy, clay, loamy skeletal and clayey skeletal. The majority of the area has clayey soil (48.90%) followed by the loamy soil (42.98%) and loamy skeletal soil (7.63%). The slope percent ranged from 0 to 404% over the study area and the largest slope category is 0–1% (nearly level) followed by 3–5% (moderate slope) covering 31.22% and 22.10% of the entire area, respectively. Slope of the Narmada River Basin pointing from east to west direction. Furthermore, the slope percent from 0 to 8% were made into a group of the entire area which is suitable for groundwater potentiality. Another slope ranges more than 8% were grouped which shows less possibility of groundwater. The study area typically receives normal monsoon rainfall between 675 and

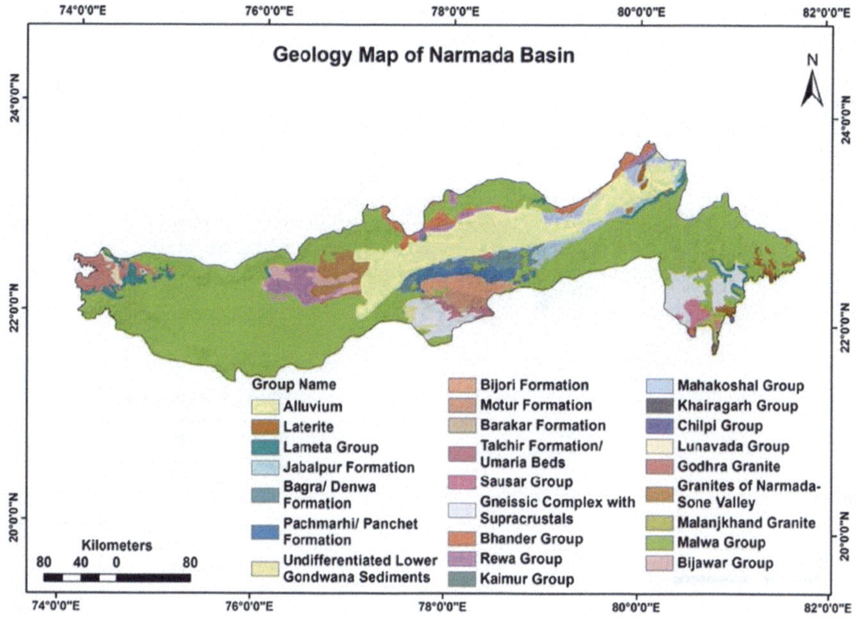

Fig. 6.2 The geology map of Narmada River basin

1422 mm. The majority of the area receives normal monsoon rainfall between 1000–1200 mm followed by the range 675–800 mm which contributes the 57.84% and 21.14% of the study area. The spatial maps on eights thematic layers were given in Figs. 6.2, 6.3, 6.4, 6.5, 6.6, 6.7, 6.8 and 6.9.

The Analytical Hierarchical Process (AHP) method, developed by Saaty in 2008, is a popular Multiple Criteria Decision Making (MCDM) technique used globally to identify potential groundwater zones on maps. In this study, we created a pairwise comparison matrix to assess the relative importance of various factors. We assigned numerical ratings from 1 to 5, and their reciprocals from 1/2 to 1/5, to represent the relative weight of each factor compared to others (Table 6.3). Then prepared the normalized pairwise comparison (Table 6.4) and developed the weighted sum matrix (Table 6.5). The AHP model was executed and yielded a consistency ratio (CR) of approximately 0.07. This CR value is below the 0.1 threshold, indicating that the model's judgments are reasonably consistent.

$$\lambda\,\max = 8.718.$$

$$\text{Consistency Index} = \text{CI} = \lambda\,\max - n/n - 1$$

$$0.102$$

$$\text{CR} = \text{CI}/\text{RI}$$

$$0.07$$

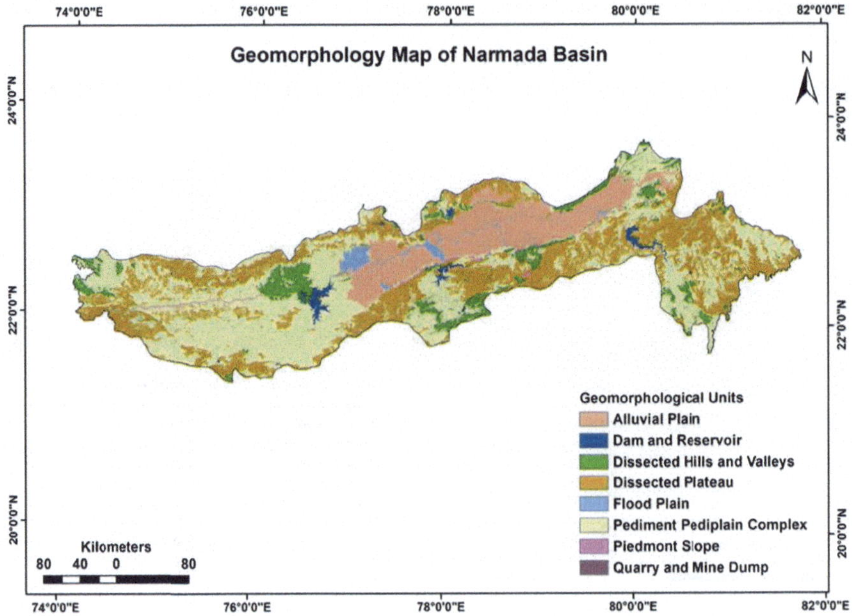

Fig. 6.3 The geomorphology map of Narmada River basin

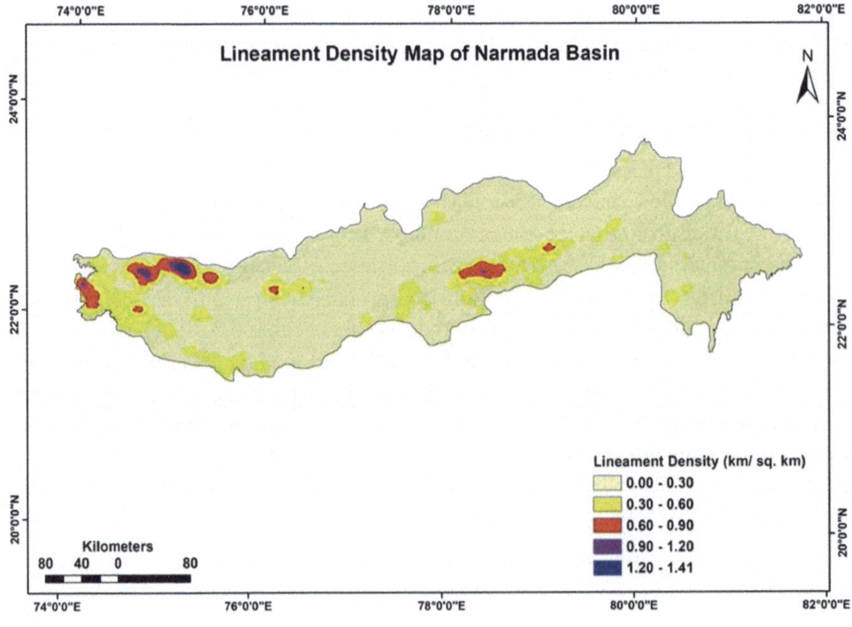

Fig. 6.4 The lineament density map of Narmada River basin

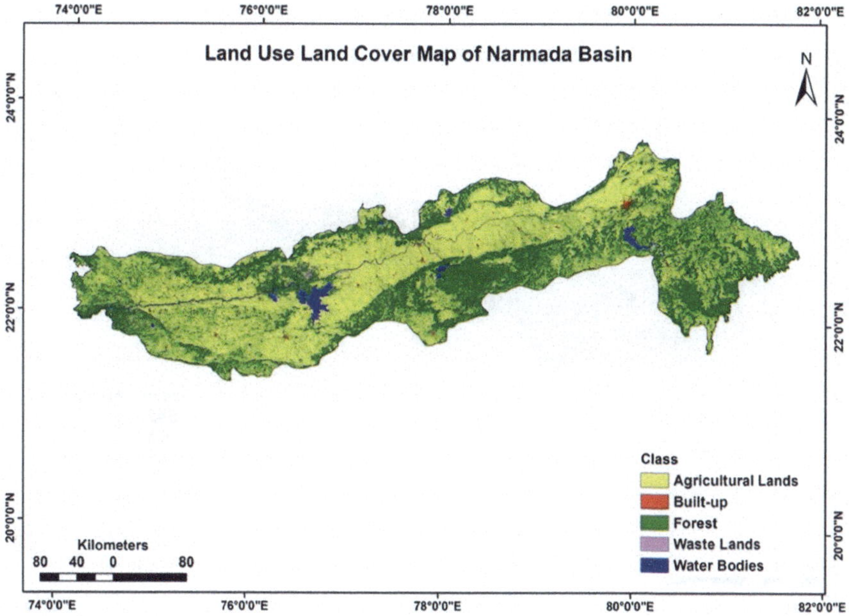

Fig. 6.5 The land use land cover map of Narmada River basin

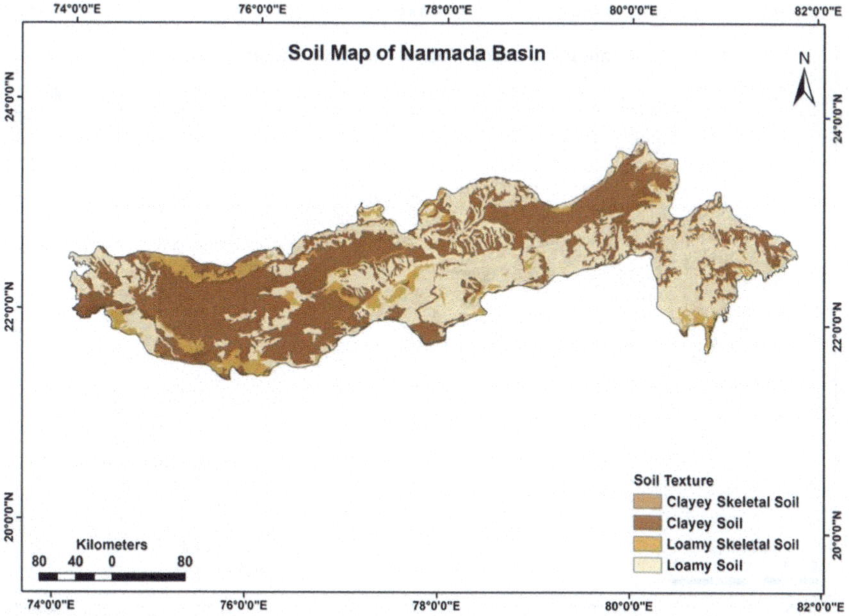

Fig. 6.6 The soil texture map of Narmada River basin

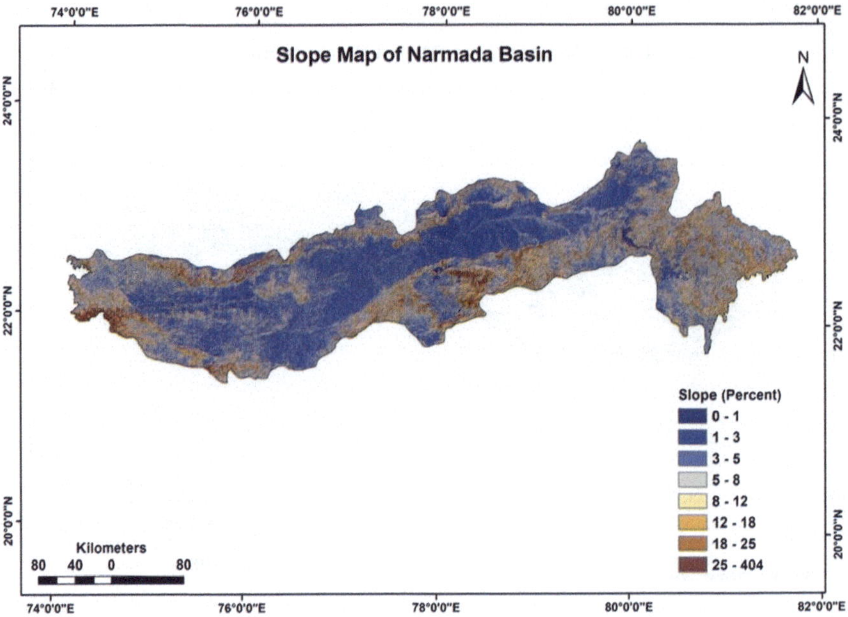

Fig. 6.7 The slope map of Narmada River basin

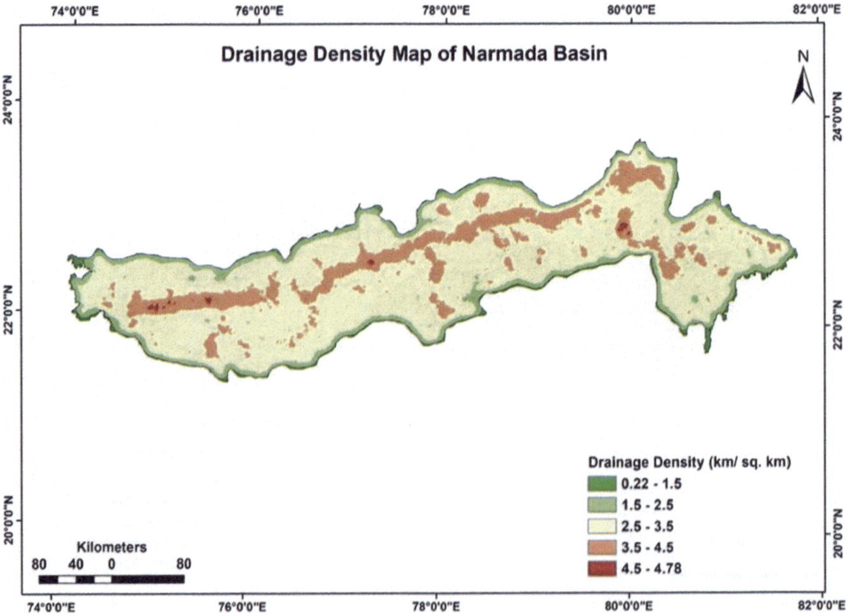

Fig. 6.8 The drainage density map of Narmada River basin

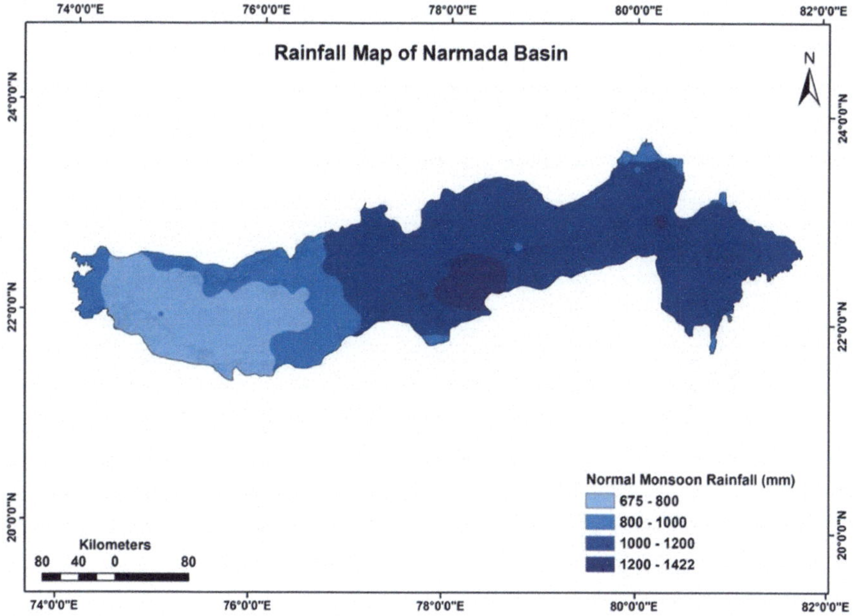

Fig. 6.9 The rainfall map of Narmada River basin

It is less than 0.1.

Finally, Weightage is given the Table 6.6.

The weights for each thematic layer were determined (see Table 6.6). Thematic factors were ranked from 5 (most influential) to 1 (least influential) based on their impact on the classes or categories. The rankings of different thematic parameters and their sub classes were assigned. Following this, a weighted sum approach was utilized to merge the layers. The integrated layer that resulted from this process exhibited a range of values between 125 and 416 which was classified into five classes viz. < 175, 175–225, 225–275, 275–325 and > 325 index value of groundwater potential map. Lower index values indicate areas with low groundwater potential, while higher values signify areas with high potential. The study area was categorized into five groundwater potential zones: very good, good, moderate, poor, and very poor, based on the range of index values. Figure 6.10 illustrates the groundwater potential map generated using the AHP method.

The area classified as having a very poor potential was 0.38% (327.52 km^2) of the total area, while the poor potential area was 10.21% (8686.20 km^2). The moderate potential area was the largest at 50.78% (43,204.94 km^2), followed by the good potential area at 26.19% (22,282.21 km^2). The very good potential area was 12.44% (10,582.12 km^2) of the total area. The central part of the Narmada Basin indicates good to very good groundwater potential zones. One of the reasons may be the widespread existence of the Alluvium and Malwa geologic group, Pediment Pediplain Complex geomorphologic unit, nearly level to moderate slope, and

Table 6.3 Pair-wise comparison matrix for all parameters

Factors	Geology	Geomorphology	Lineament density	Land use/land cover	Soil texture	Monsoon rainfall	Slope	Drainage density
Geology	1	1	2	3	3	3	3	2
Geomorphology	1	1	2	2	2	2	3	3
Lineament density	0.5	0.5	1	2	2	3	2	2
Land use/land cover	0.33	0.5	0.5	1	2	3	2	2
Soil texture	0.33	0.5	0.5	0.5	1	2	1	0.5
Monsoon rainfall	0.33	0.33	0.5	0.5	2	1	0.33	0.5
Slope	0.5	0.33	0.5	0.5	0.5	3	1	2
Drainage density	0.5	0.33	0.5	0.5	2	2	0.5	1
Sum	4.49	4.49	7.5	10	14.5	19	12.83	13

Table 6.4 Normalized pairwise comparison matrix

Factors	Geology	Geomorphology	Lineament density	Land use/land cover	Soil texture	Monsoon rainfall	Slope	Drainage density	Criteria weight
Geology	0.223	0.223	0.267	0.300	0.207	0.158	0.234	0.154	0.221
Geomorphology	0.223	0.223	0.267	0.200	0.138	0.105	0.234	0.231	0.202
Lineament density	0.111	0.111	0.133	0.200	0.138	0.158	0.156	0.154	0.145
Land use/land cover	0.073	0.111	0.067	0.100	0.138	0.158	0.156	0.154	0.120
Soil texture	0.073	0.111	0.067	0.050	0.069	0.105	0.078	0.038	0.074
Monsoon rainfall	0.073	0.073	0.067	0.050	0.138	0.053	0.026	0.038	0.065
Slope	0.111	0.073	0.067	0.050	0.034	0.158	0.078	0.154	0.091
Drainage density	0.111	0.073	0.067	0.050	0.138	0.105	0.039	0.077	0.083
Sum	1.000	1.000	1.000	1.000	1.000	1.000	1.000	1.000	1.000

Table 6.5 Weighted sum (WS) matrix

Factors	Geology	Geomorphology	Lineament density	Land use/ land cover	Soil texture	Monsoon rainfall	Slope	Drainage density	Weighted sum	CW	Ratio of WS/CW
Geology	0.221	0.202	0.290	0.359	0.222	0.194	0.272	0.165	1.926	0.221	8.732
Geomorphology	0.221	0.202	0.290	0.239	0.148	0.130	0.272	0.248	1.750	0.202	8.644
Lineament density	0.110	0.101	0.145	0.239	0.148	0.194	0.181	0.165	1.285	0.145	8.850
Land use/land cover	0.073	0.101	0.073	0.120	0.148	0.194	0.181	0.165	1.055	0.120	8.821
Soil texture	0.073	0.101	0.073	0.060	0.074	0.130	0.091	0.041	0.642	0.074	8.674
Monsoon rainfall	0.073	0.067	0.073	0.060	0.148	0.065	0.030	0.041	0.556	0.065	8.582
Slope	0.110	0.067	0.073	0.060	0.037	0.194	0.091	0.165	0.797	0.091	8.784
Drainage density	0.110	0.067	0.073	0.060	0.148	0.130	0.045	0.083	0.715	0.083	8.660

Table 6.6 Criteria weight estimated for different themes

S. No.	Thematic factor	Criteria weight	Weight (%)
1	Geology	0.221	22
2	Geomorphology	0.202	20
3	Lineament density	0.145	15
4	Land use/land cover	0.120	12
5	Soil texture	0.074	7
6	Monsoon rainfall	0.065	7
7	Slope	0.091	9
8	Drainage density	0.083	8
	Sum	1.000	100

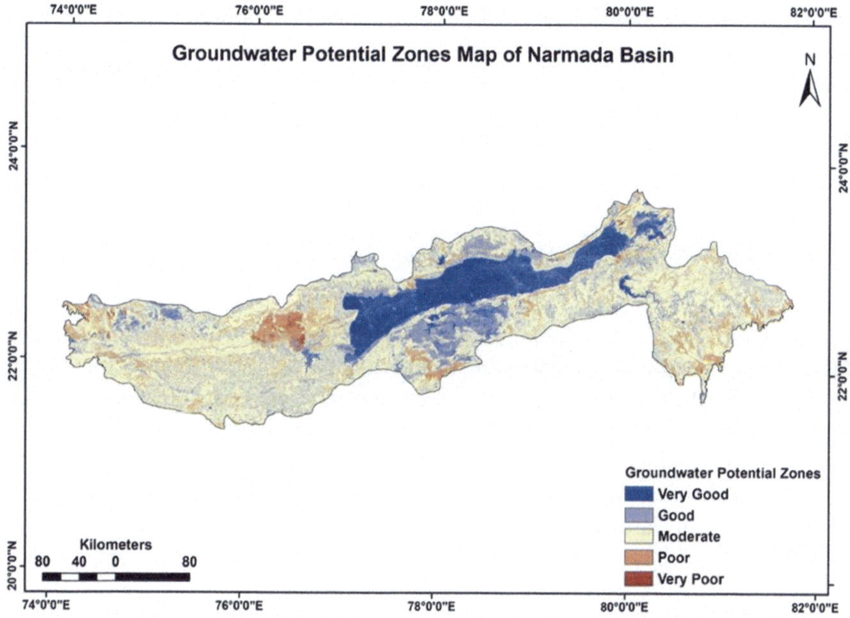

Fig. 6.10 The groundwater potential zones map of Narmada River basin

agricultural lands. Other factors like soil texture, slope, drainage density, and rainfall impacted low influence due to less weightage assigned. The majority of the Narmada basin is capable of yielding moderate to good amounts of groundwater due to varying geologic condition and other groundwater influencing factors.

6.3.2 Validation of Groundwater Potential Map

As reported by Patle et al. (2024a, b), the validation of resulted data is one of the most important works after making any model to check the proficiency of the predicted results. Various methods are extensively used to validate groundwater potential zones maps such as receiver operating characteristics (ROC) approach, groundwater well yield data, net availability of groundwater and groundwater level fluctuation data etc. (Basavarajappa et al. 2016; Arulbalaji et al. 2019; Elubid et al. 2020; Patle et al. 2022, 2024a, b). To verify the accuracy of the delineated groundwater potential zones, well yield data was cross-referenced with Table 6.7.

According to the above shown graph, 34% of points were classified under the agreement condition. About 41% and 6% of random points were found under the agreement with excess and less yield respectively. In the disagreement category, about 19% of points were found. Figure 6.11 demonstrated that the well yield points with the agreement (including agree, excess agree, and less agree) are mostly detected in entire zones of the groundwater potential zones. The overall accuracy of delineated groundwater potential zones within the Narmada Basin was found 81% which symbolizes very good accuracy.

6.3.3 Identification of Critical Area

Following the successful mapping of groundwater potential zones in the Narmada Basin, the next step was to identify the most critical areas for artificial groundwater replenishment. To this end, a groundwater level trend map was developed for the post-monsoon period between 2011 and 2020, as illustrated in Fig. 6.12.

Areas experiencing significant subsidence, exceeding 0.10 m per year, were identified as critical zones. About 48,335 km^2 area out of the 85,035 km^2 area is identified as a critical area in Narmada River Basin for Artificial Groundwater Recharge shown in Fig. 6.13.

6.3.4 Identification of Suitable Site for Artificial Recharge in Critical Areas

To increase groundwater levels, we needed to identify suitable locations for various artificial recharge structures like check dams, percolation tanks, nala bunds, and staggered contour trenches. We used several guidelines (IMSD 1995, CGWB 2007, NRSA 2011, and SAKSHAM 2017) to develop criteria for site selection. These criteria were based on factors like slope, drainage orders, soil permeability, and lineaments. This study identified potential locations for different groundwater recharge structures, including check dams, percolation tanks, nala bunds, and staggered

Table 6.7 Comparison analysis of GPZ and well yield data of Narmada Basin

Point	Longitude	Latitude	GPZ	Yield	Agreements
1	80.0683	23.7640	Moderate	Very poor	Disagree
2	80.0669	23.7509	Poor	Very poor	Agree—excess
3	80.0294	23.7189	Good	Very poor	Disagree
4	79.9971	23.6277	Very poor	Very poor	Agree
5	80.1706	23.6704	Very good	Moderate	Disagree
6	79.9254	23.4650	Very poor	Very poor	Agree
7	78.2703	23.3383	Moderate	Poor	Agree—excess
8	80.1796	23.3728	Good	Moderate	Agree—excess
9	78.3961	23.3005	Moderate	Moderate	Agree
10	79.9045	23.2905	Very good	Moderate	Disagree
11	77.7631	23.1255	Good	Moderate	Agree
12	78.1421	23.1754	Moderate	Moderate	Agree
13	80.3636	23.2110	Good	Poor	Disagree
14	80.9363	23.2266	Moderate	Poor	Agree—excess
15	78.7678	23.1161	Very good	Good	Agree—excess
16	79.6732	23.1095	Good	Good	Agree
17	79.8352	23.1115	Poor	Poor	Agree
18	80.0822	23.1336	Moderate	Moderate	Agree
19	77.7251	22.9542	Moderate	Very poor	Disagree
20	78.5999	23.0246	Very good	Very good	Agree
21	79.3730	23.0175	Good	Moderate	Agree—excess
22	80.3644	23.0202	Moderate	Very poor	Disagree
23	77.0551	22.8855	Moderate	Poor	Agree—excess
24	78.1255	22.9218	Very good	Very good	Agree
25	79.0731	22.8539	Very good	Very good	Agree
26	79.1051	22.8729	Good	Very good	Agree—less
27	79.8761	22.8646	Moderate	Poor	Agree—excess
28	80.9969	22.9390	Moderate	Poor	Agree—excess
29	77.3531	22.7810	Good	Moderate	Agree—excess
30	77.4354	22.7390	Moderate	Good	Agree—less
31	78.3114	22.8381	Very good	Very good	Agree
32	80.1880	22.8208	Moderate	Poor	Agree—excess
33	81.2878	22.8150	Moderate	Poor	Agree—excess
34	81.4721	22.8320	Good	Poor	Disagree
35	81.5809	22.8502	Poor	Very poor	Agree—excess
36	76.9222	22.6824	Moderate	Poor	Agree—excess
37	77.2248	22.6923	Very good	Very good	Agree

(continued)

Table 6.7 (continued)

Point	Longitude	Latitude	GPZ	Yield	Agreements
38	78.2307	22.7083	Very good	Very good	Agree
39	79.0834	22.7298	Good	Moderate	Agree
40	79.6236	22.7095	Moderate	Moderate	Agree
41	80.8946	22.7687	Moderate	Poor	Agree—excess
42	74.7032	22.5212	Moderate	Very poor	Disagree
43	74.6996	22.4884	Very poor	Very poor	Agree
44	74.9743	22.4764	Good	Very poor	Disagree
45	76.7055	22.5874	Moderate	Poor	Agree—excess
46	77.7770	22.5698	Very good	Very good	Agree
47	78.3957	22.6208	Good	Very poor	Disagree
48	78.8083	22.6512	Poor	Very poor	Agree—excess
49	79.3145	22.6508	Moderate	Poor	Agree—excess
50	80.8844	22.6183	Good	Very poor	Disagree
51	80.9083	22.6316	Moderate	Very poor	Disagree
52	75.0631	22.4435	Moderate	Moderate	Agree
53	76.2021	22.4595	Very poor	Very poor	Agree
54	76.4869	22.4958	Poor	Very poor	Agree—excess
55	76.4484	22.4467	Very poor	Very poor	Agree
56	76.8885	22.4827	Moderate	Poor	Agree—excess
57	77.5603	22.5363	Good	Very good	Agree—less
58	77.8916	22.4677	Very good	Good	Agree—excess
59	79.3094	22.5691	Moderate	Poor	Agree—excess
60	79.4947	22.5373	Good	Very poor	Disagree
61	81.3150	22.5324	Moderate	Poor	Agree—excess
62	75.2680	22.3525	Moderate	Poor	Agree—excess
63	75.9016	22.3381	Good	Moderate	Agree—excess
64	76.1590	22.3713	Poor	Very poor	Agree—excess
65	76.2777	22.3912	Very poor	Very poor	Agree
66	76.9165	22.3502	Moderate	Poor	Agree—excess
67	78.4853	22.3951	Good	Very poor	Disagree
68	80.4656	22.4436	Moderate	Poor	Agree—excess
69	74.5238	22.2367	Very poor	Very poor	Agree
70	74.7810	22.2651	Moderate	Poor	Agree—excess
71	75.0602	22.2121	Poor	Very poor	Agree—excess
72	75.9592	22.2907	Moderate	Moderate	Agree
73	76.0487	22.2890	Very poor	Very poor	Agree
74	76.3811	22.2450	Good	Moderate	Agree—excess

(continued)

Table 6.7 (continued)

Point	Longitude	Latitude	GPZ	Yield	Agreements
75	76.5979	22.3042	Very poor	Poor	Agree—less
76	77.0848	22.3019	Very good	Good	Agree—excess
77	78.3368	22.3022	Moderate	Very poor	Disagree
78	78.4500	22.3452	Good	Good	Agree
79	74.2269	22.1113	Moderate	Good	Agree—less
80	75.2925	22.1470	Moderate	Moderate	Agree
81	75.4625	22.1585	Poor	Good	Disagree
82	76.2571	22.2060	Very poor	Very poor	Agree
83	76.3376	22.1986	Moderate	Moderate	Agree
84	76.9904	22.2303	Good	Very good	Agree—excess
85	78.3444	22.2343	Moderate	Very poor	Disagree
86	74.4039	22.0329	Moderate	Good	Agree—less
87	74.9219	22.0833	Good	Good	Agree
88	75.7085	22.0859	Moderate	Moderate	Agree
89	77.2799	22.1619	Moderate	Very poor	Disagree
90	77.7555	22.1399	Poor	Very poor	Agree—excess
91	78.0788	22.1783	Good	Good	Agree
92	80.5684	22.1295	Moderate	Poor	Agree—excess
93	75.2133	21.9499	Moderate	Poor	Agree—excess
94	76.3517	21.9796	Moderate	Poor	Agree—excess
95	76.6754	22.0437	Good	Good	Agree
96	77.7968	22.0831	Moderate	Poor	Agree—excess
97	80.7233	22.0878	Poor	Very poor	Agree—excess
98	74.7994	21.8757	Moderate	Moderate	Agree
99	75.9490	21.9482	Moderate	Poor	Agree—excess
100	76.0497	21.9382	Good	Poor	Disagree
101	77.1736	21.9728	Moderate	Poor	Agree—excess
102	75.4355	21.7998	Moderate	Poor	Agree—excess
103	76.2769	21.8376	Good	Moderate	Agree—excess
104	76.5846	21.8433	Moderate	Moderate	Agree
105	75.2189	21.6567	Moderate	Poor	Agree—excess
106	76.4380	21.6976	Moderate	Moderate	Agree
107	76.8078	21.7907	Good	Good	Agree
108	75.7251	21.6406	Moderate	Very poor	Disagree
109	75.6540	21.5512	Moderate	Poor	Agree—excess

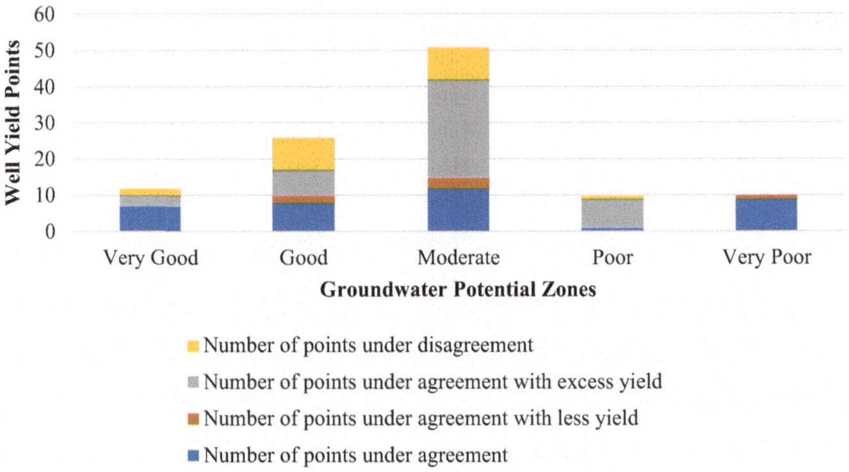

Fig. 6.11 Comparison of GPZ with well yield points of Narmada Basin

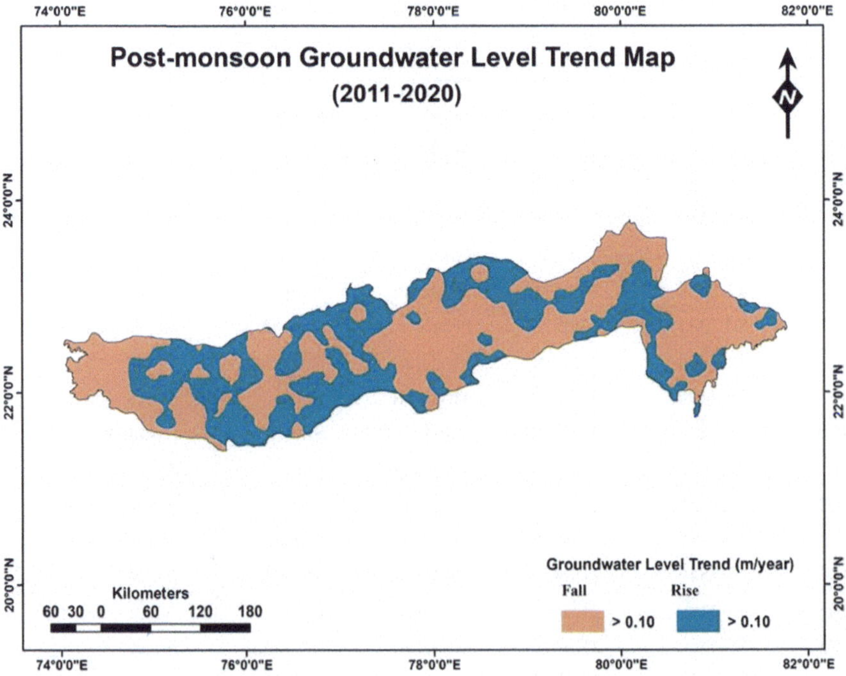

Fig. 6.12 The post-monsoon groundwater level trend map of Narmada River basin

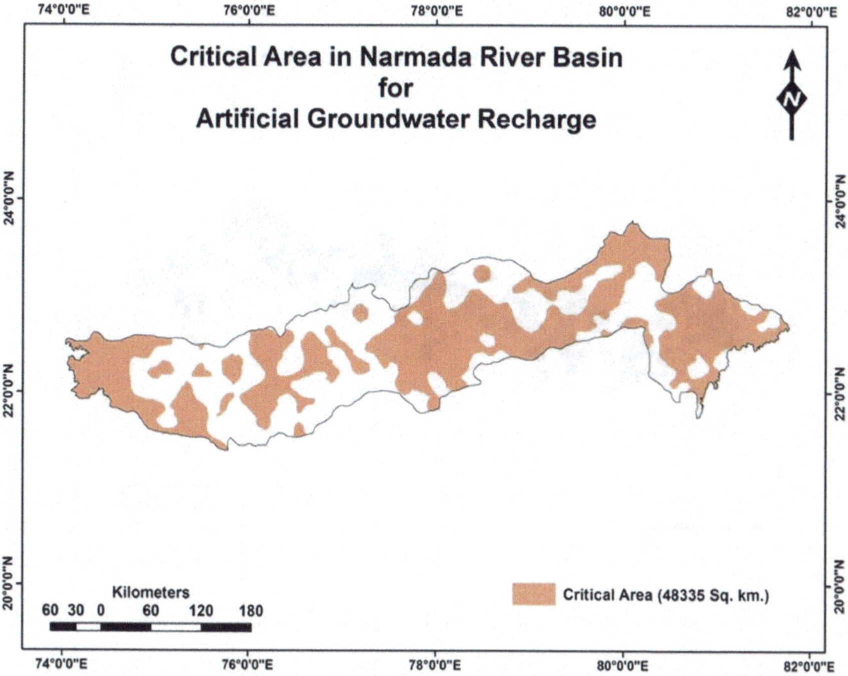

Fig. 6.13 The critical area map of Narmada River basin

contour trenches. The selection of these sites was based on the criteria outlined in Table 6.2. Using a GIS-based model in ArcGIS 10.8, we analyzed the area and identified potential sites for each type of recharge structure. The results of this analysis are shown in Figs. 6.14, 6.15, 6.16 and 6.17. After the identification of suitable sites, verification of the identified sites was conducted by visual interpretation with 100 random samples imported on Google Earth Pro imagery. It was found the good validation accuracy for the different artificial groundwater recharge structures.

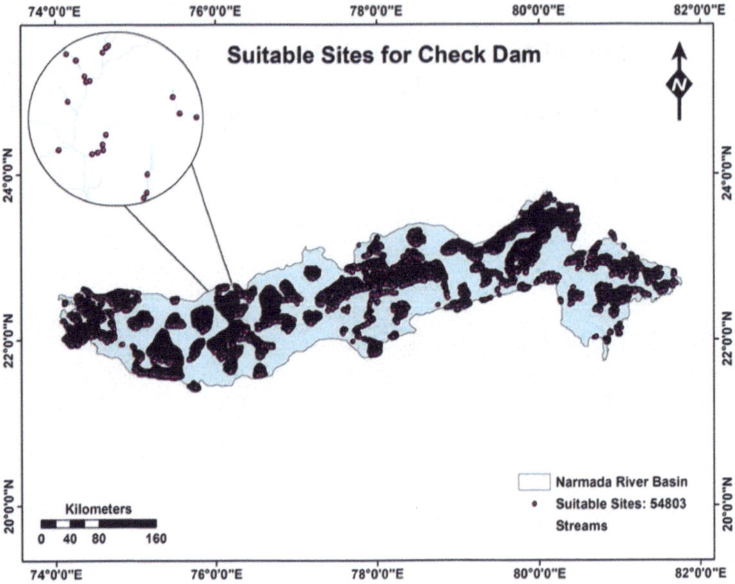

Fig. 6.14 The suitable sites for check dams in critical areas of Narmada River basin

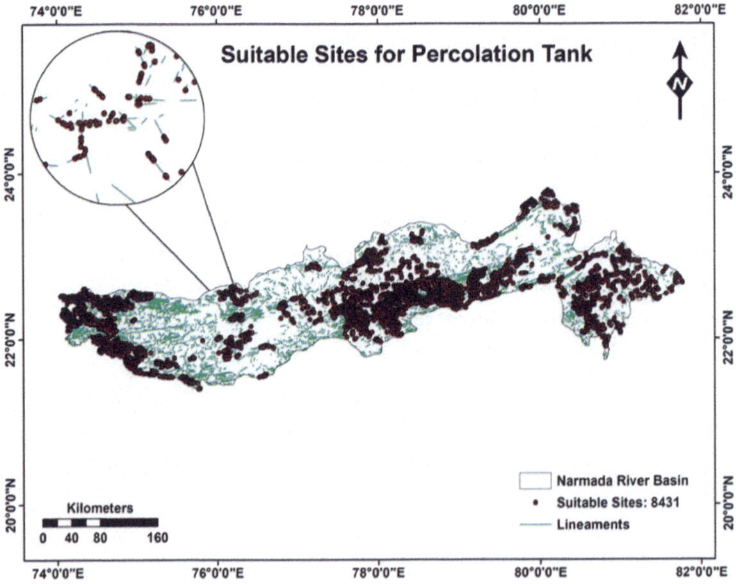

Fig. 6.15 The suitable sites for percolation tanks in critical areas of Narmada River basin

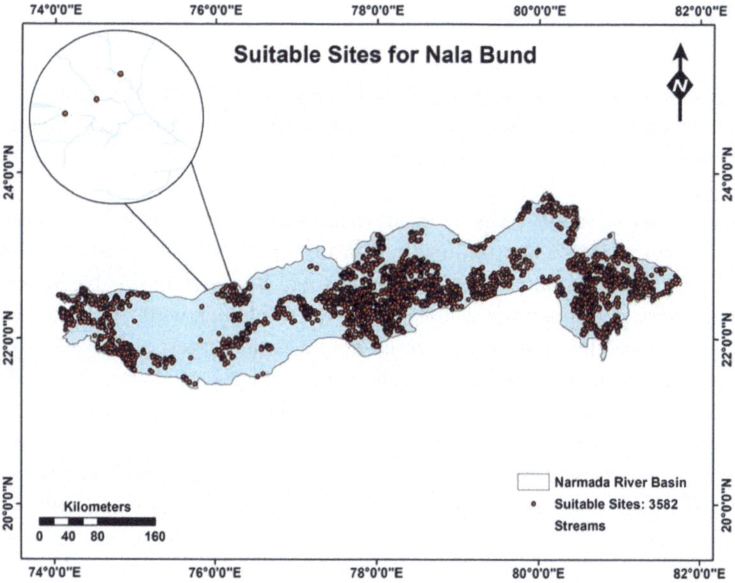

Fig. 6.16 The suitable sites for nala bunds in critical areas of Narmada River basin

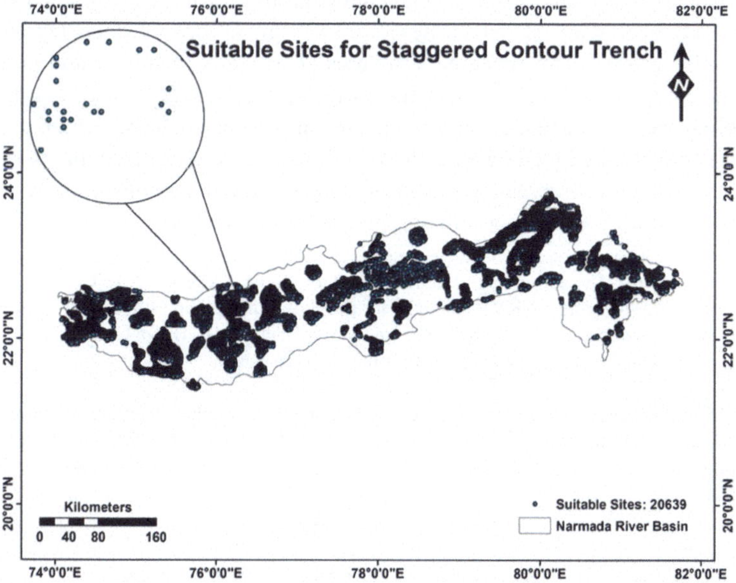

Fig. 6.17 The suitable sites for staggered contour trenches in critical areas of Narmada River basin

6.4 Conclusion

This research was conducted in Madhya Pradesh's Narmada River Basin, India's largest. Groundwater potential mapping was performed using various thematic influencing parameters, including geology, geomorphology, lineament density, land use/land cover, soil texture, rainfall, slope, and drainage density. The majority of data pertaining to these factors can be easily accessed on open platforms maintained by government sectors. The multi criterion decision making approach AHP was employed for the potential zoning of groundwater. The area in this basin classified as having a very poor potential was 0.38% of the total area, while the poor potential area was 10.21%. The moderate potential area was the largest at 50.78%, followed by the good potential area at 26.19%. The very good potential area was 12.44% of the whole area. The central part of the Narmada Basin indicates good to very good groundwater potential zones.

The main aim was to recognize the critical area in Narmada River Basin. The decadal post-monsoon season groundwater level trend map (where depletion is > 0.10 m/yr) was considered as critical area. About 56% area found critical for artificial groundwater recharge structures. By combining remote sensing and GIS techniques with a strong multi-criterion decision-making process, we were able to accurately locate optimal sites for various artificial groundwater recharge structures. This included percolation tanks (8431), staggered contour trenches (20,639), nala bunds (3582), and check dams (54,803). Reviewed key guidelines for choosing suitable locations for groundwater recharge structures. This study identified and proposed a specific number of suitable spots for various groundwater recharge structures, pinpointing their exact places on a map. The sites were confirmed to be accurate through verification with Google Earth Pro software. It is recognized that the geoinformatics presents a comprehensive and agile solution for mapping groundwater and managing natural resources, optimizing both time and budget.

Acknowledgements The authors express their gratitude to the NAHEP-CAAST-CSDA, College of Agricultural Engineering, JNKVV Jabalpur for providing the resources needed to conduct this research. They also acknowledge the State Ground Water Data Center, Bhopal and CGWB, Bhopal for sharing essential data and insights that contributed to this study.

Authors' Contributions Dr. Deepak Patle carried out the study, analyzed and performed different approaches under the supervision of Dr. M. K. Awasthi. Dr. Patle wrote the draft manuscript, which was modified by Dr. M. K. Awasthi. Dr. S. K. Sharma facilitated to modification of Remote Sensing GIS figures.

Competing Interests The authors declare no potential conflicts of interest.

References

Arulbalaji P, Padmalal D, Sreelash K (2019) GIS and AHP techniques based delineation of groundwater potential zones: a case study from southern Western Ghats, India. Sci Repo 91:1–17. https://doi.org/10.1038/s41598-019-38567-x

Awasthi MK, Patle D (2019) Water harvesting in Kharif fallow for augmenting ground water recharge. In: 4th International conference on soil and water resources management for climate smart agriculture, global food and livelihood security. SCSI, New Delhi at NASC, New Delhi, India, p 94

Awasthi MK, Patle D (2020) Trend analysis of ground water recharge in Tikamgarh district of Bundelkhand using geospatial technology. Int J Chemi Stud 8(4):417–420. https://doi.org/10.22271/chemi.2020.v8.i4g.10181

Basavarajappa HT, Dinakar S, Manjunatha MC (2016) Validation of derived groundwater potential zones (GWPZ) using geoinformatics and actual yield from well points in parts of upper Cauvery basin of Mysuru and Chamarajanagara districts, Karnataka, India. Int J Civil Engi Tech 7(1):140–161.. https://iaeme.com/MasterAdmin/Journal_uploads/IJCIET/VOLUME_7_ISSUE_1/IJCIET_07_01_012.pdf

CGWB (2007) Manual on artificial recharge of ground water. Central Ground Water Board (CGWB), Govt of India 2007

Das S, Pardeshi SD (2018) Integration of different influencing factors in GIS to delineate groundwater potential areas using IF and FR techniques: a study of Pravara basin, Maharashtra, India. App Water Sci 8:1–16. https://doi.org/10.1007/s13201-018-0848-x

DGWR (2020) Dynamic ground water resources of Madhya Pradesh. CGWB, NCR, Govt of India and GWS, WRD, Govt. of MP. http://mpwrd.gov.in/wp-content/uploads/2018/03/07-dynamic-gw-assessment-2011.pdf

Dubey S, Rao JH, Patle D (2020) Morphometric analysis and prioritization of sub watersheds of Umar Nala Watershed, Madhya Pradesh using geospatial technique. Int J Agri, Envi Biotech 13(3):269–274. https://doi.org/10.30954/0974-1712.03.2020.2

Elubid BA, Huang T, Peng DP, Ahmed EH, Babiker MM (2020) Delineation of groundwater potential zones using integrated remote sensing, GIS and multi-criteria decision making MCDM. Desalin Water Treat 192:248–258. https://doi.org/10.5004/dwt.2020.25761

Hosmer DW, Lemeshow S (2000) Applied logistic regression, 2nd edn. Wiley, New York, pp 373

Hussein TAA, Govindu V, Nigusse AG (2017) Evaluation of groundwater potential using geospatial techniques. App Water Sci 75:2447–2461. https://doi.org/10.1007/s13201-016-0433-0

IMD (2021) IMD rainfall gridded data. https://cdsp.imdpune.gov.in/

IMSD (1995) Integrated mission for sustainable development IMSD. National Remote Sensing Agency, Department of Space, Govt. of India

NRSA (2011) Groundwater prospects map study. National Remote Sensing Agency. Dept. of Space. Govt. of India, Balanagar, Hyderabad 2011

NRSC (2019) Land use/land cover database on 1:50,000 scale, Natural Resources Census Project, LUCMD, LRUMG, RSAA, National Remote Sensing Centre, ISRO, Hyderabad 2019

Patle D, Awasthi MK (2019) Groundwater potential zoning in Tikamgarh District of Bundelkhand using remote sensing and GIS. Int J Agri, Envi Biotech 124:311–318. https://doi.org/10.30954/0974-1712.12.2019.3

Patle D, Rao JH, Dubey S (2020) Morphometric analysis and prioritization of sub-watersheds in Nahra watershed of Balaghat district, Madhya Pradesh: a remote sensing and GIS perspective. J Exp Bio Agri Sci 8(4):447–455. https://doi.org/10.18006/2020.8(4).447.455

Patle D, Awasthi MK, Sharma SK, Tiwari YK (2022) Application of geoinformatics with frequency ratio FR model to delineate different groundwater potential zones in Ken Basin, India. Indi J Eco 492:313–323. https://doi.org/10.55362/IJE/2022/3523

Patle D, Awasthi MK, Nema S (2024a) Validation of derived groundwater potential zones using well yield data through agreement scheme approach with geoinformatics in Ken River Basin. Indi J Eco 51(1):118–125. https://doi.org/10.55362/IJE/2024/4204

Patle D, Nema S, Awasthi MK, Sharma SK (2024b) Identification of suitable sites for different artificial groundwater recharge structures for sustainable water resources management in Ken River Basin. Envi Dev Sustain 26(10):1–49. https://doi.org/10.1007/s10668-024-05484-7

Rao JH, Patle D, Sharma SK (2020) Remote sensing and GIS technique for mapping land use/land cover of Kiknari Watershed, India. Ind J Pure App Biosci 8(6):455–463. https://doi.org/10.18782/2582-2845.8458

Saaty TL (2008) The analytic hierarchy process: planning, priority settling. Resource Allocation, New York

Sahu H, Purohit P, Srivastava A, Singh R, Mishra AP, Arunachalam K, Kumar U (2024) Comparative assessment of soil parameters and ecological dynamics in the western Himalayan wetland and its surrounding periphery. Environ Qual Manage 34(1):e22283

SAKSHAM (2017) Natural resources management through integrated watershed development under Mahatma Gandhi national rural employment guarantee scheme (MGNREGS). National Ground Water Training and Research Institute, India

Singh R, Saritha V, Pande CB (2024) Dynamics of LULC changes, LST, vegetation health and climate interactions in wetland buffer zone: a remote sensing perspective. Phys Chem Earth, Parts A/B/C 135:103660. https://doi.org/10.1016/j.pce.2024.103660

Singh R, Saritha V, Mishra AP, Pande CB, Sahu H (2025) A comprehensive analysis of water quality index in a wetland ecosystem supporting drinking water to major cities in Rajasthan, India. J Clean Prod 487:144593. https://doi.org/10.1016/j.jclepro.2024.144593

Chapter 7
Remote Sensing of Freshwater Algal Blooms: A Spectral Index Approach

Krishna Patil, Kajal Rathod, Ashwin Gujrati, and Ravindra Pawar

Abstract Climate change and anthropogenic activities have intensified freshwater algal blooms, creating substantial risks to aquatic ecosystems and human health. Although traditional in-situ monitoring methods deliver accurate measurements, their spatial and temporal limitations have driven a shift towards remote sensing as a viable alternative with extensive monitoring potential. This chapter reviews recent advancements in spectral indices for detecting and monitoring freshwater algal blooms, tracing their evolution from primary vegetation indices to specialised algorithms. Key indices, including the Normalized Difference Chlorophyll Index (NDCI), Floating Algae Index (FAI), and Algal Bloom Detection Index (ABDI), along with new approaches for macroalgal detection, are analysed for their strengths and limitations across different remote sensing platforms. While these indices enhance the capability to track algal blooms across diverse spatial and temporal scales, challenges remain regarding universal applicability and species-level discrimination. The chapter identifies research gaps and proposes future directions, emphasising the need for integrated monitoring frameworks and standardised methods across satellite platforms to improve algal bloom detection and risk management.

Keywords Freshwater algal bloom · Remote Sensing · Spectral Index

---

K. Patil · K. Rathod · R. Pawar (✉)
Department of Fisheries Biology, College of Fisheries, Dr. Balasaheb Sawant Konkan Krishi Vidyapeeth, Ratnagiri, Maharashtra, India
e-mail: rapawar@dbskkv.ac.in; ravindra.fisheries@gmail.com

K. Patil
e-mail: pkrishna2405@gmail.com

K. Rathod
e-mail: kajalbr2024@gmail.com

A. Gujrati
Space Applications Centre, ISRO, Ahmedabad, India
e-mail: ashwin8199@sac.isro.gov.in

V. Saritha et al. (eds.), *Remote Sensing for Environmental Monitoring*,
https://doi.org/10.1007/978-981-96-5546-5_7

7.1 Introduction

Freshwater resources, including lakes, rivers, and wetlands, play an essential role in the environment by providing critical resources for human use and supporting a rich diversity of habitats and ecosystem services (Vári et al. 2022). Despite their relatively small coverage on Earth's surface, freshwater resources are indispensable for sustaining biodiversity and enabling various species to thrive (Musie and Gonfa 2023). These ecosystems contribute significantly to global carbon and nutrient cycles, regulating processes that maintain ecological balance and climate stability. Anthropogenic activities and climate change have intensified the impact on freshwater resources, pushing them to the brink of environmental crisis (Bănăduc et al. 2022; IPCC 2022). These ecosystems are now emerging as critical but limited resources, both in terms of water quality and availability, for human development and long-term ecological stability. Among the most significant challenges are nutrient enrichment from agricultural runoff, industrial pollution, climate change-induced alterations, acidification, and the spread of invasive species (Palmer et al. 2015; Singh et al. 2025).

The increasing frequency and intensity of algal blooms are among the most severe consequences of these cumulative environmental stresses. These harmful blooms are driven mainly by eutrophication—an excess of nutrients in water bodies—coupled with rising atmospheric CO_2 concentrations and global warming (Griffith and Gobler 2020; Igwaran et al. 2024). Evidence shows that algal blooms have significantly increased globally in recent decades, with projections indicating that this trend will continue (Ho et al. 2019; Singh et al. 2024). These blooms severely- threaten ecosystems ecological structure, functioning, and aesthetics, leading to a cascade of problems, including biodiversity loss, degradation of water quality, and disruptions to ecosystem services (Griffith and Gobler 2020; Amorim and do Nascimento Moura 2021; Roegner et al. 2023). Addressing these challenges is critical to safeguarding the ecological integrity of freshwater resources and the humans and wildlife that depend on them.

7.2 Freshwater Algal Blooms

Algae are diverse, primarily aquatic organisms that are essential to freshwater ecosystems. These organisms range widely in size and complexity, including microalgae (small, often single-celled organisms like filamentous cyanobacteria) and macroalgae (larger, multicellular forms like seaweed) (Bellinger and Sigee 2015). Algae are fundamental to these ecosystems, mainly because they contribute to primary production by generating organic matter, which serves as the base of the food web and supports higher trophic levels (Paerl and Otten 2013; Guo et al. 2016). Algal classification extends beyond size and structure and includes groupings based on their unique photosynthetic pigments. These pigments allow algae to capture sunlight for energy;

each group's pigments contribute to specific light absorption properties. Major pigment groups include chlorophyll-a (Chl-a), present in Phaeophyta, Rhodophyta, and Chlorophyta, as well as accessory pigments in shades of blue, yellow, and grey that enable algae to absorb different light wavelengths, thereby enhancing their photosynthetic efficiency and giving them distinct colouration (Sheath and Wehr 2015; Sahu et al. 2024).

Algae are exceptionally responsive to changes in nutrient availability, a critical factor in their growth and productivity (Burford et al. 2023). Freshwater ecosystems are sensitive to nutrient inputs, such as nitrogen and phosphorus from agricultural runoff, fertilisers, or other pollutants (Akinnawo 2023). When nutrient concentrations increase beyond natural levels, algal productivity is amplified, often leading to excessive growth—a phenomenon termed eutrophication. This nutrient-driven surge can result in high concentrations of algae in the water, leading to algal blooms (Sanseverino et al. 2016). Algal blooms are also defined as rapid increases in algal populations beyond typical background levels, often resulting in dense accumulations of cells in the water. These blooms occur primarily in surface waters exposed to sunlight, where algae can capture energy efficiently and proliferate through rapid cell division. The term "bloom" conveys the abundant growth and flourishing of algae, similar to how plants proliferate on land under favourable conditions (Huisman et al. 2018). While algal blooms are somewhat natural and contribute to oxygen production and organic matter for food webs, excessive blooms can disrupt aquatic ecosystems. Certain bloom-forming algae, particularly within the phytoplankton group, release toxins or create oxygen-depleted zones as they decompose, posing risks to aquatic life and human health (Paerl and Otten 2013).

Microscopic algae, or phytoplankton, represent the drifting, single-celled forms that are especially critical to freshwater ecosystems. These non-motile algae rely on water movements for distribution and play an essential role in producing oxygen and organic matter. Algal blooms, however, can have both beneficial and detrimental impacts. While blooms provide oxygen and are foundational to aquatic food webs, certain species within these blooms can release toxins or otherwise disrupt ecosystem balance (Hallegraeff 2003; Huisman et al. 2018; Sanseverino et al. 2016). Though few, harmful algal species are widely distributed across various phytoplankton groups. Some harmful species are eukaryotes—organisms with a nucleus and organelles, such as dinoflagellates, diatoms, and chlorophytes—while others are prokaryotes, such as cyanobacteria, which lack membrane-bound organelles (Watson et al. 2015; Brooks et al. 2016).

Thus, while algae are essential to life on Earth—producing roughly half of the oxygen we breathe and supporting aquatic food chains—uncontrolled growth through nutrient overloading poses ecological and health risks, marking the complex and significant role of algae in freshwater ecosystems. Consequently, continuous monitoring of algal populations is essential to detect and manage blooms effectively, ensuring the stability and health of freshwater resources.

7.3 Remote Sensing of Freshwater Algal Blooms

Monitoring freshwater algal blooms is essential to protect aquatic life, ensure water quality, and safeguard public health. Traditional in-situ approaches primarily relied on field sampling and laboratory analyses. It mainly measures toxic cell concentrations in water and additional parameters such as water temperature, turbidity, phosphates, nitrates, pH, and dissolved oxygen. While effective at localised scales, it presents challenges for widespread monitoring, being labour-intensive, time-consuming, and costly. It also requires regular sample collection, which is impractical at regional or national scales. Freshwater algal blooms exhibit high spatiotemporal variability, complicating comprehensive long-term monitoring efforts through traditional approaches alone (Shi et al. 2019). In recent years, various methods for real-time detection, spatiotemporal monitoring and prediction of algal blooms are available. Remote sensing significantly improves algal bloom monitoring in freshwaters by providing large-scale, simultaneous data collection with frequent temporal updates (Rolim et al. 2023).

Remote sensing collects information about Earth's surface without direct contact, primarily through satellite or aerial sensors. It involves detecting and measuring reflected or emitted energy from the Earth in various wavelengths of the electromagnetic spectrum. Remote sensing technology has been widely used to analyse water quality (Huang et al. 2018; Chen et al. 2022). Key biogeochemical parameters investigated include temperature, colour, suspended matter, dissolved carbon, nutrients, trophic status and aquatic vegetation (Gholizadeh et al. 2016a). Various remote sensing platforms are utilised for data collection, including groundborne, spaceborne, and airborne systems.

Groundborne platforms include sensors on the Earth's surface, such as fixed stations or mobile units that capture data directly from the environment. These are typically used for real-time, high-resolution, localised studies and can provide validation for satellite data. In China, a ground-based remote sensing system (GRSS) has been used to capture rapid changes in phytoplankton bloom (Wang et al. 2022). Airborne platforms include balloons, aircraft, and Unmanned Aerial Vehicles (UAVs). These can cover large areas at higher resolutions than satellites, making them useful for detailed studies. UAVs, also known as drones, are increasingly popular for remote sensing due to their flexibility, cost-effectiveness, and ability to access remote areas. It can be equipped with various sensors, including multispectral and hyperspectral cameras, to gather detailed spatial information. Wu et al. (2023), used multispectral UAV-derived vegetation indices and chlorophyll-a concentrations to monitor algal blooms in Southern Illinois. Satellites orbiting the Earth are spaceborne platforms with sensors that collect data over large areas. These platforms provide comprehensive and consistent observations of the Earth's surface, enabling global monitoring of environmental changes, land use, and water quality (Tripathi and Tiwari 2021; Gholizadeh et al. 2016b). Satellite remote sensing greatly enhances algal bloom monitoring in freshwaters by enabling simultaneous, large-scale data acquisition with frequent temporal updates (Caballero et al. 2020; Nazeer et al. 2023).

Satellite remote sensing data is mainly optical, radar, and LiDAR, each offering distinct capabilities for monitoring freshwater algal bloom. Optical sensors, capturing data in visible and infrared wavelengths, are widely applied in monitoring vegetation health and water quality. Optical data includes multispectral sensors (e.g., Sentinel-2 MSI, Landsat 8 OLI) for broad spectral bands and hyperspectral sensors (e.g., AVIRIS) for detailed analysis across numerous narrow bands. It has been widely utilised in freshwater algal bloom studies. Radar technology in the microwave spectrum facilitates cloud-penetrating, all-weather data collection, enabling detailed assessments. The effectiveness of radar remote sensing techniques in monitoring and controlling algal blooms is still under study. LiDAR, through laser pulse emissions, provides high-resolution, three-dimensional elevation data, supporting precision. Attempts have been made in recent years to conduct LiDAR studies on freshwater algal blooms (Rolim et al. 2023). These remote sensing modalities collectively enhance our ability to conduct large-scale, temporally consistent environmental monitoring and contribute critical insights for sustainable resource management.

Algal bloom monitoring is based on optical properties, categorised into inherent optical properties (IOPs) and apparent optical properties (AOPs). IOPs, such as absorption and scattering coefficients, depend solely on the composition and concentration of the medium, independent of the ambient light. In contrast, AOPs, including remote sensing reflectance, rely on the medium's characteristics and the light field (Shen et al. 2012). Blooms affect IOPs through their biological properties, altering AOPs (remote sensing reflectance) to allow the extraction of biological information. Current remote sensing of algal blooms employs various algorithms based on these optical characteristics of primarily absorption and fluorescence of pigments (Shi et al. 2019). Over the past several decades, satellite remote sensing of freshwater algal blooms has advanced significantly. Furthermore, developing algorithms for remotely estimating algal biomass facilitates the satellite-based quantification of algal bloom phenology.

7.4 Spectral Indices for Freshwater Algal Blooms Monitoring

Spectral indices are widely used mathematical combinations of various spectral bands (Montero et al. 2023). Various spectral indices have been developed based on normalized band ratio and spectral shape measuring peaks at specific wavelengths to detect and characterise algal blooms (Nazeer et al. 2023; Rolim et al. 2023). By analysing the spectral signatures of water bodies, these indices provide insights into the biochemical and biophysical properties associated with algal biomass. Chlorophyll-a and phycocyanin concentrations have been utilised as proxies for monitoring freshwater algal blooms (Liu et al. 2022; Beal et al. 2024). These pigments exhibit distinct spectral characteristics, with noticeable peaks in remote sensing reflectance at nearly 440 nm in the blue and nearly 675 nm in the red wavelengths (Shi et al. 2019). This chapter investigated various recent spectral indices for detecting, mapping, and monitoring algal blooms in freshwaters.

7.4.1 Normalized Difference Vegetation Index (NDVI) and Enhanced Vegetation Index (EVI)

The NDVI (Rouse et al. 1974) is based on comparing the amount of near-infrared light reflected by vegetation to the amount of visible light reflected. It is the most widely used spectral index in remote sensing to assess vegetation health and chlorophyll concentrations. The NDVI has also been utilised to detect freshwater algal blooms. The NDVI values range from − 1 to + 1, where values close to zero indicate non-vegetated surfaces (like water), while positive values indicate vegetation (Rodríguez-López et al. 2020). In algal bloom conditions, NDVI can show increased values due to the high chlorophyll content in algae, especially during peak bloom periods. It can be used for temporal analysis of algal blooms to track changes in bloom extent over time (Choi et al. 2023). Hu and He (2008) found that while NDVI effectively detected floating algae, its values were sensitive to changing environmental conditions, such as aerosols and solar/viewing geometry, affecting measurements' accuracy in nearby waters. In very dense algal blooms, NDVI may saturate and fail to effectively distinguish between high biomass areas (Colkesen et al. 2024). This can lead to underestimations of bloom extent. Atmospheric conditions, cloud cover, and turbid surface waters can significantly impact the detection of algal blooms using the NDVI (Cao et al. 2021).

$$NDVI = (NIR - Red)/(NIR + Red)$$

Huete et al. (1999) developed the Enhanced Vegetation Index (EVI) to improve the NDVI. It is designed to optimise the vegetation signal while minimising the influence of atmospheric conditions and soil background. It is more sensitive to dense vegetation and has also been used in studies to detect chlorophyll-a in lakes. Hence, it has been effectively utilized to detect freshwater algal blooms. It is affected by varying environmental conditions, such as turbidity and sediment load in water bodies, as high turbidity can affect the chlorophyll signals, leading to underestimations of algal bloom presence (Cao et al. 2021; Colkesen et al. 2024).

$$EVI = G \times (NIR - Red)/(NIR + C_1 \times Red - C_2 \times Blue + L)$$

where $G = 2.5$, $C_1 = 6$and $C_2 = 7.5$, and $L = 1$

7.4.2 Normalized Difference Chlorophyll Index (NDCI)

Based on the principles of the NDVI, the NDCI is calculated by taking the difference between reflectance values at 708 nm and 665 nm, normalised by their sum. This method minimises uncertainties in remote sensing reflectance estimates, accounts for seasonal solar azimuth variations, and reduces atmospheric influences (Mishra and Mishra 2012). NDCI effectively captured peak chlorophyll-a concentrations observed in water bodies with high accuracy. NDCI values can directly correlate with Chl-a concentrations based on various models derived from extensive datasets (Mishra et al. 2014; Beck et al. 2016; Karimi et al. 2024; Mpakairi et al. 2024). These models, validated against measured Chl-a, showed strong correlations across diverse water types (Neil et al. 2019). It effectively overcomes the sensitivity of NDVI to chlorophyll-a concentrations, particularly in turbid and productive waters, highlighting its ability to map algal bloom dynamics. It demonstrated reliable performance across various platforms and environments, minimising measurement uncertainty (Caballero et al. 2020). It has also been effectively utilized to assess spatiotemporal variability of algal blooms in small reservoirs (Kislik et al. 2022). The NDCI is simple and versatile, making it most suitable for multiple applications in monitoring and analysing algal blooms.

$$NDCI = (Rededge1 - Red)/(Rededge1 + Red)$$

7.4.3 Floating Algae Index (FAI) and Adjusted Floating Algae Index (AFAI)

Hu (2009) developed the FAI based on MODIS data, which calculates the difference between Near-Infrared (NIR) reflectance and a linear interpolation of red and Shortwave Infrared (SWIR) reflectance values. Positive FAI values generally signal the presence of floating vegetation (Dogliotti et al. 2018). The FAI has been widely applied to monitor macro- and micro-algal blooms in various regions (Oyama et al. 2015). Algal bloom mapping is limited by detection thresholds, such as FAI > − 0.004, initially developed for MODIS but potentially less adaptable to other sensors, leading to possible inaccuracies in detection across platforms. FAI also struggles to distinguish cyanobacteria from aquatic macrophytes, requiring post-classification adjustments (Li et al. 2022).

$$FAI = NIR - [Red + (SWIR - Red) \times (\lambda NIR - \lambda Red)/(\lambda SWIR - \lambda Red)]$$

Fang et al. (2018) developed an adjusted floating algae index (AFAI). It enhances the near-infrared (NIR) peak, distinguishing harmful algal blooms (HABs) from clear water easier. AFAI normalises across different sensors by disregarding the central wavelength, allowing efficient algal bloom extraction and integrating multiple satellite images. Additionally, its flexible threshold enables more accurate detection. It has limitations due to vegetation interference, making distinguishing algal blooms difficult.

$$AFAI = NIR - Red + (SWIR - Red) \times 0.5$$

7.4.4 Algal Bloom Detection Index (ABDI)

The ABDI is one of the most recent indices applied for freshwater algal bloom mapping. Cao et al. (2021) employed the Algal Bloom Detection Index (ABDI) using Sentinel-2 MSI imagery to identify algal blooms in China. After applying an appropriate threshold, the resulting algal bloom maps demonstrated high consistency with visual interpretation maps. It was less sensitive to thin cloud cover and turbid waters than other spectral indices. It proves high accuracy (over 96%) in identifying algal blooms, but its detection threshold variability can lead to inconsistencies, particularly in complex ecosystems with both algal blooms and aquatic vegetation. This limitation underscores the need for adaptive approaches to enhance detection accuracy.

$$ABDI = \left[Rededge2 - Red - (NIR - Red) \times \frac{\lambda Rededge2 - \lambda Red}{\lambda NIR - \lambda Red}\right] - [Red - 0.5 \times Green]$$

7.4.5 Spectral Indices Based Algal Blooms Detection in Ukai Reservoir

The Ukai Reservoir, situated on the Tapi River, is the second-largest reservoir in Gujarat, India. It covers an area of 494.01 km^2. The catchment area of the Ukai Reservoir spans around 62,225 km. A sentinel-2 image (ID: COPERNICUS/S2_HARMONIZED/20220221T053839_20220221T054432_T43QCD), from February 21, 2022, was used to extract algal blooms using NDVI, EVI, FAI, AFAI, NDCI, and ABDI indices in google earth engine platform. Spectral features highlighted the presence of algal blooms in the Ukai reservoir. Algal blooms showed reflectance peaks in the green, red-edge, and near-infrared (NIR) bands, with a pronounced peak between Red Edge 1 and Red Edge 3, likely attributed to chlorophyll and other pigments. In contrast, water demonstrated low reflectance across all bands due to its high absorption characteristics, particularly in the NIR and shortwave infrared (SWIR) regions (Fig. 7.1). Algal blooms exhibited higher values, especially for NDVI and EVI, indicating strong vegetation-like spectral properties, while water indicated negative values for all indices (Fig. 7.2). This highlighted the effective identification of algal blooms using spectral indices as evidenced in Fig. 7.3.

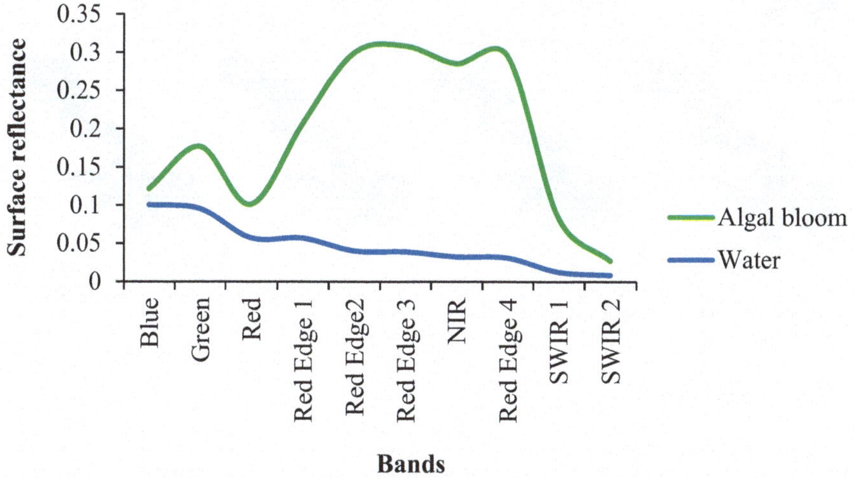

Fig. 7.1 Spectral signatures of algal bloom and water across Sentinel-2 bands

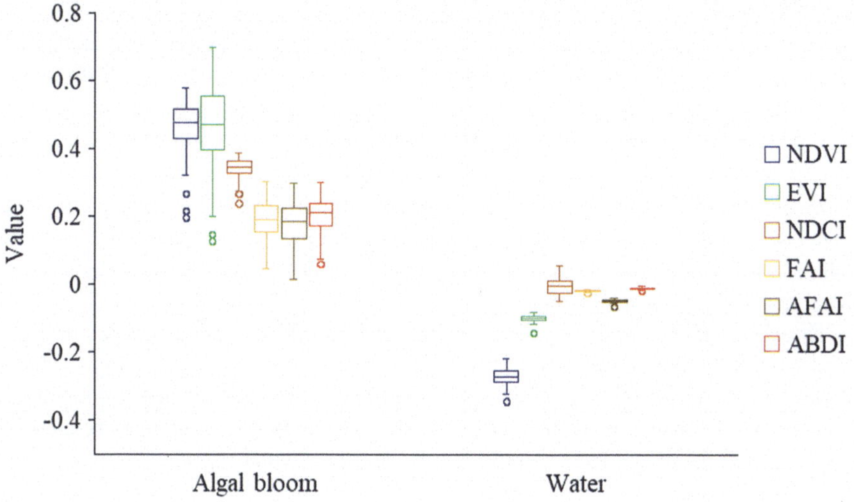

Fig. 7.2 Box-plot distribution of spectral indices values for algal bloom and water

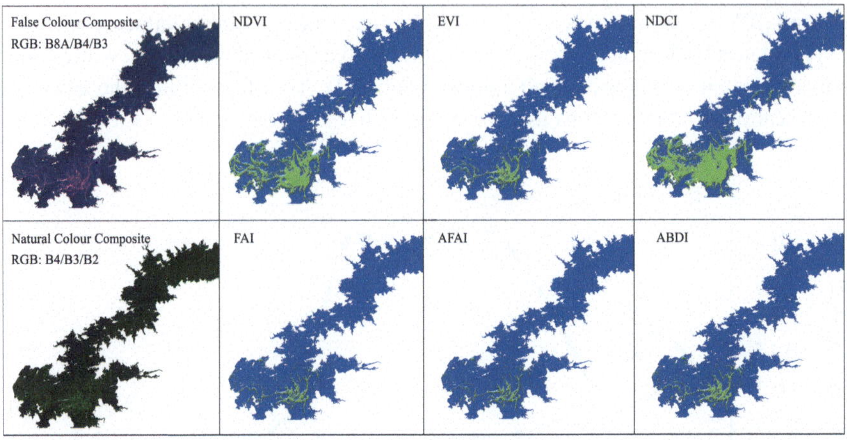

Fig. 7.3 Algal blooms detection in Ukai Reservoir using various spectral indices, highlighting the spatial distribution of algal blooms

7.4.6 Other Indices

Recent advances have led to the development of various new algorithms and spectral indices tailored for algal bloom monitoring. In addition to spectral indices, multiple studies have attempted to develop different band ratios specifically for monitoring algal blooms in freshwater systems (Beck et al. 2016; King et al. 2022; Kislik et al. 2022). Most studies have focused on using chlorophyll-a as a proxy for detecting algal blooms. Some studies, however, have explored the phycocyanin proxy for enhanced detection, particularly for cyanobacterial blooms (Qi et al. 2014; Sòria-Perpinyà et al. 2020). Studies have also concentrated on spectral shape analysis to improve detection accuracy (Maniyar et al. 2022). Microalgal bloom monitoring has been widely explored, with numerous studies establishing effective methods and indices for detection. In contrast, freshwater macroalgal monitoring is emerging as an area of interest, with recent studies developing and refining new indices (as shown in Table 7.1) to improve detection accuracy and offer novel insights into macroalgal dynamics in freshwater systems.

These indices are primarily site-specific, developed and validated according to local environmental conditions, limiting their applicability to broader areas. Many of these were initially designed for coastal and ocean waters and are now increasingly being adapted for use in freshwater environments. However, challenges remain in ensuring their accuracy and reliability across different water types.

Table 7.1 Spectral indices for freshwater macroalgal bloom monitoring

Index	Formula	Satellite	References
Surface algal bloom index (SABI)	$(NIR - Red)/(Blue + Green)$	MODIS	Alawadi (2010)
Normalized difference aquatic vegetation index (NDAVI)	$(NIR - Blue)/(NIR + Blue)$	Landsat TM/ ETM+	Villa et al. (2014)
Water adjusted vegetation index (WAVI)	$(1 + L) \times \frac{NIR-Blue}{NIR+Blue+L}$	Landsat TM/ ETM+	Villa et al. (2014)
Cyanobacteria and macrophytes index (CMI)	$Green - Blue - ((SWIR - Blue) \times (\lambda Green - \lambda Blue)/(\lambda SWIR - \lambda Red)$	Landsat OLI	Chen et al. (2020)
Macroalgae index (MAI)	$(Green + Red) - [SWIR1 + (Blue - SWIR1) \times ((\lambda SWIR1 - \lambda Green)/(\lambda SWIR1 - \lambda Blue))$	MODIS	Liang et al. (2017)
Three-band macroalgae detection index (TMI)	$((Green - Red) \times NIR/Red) \times 50$	Landsat OLI	Nazeer et al. (2023)

7.5 Conclusion

Remote sensing-based spectral indices have revolutionized the monitoring of freshwater algal blooms, offering solutions to the limitations of traditional monitoring methods. The evolution from basic vegetation indices like NDVI to specialized indices such as NDCI, FAI, and ABDI reflects significant progress in this field. These advancements have enhanced the ability to detect, quantify, and monitor algal blooms across diverse freshwater ecosystems with improved temporal and spatial resolution. Satellite remote sensing has emerged as a valuable tool for early detection and monitoring of algal blooms. The capability to provide synoptic coverage over large areas, combined with regular revisit times, enables tracking bloom dynamics and their spatiotemporal evolution. This has proven crucial for timely management interventions and public health protection. The development of specialized indices like ABDI has improved detection accuracy even under challenging conditions such as thin cloud cover and turbid waters, demonstrating the continuing advancement in this field.

However, several key research gaps and challenges remain. The site-specific nature of many indices limits their universal applicability, suggesting the need for more robust, globally applicable approaches. Atmospheric correction and sensor calibration present technical challenges, particularly in complex inland water bodies. The selection of accurate threshold limits also poses a challenge, as variability in environmental conditions can complicate this process, potentially reducing classification accuracy. Additionally, the distinction between different types of aquatic vegetation and various algal species remains problematic, highlighting the need for more sophisticated detection methods. The development of adaptive algorithms that can account for regional variations in water quality parameters and environmental conditions is essential. Integrating multiple spectral indices with machine learning approaches could enhance detection accuracy and reliability. Future research should also focus on improving the ability to distinguish between different algal species, particularly toxic varieties, and strengthening early warning capabilities by combining spectral indices with environmental parameters. Standardization of methods across different satellite platforms is crucial to ensure data continuity and compatibility.

As climate change continues to influence the frequency and intensity of algal blooms, the continued development and refinement of spectral indices will be crucial. Integrating these remote sensing approaches with traditional monitoring methods could provide comprehensive monitoring systems that support effective ecosystem management and public health protection. Furthermore, the advancement of satellite technology and the launch of upcoming sensors with improved spectral, spatial, and temporal resolutions like 3MI, GISAT1R, TRISHNA will likely open new opportunities for more accurate and timely monitoring of freshwater algal blooms. This chapter has demonstrated that while significant progress has been made in spectral index-based monitoring of freshwater algal blooms, there remains considerable scope for improvement. The future of algal bloom monitoring lies in the development of more sophisticated, integrated approaches that can provide reliable, timely, and actionable information for water resource managers and stakeholders. As we continue to face increasing environmental challenges, the role of remote sensing and spectral indices in monitoring and managing freshwater ecosystems will become increasingly important.

Competing Interests The author declares that there are no competing interests related to this chapter.

References

Akinnawo SO (2023) Eutrophication: causes, consequences, physical, chemical and biological techniques for mitigation strategies. Environ Challenges 12:100733. https://doi.org/10.1016/j.envc.2023.100733

Alawadi F (2010) Detection of surface algal blooms using the newly developed algorithm surface algal bloom index (SABI). In: Remote sensing of the ocean, sea ice, and large water regions 2010, vol 7825. SPIE, pp 45–58. https://doi.org/10.1117/12.862096

Amorim CA, do Nascimento Moura A (2021) Ecological impacts of freshwater algal blooms on water quality, plankton biodiversity, structure, and ecosystem functioning. Sci Total Environ 758:143605. https://doi.org/10.1016/j.scitotenv.2020.143605

Bănăduc D, Simić V, Cianfaglione K et al (2022) Freshwater as a sustainable resource and generator of secondary resources in the 21st century: Stressors, threats, risks, management and protection strategies, and conservation approaches. Int J Environ Res Public Health 19(24):16570. https://doi.org/10.3390/ijerph192416570

Beal MR, Özdoğan M, Block PJ (2024) A machine learning and remote sensing-based model for algae pigment and dissolved oxygen retrieval on a small inland lake. Water Resour Res 60(3):e2023WR035744. https://doi.org/10.1029/2023WR035744

Beck R, Zhan S, Liu H et al (2016) Comparison of satellite reflectance algorithms for estimating chlorophyll-a in a temperate reservoir using coincident hyperspectral aircraft imagery and dense coincident surface observations. Remote Sens Environ 178:15–30. https://doi.org/10.1016/j.rse.2016.03.002

Bellinger EG, Sigee DC (2015) Freshwater algae: identification, enumeration and use as bioindicators. Wiley, London. https://doi.org/10.1002/9780470689554

Brooks BW, Lazorchak JM, Howard MD et al (2016) Are harmful algal blooms becoming the greatest inland water quality threat to public health and aquatic ecosystems? Environ Toxicol Chem 35(1):6–13. https://doi.org/10.1002/etc.3220

Burford MA, Willis A, Xiao M et al (2023) Understanding the relationship between nutrient availability and freshwater cyanobacterial growth and abundance. Inland Waters 13(2):143–152. https://doi.org/10.1080/20442041.2023.2204050

Caballero I, Fernández R, Escalante OM et al (2020) New capabilities of Sentinel-2A/B satellites combined with in situ data for monitoring small harmful algal blooms in complex coastal waters. Sci Rep 10:8743. https://doi.org/10.1038/s41598-020-65600-1

Cao M, Qing S, Jin E et al (2021) A spectral index for the detection of algal blooms using Sentinel-2 Multispectral Instrument (MSI) imagery: a case study of Hulun Lake, China. Int J Remote Sens 42(12):4514–4535. https://doi.org/10.1080/01431161.2021.1897186

Chen N, Wang S, Zhang X, Yang S (2020) A risk assessment method for remote sensing of cyanobacterial blooms in inland waters. Sci Total Environ 740:140012. https://doi.org/10.1016/j.scitotenv.2020.140012

Chen J, Chen S, Fu R et al (2022) Remote sensing big data for water environment monitoring: current status, challenges, and future prospects. Earth's Future 10(2):e2021EF002289. https://doi.org/10.1029/2021EF002289

Choi B, Lee J, Park B, Sungjong L (2023) A study of cyanobacterial bloom monitoring using unmanned aerial vehicles, spectral indices, and image processing techniques. Heliyon 9(5). https://doi.org/10.1016/j.heliyon.2023.e16343

Colkesen I, Ozturk MY, Altuntas OY (2024) Comparative evaluation of performances of algae indices, pixel-and object-based machine learning algorithms in mapping floating algal blooms using Sentinel-2 imagery. Stochastic Environ Res Risk Assess 38(4):1613–1634. https://doi.org/10.1007/s00477-023-02648-1

Dogliotti AI, Gossn JI, Vanhellemont Q, Ruddick KG (2018) Detecting and quantifying a massive invasion of floating aquatic plants in the Río de la Plata turbid waters using high spatial resolution ocean color imagery. Remote Sens 10(7):1140. https://doi.org/10.3390/rs10071140

Fang C, Song KS, Shang YX et al (2018) Remote sensing of harmful algal blooms variability for Lake Hulun using adjusted FAI (AFAI) algorithm. J Environ Inform 34(2):108–122

Gholizadeh MH, Melesse AM, Reddi L (2016) A comprehensive review on water quality parameters estimation using remote sensing techniques. Sensors 16(8):1298. https://doi.org/10.3390/s16081298

Gholizadeh MH, Melesse AM, Reddi L (2016) Spaceborne and airborne sensors in water quality assessment. Int J Remote Sens 37(14):3143–3180. https://doi.org/10.1080/01431161.2016.1190477

Griffith AW, Gobler CJ (2020) Harmful algal blooms: a climate change co-stressor in marine and freshwater ecosystems. Harmful Algae 91:101590. https://doi.org/10.1016/j.hal.2019.03.008

Guo F, Kainz MJ, Sheldon F, Bunn SE (2016) The importance of high-quality algal food sources in stream food webs–current status and future perspectives. Freshwater Biol 61(6):815–831. https://doi.org/10.1111/fwb.12755

Hallegraeff GM (2003) Harmful algal blooms: a global overview. In: Manual on Harmful Marine Microalgae. Monographs on Oceanographic Methodology, 2nd edn, IOC-UNE-SCO, Paris, 25–49

Ho JC, Michalak AM, Pahlevan N (2019) Widespread global increase in intense lake phytoplankton blooms since the 1980s. Nature 574(7780):667–670. https://doi.org/10.1038/s41586-019-1648-7

Hu C (2009) A novel ocean color index to detect floating algae in the global oceans. Remote Sens Environ 113(10):2118–2129. https://doi.org/10.1016/j.rse.2009.05.012

Hu C, He MX (2008) Origin and offshore extent of floating algae in Olympic sailing area. Eos Trans Am Geophys Union 89(33):302–303. https://doi.org/10.1029/2008EO330002

Huang C, Chen Y, Zhang S, Wu J (2018) Detecting, extracting, and monitoring surface water from space using optical sensors: a review. Rev Geophys 56(2):333–360. https://doi.org/10.1029/2018RG000598

Huete A, Justice C, Van Leeuwen W (1999) MODIS vegetation index (MOD13). Algorithm Theor Basis Doc 3(213):295–309

Huisman J, Codd GA, Paerl HW et al (2018) Cyanobacterial blooms. Nat Rev Microbiol 16(8):471–483. https://doi.org/10.1038/s41579-018-0040-1

Igwaran A, Kayode AJ, Moloantoa KM et al (2024) Cyanobacteria harmful algae blooms: causes, impacts, and risk management. Water Air Soil Pollut 235(1):71. https://doi.org/10.1007/s11270-023-06782-y

IPCC (2022) Climate change 2022: impacts, adaptation and vulnerability. Contribution of working group II to the sixth assessment report of the intergovernmental panel on climate change. Cambridge University Press, Cambridge. https://doi.org/10.1017/9781009325844.033

Karimi B, Hashemi SH, Aghighi H (2024) Application of Landsat-8 and Sentinel-2 for retrieval of chlorophyll-a in a shallow freshwater lake. Adv Space Res 74(1):117–129. https://doi.org/10.1016/j.asr.2024.03.056

King T, Hundt S, Hafen K et al (2022) Mapping the probability of freshwater algal blooms with various spectral indices and sources of training data. J Appl Remote Sens 16(4):044522. https://doi.org/10.1117/1.JRS.16.044522

Kislik C, Dronova I, Grantham TE, Kelly M (2022) Mapping algal bloom dynamics in small reservoirs using Sentinel-2 imagery in Google Earth Engine. Ecol Indic 140:109041. https://doi.org/10.1016/j.ecolind.2022.109041

Li J, Liu Y, Xie S et al (2022) Landsat-satellite-based analysis of long-term temporal spatial dynamics of cyanobacterial blooms: a case study in Taihu Lake. Land 11(12):2197. https://doi.org/10.3390/land11122197

Liang Q, Zhang Y, Ma R, Loiselle S, Li J, Hu M (2017) A MODIS-based novel method to distinguish surface cyanobacterial scums and aquatic macrophytes in Lake Taihu. Remote Sens 9(2):133. https://doi.org/10.3390/rs9020133

Liu S, Glamore W, Tamburic B et al (2022) Remote sensing to detect harmful algal blooms in inland waterbodies. Sci Total Environ 851:158096. https://doi.org/10.1016/j.scitotenv.2022.158096

Maniyar CB, Kumar A, Mishra DR (2022) Continuous and synoptic assessment of Indian inland waters for harmful algae blooms. Harmful Algae 111:102160. https://doi.org/10.1016/j.hal.2021.102160

Mishra S, Mishra DR (2012) Normalized difference chlorophyll index: a novel model for remote estimation of chlorophyll-a concentration in turbid productive waters. Remote Sens Environ 117:394–406. https://doi.org/10.1016/j.rse.2011.10.016

Mishra DR, Schaeffer BA, Keith D (2014) Performance evaluation of normalized difference chlorophyll index in northern Gulf of Mexico estuaries using the Hyperspectral Imager for the Coastal Ocean. Giscience Remote Sens 51(2):175–198. https://doi.org/10.1080/15481603.2014.895581

Montero D, Aybar C, Mahecha MD et al (2023) A standardized catalogue of spectral indices to advance the use of remote sensing in Earth system research. Sci Data 10(1):1–20. https://doi.org/10.1038/s41597-023-02096-0

Mpakairi KS, Muthivhi FF, Dondofema F et al (2024) Chlorophyll-a unveiled: unlocking reservoir insights through remote sensing in a subtropical reservoir. Environ Monit Assess 196(4):401. https://doi.org/10.1007/s10661-024-12554-w

Musie W, Gonfa G (2023) Fresh water resource, scarcity, water salinity challenges and possible remedies: A review. Heliyon. https://doi.org/10.1016/j.heliyon.2023.e18685

Nazeer M, Alsahli MM, Nichol JE et al (2023) A novel three-band macroalgae detection index (TMI) for aquatic environments. Int J Remote Sens 44(7):2359–2381. https://doi.org/10.1080/01431161.2023.2202339

Neil C, Spyrakos E, Hunter PD, Tyler AN (2019) A global approach for chlorophyll-a retrieval across optically complex inland waters based on optical water types. Remote Sens Environ 229:159–178. https://doi.org/10.1016/j.rse.2019.04.027

Oyama Y, Matsushita B, Fukushima T (2015) Distinguishing surface cyanobacterial blooms and aquatic macrophytes using Landsat/TM and ETM+ shortwave infrared bands. Remote Sens Environ 157:35–47. https://doi.org/10.1016/j.rse.2014.04.031

Paerl HW, Otten TG (2013) Harmful cyanobacterial blooms: causes, consequences, and controls. Microb Ecol 65:995–1010. https://doi.org/10.1007/s00248-012-0159-y

Palmer SC, Kutser T, Hunter PD (2015) Remote sensing of inland waters: challenges, progress and future directions. Remote Sens Environ 157:1–8. https://doi.org/10.1016/j.rse.2014.09.021

Qi L, Hu C, Duan H et al (2014) A novel MERIS algorithm to derive cyanobacterial phycocyanin pigment concentrations in a eutrophic lake: theoretical basis and practical considerations. Remote Sens Environ 154:298–317. https://doi.org/10.1016/j.rse.2014.08.026

Rodríguez-López L, Duran-Llacer I, González-Rodríguez L et al (2020) Spectral analysis using LANDSAT images to monitor the chlorophyll-a concentration in Lake Laja in Chile. Ecol Inform 60:101183. https://doi.org/10.1016/j.ecoinf.2020.101183

Roegner AF, Corman JR, Sitoki LM et al (2023) Impacts of algal blooms and microcystins in fish on small-scale fishers in Winam Gulf, Lake Victoria: implications for health and livelihood. Ecol Soc 28(1):49. https://doi.org/10.5751/es-13860-280149

Rolim SBA, Veettil BK, Vieiro AP, Kessler AB, Gonzatti C (2023) Remote sensing for mapping algal blooms in freshwater lakes: a review. Environ Sci Pollut Res 30(8):19602–19616. https://doi.org/10.1007/s11356-023-25230-2

Rouse JW, Haas RH, Schell JA, Deering DW (1974) Monitoring vegetation systems in the Great Plains with ERTS. NASA Spec Publ 351(1):309

Sanseverino I, Conduto D, Pozzoli L et al (2016) Algal bloom and its economic impact. EUR 27905 EN. https://doi.org/10.2788/660478

Sahu H, Purohit P, Srivastava A, Singh R, Mishra AP, Arunachalam K, Kumar U (2024) Comparative assessment of soil parameters and ecological dynamics in the Western Himalayan wetland and its surrounding periphery. Environ Qual Manage 34(1):e22283. https://doi.org/10.1002/tqem.22283

Sheath RG, Wehr JD (2015) Introduction to the freshwater algae. In: Freshwater algae of North America. Academic Press, pp 1–11. https://doi.org/10.1016/B978-0-12-385876-4.00001-3

Shen L, Xu H, Guo X (2012) Satellite remote sensing of harmful algal blooms (HABs) and a potential synthesized framework. Sensors 12(6):7778–7803. https://doi.org/10.3390/s120607778

Shi K, Zhang Y, Qin B, Zhou B (2019) Remote sensing of cyanobacterial blooms in inland waters: present knowledge and future challenges. Sci Bull 64(20):1540–1556. https://doi.org/10.1016/j.scib.2019.07.002

Singh R, Saritha V, Pande CB (2024) Dynamics of LULC changes, LST, vegetation health and climate interactions in wetland buffer zone: a remote sensing perspective. Phys Chem Earth, Parts A/B/C 135:103660. https://doi.org/10.1016/j.pce.2024.103660

Singh R, Saritha V, Mishra AP, Pande CB, Sahu H (2025) A comprehensive analysis of water quality index in a wetland ecosystem supporting drinking water to major cities in Rajasthan, India. J Clean Prod 487:144593. https://doi.org/10.1016/j.jclepro.2024.144593

Sòria-Perpinyà X, Vicente E, Urrego P et al (2020) Remote sensing of cyanobacterial blooms in a hypertrophic lagoon (Albufera of València, Eastern Iberian Peninsula) using multitemporal Sentinel-2 images. Sci Total Environ 698:134305. https://doi.org/10.1016/j.scitotenv.2019.134305

Tripathi A, Tiwari RK (2021) Role of space-borne remote sensing technology for monitoring of urban and environmental hazards. In: Recent technologies for disaster management and risk reduction: sustainable community resilience & responses, pp 295–317. https://doi.org/10.1007/978-3-030-76116-5_18

Vári Á, Podschun SA, Erős T et al (2022) Freshwater systems and ecosystem services: challenges and chances for cross-fertilization of disciplines. Ambio 51(1):135–1511. https://doi.org/10.1007/s13280-021-01556-4

Villa P, Mousivand A, Bresciani M (2014) Aquatic vegetation indices assessment through radiative transfer modeling and linear mixture simulation. Int J Appl Earth Obs Geoinf 30:113–127. https://doi.org/10.1016/j.jag.2014.01.017

Wang W, Shi K, Zhang Y, Li N, Sun X, Zhang D et al (2022) A ground-based remote sensing system for high-frequency and real-time monitoring of phytoplankton blooms. J Hazard Mater 439:129623. https://doi.org/10.1016/j.jhazmat.2022.129623

Watson SB, Whitton BA, Higgins SN, Paerl HW, Brooks BW, Wehr JD (2015) Harmful algal blooms. In: Wehr JD, Sheath RG, Kociolek JP (eds) Freshwater algae of North America. Academic Press, pp 873–920

Wu D, Li R, Liu J, Khan N (2023) Monitoring algal blooms in small lakes using drones: a case study in southern illinois. J Contemp Water Res Educ 177(1):83–93. https://doi.org/10.1111/j.1936-704X.2022.3383.x

Chapter 8
Climate Change Adaptation and Mitigation: A Short Review

Prerana Badoni and Rekha Dhanai

Abstract Climate change affects people's livelihoods, health, and access to natural resources worldwide, with vulnerable communities disproportionately impacted due to factors like poverty and reliance on subsistence farming. This review examines the climate change concept with respect to the agriculture field. Mountain regions serve as early indicators of climate change. Changes in the strength and timing of monsoons, driven by climate change, have significant consequences for river flows, groundwater recharge, natural hazards, ecosystems, and human livelihoods. While researchers struggle to grasp the diverse impacts of climate change fully, it is clear that it increasingly affects people's lives, particularly in developing nations. Men and women experience climate change differently, with men often compelled to migrate while climate change exacerbates the feminization of agriculture. Developing effective adaptation strategies is crucial to address these challenges. To address emissions of greenhouse gas (GHG) and climate change in food and agriculture, prioritizing efficiency of energy use and ecologically sustainable production practices are crucial. Increasing soil carbon levels and implementing energy-efficient systems can positively impact greenhouse gas levels and climate change outcomes. However, fragmented agriculture and localized mitigation efforts pose challenges, especially for resource-intensive implementation. Disparities exist between developed nations required to reduce emissions and developing nations experiencing climate change effects, complicating mitigation efforts. International aid mechanisms could assist vulnerable smallholders in developing nations, but current opportunities are limited. Some mitigation strategies may disrupt conventional food production, risking food security. Ultimately, the most effective and economical approach is to avoid activities contributing to global warming.

Keywords Climate change · Mitigation · Global warming · Agriculture

---

P. Badoni
Department of Microbiology, Combined PG Institute of Medical Sciences and Research, Kuanwala, Dehradun, India

R. Dhanai (✉)
Department of Agriculture, Tula's Institute, Dehradun, India
e-mail: dhanai@tulas.edu.in

V. Saritha et al. (eds.), *Remote Sensing for Environmental Monitoring*,
https://doi.org/10.1007/978-981-96-5546-5_8

8.1 Introduction

As a result of utilizing fossil fuels for energy and other human activities, greenhouse gases have accumulated in the atmosphere, which has led to climate change. Peoples are vulnerable to climate change directly through variations in water, air, and food quality and indirectly through altered weather patterns like temperature, precipitation, and the severity of life-threatening events.

Therefore, Climate change refers to a shift in the climate that is caused by either natural or human-induced variations and that lasts for a long period of time, typically decades or longer, as evidenced by variations in the mean and/or variability of its attributes (Shah et al. 2014). In addition to unpredictability in natural climate has been observed over comparable periods, United Nations Framework Convention on Climate Change (UNFCCC) defined climate change as an alteration in the climate brought on by different human activities that changes the composition of the atmosphere globally (Singh et al. 2010). Variations in the average condition and other climate data across all time and space scales are referred to as climate variability, going beyond the size of a single weather event.

Variability may result from changes in external forcing, whether man-made or natural, as well as from the internal natural processes that take place within the climate system. Mountain ranges are primary signs of a changing climate (Dhanai et al. 2014; Singh et al. 2010). Although there is not enough particular information and data on human well-being, climatic changes will also have an influence on people's health, livelihoods, and security of natural resources everywhere (World Health Organisation 2008). People are at greater risk from climate change due to poverty, inadequate infrastructure, dependency on subsistence farming, and usage of forest products as a means of livelihood. The Climate change consequences are compounded by numerous additional societal and environmental pressures, many of which are already known to be severe (Haigh 1989). As a result, it is necessary to strengthen community capacities and develop strategies for climate change adaptation.

The monsoon's timing and intensity are drastically changing as an effect of ongoing climate change. While not uniform with relation to direction, intensity, or rate across the region, the consequences could be profound for people and their means of subsistence, as well as for the ecosystem, natural hazards, groundwater recharge, and river flows (Treidel et al. 2011). Researchers are having difficulty identifying the variety of impacts due to the present condition of our understanding of risk evaluation and climate change is required to direct future action (Hurlbert et al. 2019). People's lives are being impacted by climate change more and more, especially the poor in developing nations. The susceptibility of men and women to climate change is not equal, they will experience different consequences as a result. The case studies demonstrate that while men are compelled to migrate, climate change exacerbates the trend of feminization in agriculture (Agrawal et al. 2015; Tumbe 2015). Because their livelihoods depend on agriculture and availability or accessibility to natural resources (Rautela and Karki 2015), in developing countries both men and women are already dealing with problems that climate change exacerbates.

8.2 Vulnerability to Climate Change

The term "vulnerability" describes how susceptible a system or society is to adverse impact of changing climate that includes climate extremes and variability, in addition to how well they able to resist these impacts. The degree of vulnerability of a system is determined by its sensitivity and adaptability along with the kind, intensity, and pace of changing climate and the differences to which it is exposed (Change 2007). Vulnerability can be highly dynamic in both space and time due to the system consequences and adaptive capacity can vary greatly over decades and within nations (Change 2007).

It is consequently essential to create resilient agricultural systems that can absorb disruptions and have a significant capacity for stress and change adaptation (Lipper et al. 2014). Climate change has both local and global effects on food security (Betts et al. 2018). Climate change will impact agricultural food security systems worldwide, including subsistence-level, import and export-based systems. Affected industries including agriculture, forestry, livestock, and fisheries will be affected by variations in average temperature and precipitation as well as an increase in severe events. Numerous effects need to be addressed across sectors, including increased events of soil erosion and land degradation, changes in availability of water, loss of biodiversity, more recurring and severe disease and pest outbreaks, as well as natural disasters (Rosenzweig and Liverman 1992).

8.3 Adaptation to Climate Change

The Intergovernmental Panel on Climate Change (IPCC) characterizes adaptation as a modification of human or natural systems in reaction to real or predicted climatic stressors or their consequences, which minimizes damage or exploits of advantageous opportunities (Change 2007). The act of changing to better fit a circumstance or environment is called adaptation. It is a process that is perpetual. It entails managing both abrupt events and gradual, long-term changes that occur in a region over time. These changes can be caused by both climate and weather, in addition to other factors like economic and societal pressures, market volatility or connect, ecosystem dynamics, regulations and laws, infrastructural facilities, etc.

It is possible to distinguish between anticipatory, autonomous, and planned adaptations. People whose livelihoods depend on agriculture have historically created means of coping with climate change independently. The rate of climate change occurring today will affect known variability patterns to the extent that individuals will face challenges they are ill-prepared to handle (Luers and Moser 2006). Deliberate and proactive adaptation is therefore a contemporary problem. However, since major vulnerability is regional, hence site-specific adaptation is necessary. It has been noted that the Himalayan Mountains' traditional agriculture offers a wealth of agricultural biodiversity and has proven resistive to crop diseases. The crops have the innate potentials to resist environmental risks and other natural threats because they have been adapted to the environmental circumstances in the area (Dhanai et al. 2014).

8.4 Effective Adaptation Strategies

(a) Creation of more economical strategies with a number of benefits. Implementing substantial financial incentives such as microcredit, covering environmental services, and reducing the agriculture supply sector's marketing power may be necessary to achieve this (Turral et al. 2011).
(b) Employing regional coping mechanisms and indigenous knowledge as a springboard for adaptation planning. Although local communities have a tremendous amount of information regarding their ability to cope with climate extremes and severe weather occurrences, rapidly shifting environmental conditions will necessitate updating regional knowledge with additional scientific observations and creating association between neighboring regions and nations to exchange know-how with regions currently undergoing these transformations (Intergovernmental Panel on Climate Change 2023).
(c) Support for pertinent domestic agricultural research. Research should concentrate on cultivars resistance to emerging diseases and pests as well as heat, salinity, and drought (Change 2007).
(d) The use of gender-responsive tactics. When creating strategies, it is important to take into account the distinct roles, responsibilities, rights, and resources of boys and girls as well as men and women (Agrawal et al. 2015).
(e) Supporting multidisciplinary institutions and procedures that can ease changes in the use and access to resource, resolve differences, and defend the right of both individuals' and groups' to natural assets and land.

8.5 Adaptation Planning

Any planning for adaptation must consider the degree of uncertainty in scenarios for environmental changes, and the actual plans must be adaptable. The adaptation planning may encounter some issues, which must be considered when seeking a solution. Although urgent, adaptation calls for significant financial investment. Developing countries are unlikely to, especially the least developed ones, will have the funds and technical competence necessary for a foreseen and planned intervention. To cover the additional expenses of designing and putting interventions into place, technical and financial assistance will be needed (Matsa and Matsa 2021).

Climate change is regionally and locally dependent. In-depth analyses of local impacts must be included in adaptation assessment methodologies to comprehend and plan interventions. These interventions must be incorporated into more comprehensive, coherent adaptation programmes. As the effects of climate change evolve, so must each component of adaptation. Work on adaptation calls for a variety of technical techniques that can be used at various rates and times. This also implies that any necessary inputs must be planned and maintained throughout the entire adaptation period.

8.6 Mitigation to Climate Change

Some measures, which are characterized as any human (anthropogenic) action that can improve the sinks of greenhouse gas emissions or lessen their sources (abatement), have been adopted to lessen the negative effects of climate change (sequestration). A long-term mitigation scenario or several are typically developed as part of mitigation assessments. A quantified projection of how future greenhouse gas emissions can be decreased in comparison to one or more baseline scenarios is known as a mitigation scenario. An evaluation of technologies and procedures that can both mitigate climate change and advance governmental development goals is provided to policymakers by a mitigation assessment.

The assessment also identifies potential investments for projects and programmes and gives a general idea of how much would it cost to stop climatic changes. Climate change is largely caused by changes in land use and agriculture. The Fourth Assessment Report of the IPCC states that agriculture, which includes the production of crops, pastures, and cattle, and forestry account for 13 and 17 percent of total anthropogenic greenhouse gas emissions, respectively (Change 2007). Other emissions related to agriculture such as fertilizer production, food supply, packaging (waste), and cooling and heating are not included in this contribution (energy supply).

A human intervention intended to lessen greenhouse gas sources or improve sinks is known as climate change mitigation (Change 2007). Climate change mitigation is a global responsibility. A significant amount of potential for reducing GHGs is offered by forestry and agriculture. By 2030, the IPCC predicts that agriculture will have a global technical mitigation potential of between 5500 and 6000 Mt CO_2 equivalent annually, with 89 percent of that potential coming from soil carbon sequestration (Change 2007).

At the national level, the evaluation of mitigation potential continues to be a crucial tool for setting priorities. The focus of mitigation activities should be on the five major natural resources sector, which includes cattle, forestry, rangeland, agriculture, and fisheries. In the agricultural industry as a whole, the traditional mitigation options include forest-related measures to decrease degradation of forest and deforestation and increase reforestation and afforestation, as well as interventions in forest management to preserve or raise the carbon density of forests, and also take initiatives to increase carbon stocks and improve fuel substitution.

Even though there are many adaptation options, there are still no established mitigation strategies for cropland. These actions include increasing biodiversity, encouraging production of legumes in crop rotations, making quality seeds more accessible, implementing integrated livestock and agriculture systems, avoiding the burning of crop residue, encouraging systems of low-energy production, improving forest fire control and increasing energy efficiency in commercial agriculture and agro based industries. Among the most promising choices with a variety of synergies is soil carbon sequestration. This option offers advantages for soil fertility and productivity, biodiversity, and water storage capacity of soil by raising concentrations of carbon in the soil through more suitable management practices.

8.7 Remote Sensing Applications in Climate Mitigation

Remote sensing and GIS (Geographic Information System) technologies are crucial for studying climate change and agriculture. They provide a wide range of environmental data to create precision maps, offer crop information, estimate yields, and more. This is highly beneficial not only for farmers but also for governments, enabling them to implement necessary projects and support initiatives to aid farmers effectively. It helps in monitoring of natural disasters, evaluating alterations in land use pattern and also in tracking deforestation.

The abundance of data produced facilitates well-informed resource management and environmental conservation decision-making (Machireddy 2023). There are many ways that remote sensing is used in carbon-related tasks. Remote sensing plays a vital role in understanding climate change dynamics by providing researchers with essential data on Earth's climate system. It captures temperature, precipitation, sea level, ice cover, and more information. The high spatial and temporal resolution of remote sensing data enables more accurate climate modeling and promptly enhances decision-making for addressing climate change challenges.

8.8 Recommendation and Conclusion

Energy efficiency and ecologically sustainable production methods must be prioritized more in order to deal with GHGs and climate change concerns in the food and agricultural sector (Machireddy 2023). Enhancing soil carbon levels and developing energy-efficient food and agricultural systems has the ability to enhance climate change and greenhouse gas situations. Because agriculture is fragmented and mitigation is localized, implementation requires a lot of resources (Vermeulen et al. 2012), which may cause mitigation implementation to suffer. Additionally, if developing nations' extensive systems are to play a significant role. In this regard, countries that are required to reduce the emissions of GHGs (mainly developed) and those that are more affected by the ways that climate change is altering patterns of climate variability have very different perceptions of mitigation. However, there is a new opportunity—international procedures that may direct financial aid to the most disadvantaged in developing countries—that is yet a long way from improving the lives of smallholders. Some mitigation strategies may even interfere with conventional food production processes, endangering the security of those systems' supply of food. Avoiding actions that are mostly contribute to the global warming is the most cost-effective and simplest strategy to stop harm from the effects of human activity on the food production systems and climate.

Competing Interests The authors declare no potential conflicts of interest.

References

Agrawal T, Chandrasekhar S, Gandhi I (2015) Short term migrants in India: Characteristics, wages and work transition. Retrieved from Mumbai, India: https://EconPapers.repec.org/RePEc:ind:igiwpp

Betts RA, Alfieri L, Bradshaw C, Caesar J, Feyen L, Friedlingstein P, Morfopoulos C et al (2018) Changes in climate extremes, fresh water availability and vulnerability to food insecurity projected at 1.5 C and 2 C global warming with a higher-resolution global climate model. Philos Trans R Soc A: Math Phys Eng Sci 376(2119):20160452

Change OC (2007) Intergovernmental panel on climate change. World Meteor Org 52:1–43

Dhanai R, Negi RS, Singh S, Parmar MK (2014) The effects of climate change on natural resources and socio-economic condition of Himalayan communities of Uttarakhand, India. Int J Modern Commun Technol Res 2(11):265749

Haigh MJ (1989) The Himalayan dilemma: reconciling development and conservation. Routledge

Hurlbert M et al. (2019) Risk management and decision making in relation to sustainable development. IPCC (Intergovernmental Panel on Climate Change). https://www.ipcc.ch/site/assets/uploads/sites/4/2019/11/10_Chapter-7.pdf

Intergovernmental Panel on Climate Change (2023) Climate Change 2022—impacts, adaptation and vulnerability: working group II contribution to the sixth assessment report of the Intergovernmental Panel on Climate Change. Cambridge University Press, Cambridge

Lipper L, Thornton P, Campbell BM, Baedeker T, Braimoh A, Bwalya M, Henry K et al (2014) Climate-smart agriculture for food security. Nat Clim Chang 4(12):1068–1072

Luers AL, Moser SC (2006) Preparing for the impacts of climate change in California: opportunities and constraints for adaptation. UC Berkeley: California Institute for Energy and Environment (CIEE). Retrieved from https://escholarship.org/uc/item/28r0c56d

Machireddy SR (2023) Natural resource management using remote sensing and geographic information systems. Environ Sci Eng 2(2):73–82

Matsa M, Matsa M (2021) Impact of climate change in Zimbabwe. In: Climate change and agriculture in Zimbabwe: sustainability in minority farming communities, pp 21–30

Rautela P, Karki B (2015) Impact of climate change on life and livelihood of indigenous people of higher Himalaya in Uttarakhand, India. Am J Environ Protect 3(4):112–124

Rosenzweig C, Liverman D (1992) Predicted effects of climate change on agriculture: a comparison of temperate and tropical regions. In: Global climate change: implications, challenges, and mitigation measures, pp 342

Shah F, Nie L, Cui K, Shah T, Wu W, Chen C, Huang J et al (2014) Rice grain yield and component responses to near 2 C of warming. Field Crop Res 157:98–110

Singh SP, Singh V, Skutsch M (2010) Rapid warming in the Himalayas: ecosystem responses and development options. Climate Dev 2(3):221–232

Treidel H, Martin-Bordes JL, Gurdak JJ (2011) Climate change effects on groundwater resources: a global synthesis of findings and recommendations. CRC Press

Tumbe C (2015) Missing men, migration and labour markets: evidence from India. Indian J Labour Econ 58(2):245–267

Turral H, Burke J, Faurès J-M (2011) Climate change, water and food security. https://openknowledge.fao.org/server/api/core/bitstreams/6abba49b-2200-41e0-b49c-53ea7c904eab/content/i2096e.htm

Vermeulen SJ, Campbell BM, Ingram JSI (2012) Climate change and food systems. Annu Rev Environ Resour 37(1):195–222

World Health Organisation (2008) Climate change and health. https://unfccc.int/files/adaptation/sbi_agenda_item_adaptation/application/pdf/who.pdf

Chapter 9
Satellite-Based Monitoring of Trophic State: Assessing Water Quality in Lake Llanquihue

Neftalí Flores Betansson, Lien Rodríguez-López, and Santiago Yépez

Abstract Lake Llanquihue is a body of water of significant economic and ecological importance for Chile, being the second largest lake in the country and a hub for essential tourist activities. Consequently, this study aims to assess its water quality by estimating chlorophyll-a concentrations using algorithms derived from Sentinel-2 satellite data and three atmospheric correction methods: Sen2Cor, C2X, and Acolite. The results suggest that Acolite, in combination with the B4/B5 algorithm, yields the most efficient simulations. However, the accuracy of the results is limited due to the large surface area of the lake and the small number of monitoring stations. Therefore, greater spatial coverage of in-situ measurements is needed to refine the model and improve its accuracy. This study highlights the importance and potential of satellite remote sensing in monitoring and predicting environmental changes in aquatic ecosystems.

Keywords Water quality · Sentinel 2 · Chlorophyll-a · Atmospheric correction

9.1 Introduction

Eutrophication is a natural process that occurs in water bodies, characterized by increased nutrient concentrations, such as phosphorus and nitrogen, which negatively affect aquatic ecosystems (Glibert et al. 2005). However, human activities, such as urbanization, pollution, agriculture, and tourism, have accelerated this process (Glibert et al. 2005).

To assess the trophic status of a water body, indices have been developed that relate factors such as phosphorus, nitrogen, and chlorophyll-a concentrations, as

N. F. Betansson · S. Yépez
Departamento Manejo de Bosques y Medio Ambiente, Facultad de Ciencias Forestales, Universidad de Concepción, Concepción, Chile

L. Rodríguez-López (✉)
Facultad de Ingeniería, Universidad San Sebastián, Concepción, Chile
e-mail: lien.rodriguez@uss.cl

V. Saritha et al. (eds.), *Remote Sensing for Environmental Monitoring*,
https://doi.org/10.1007/978-981-96-5546-5_9

well as Secchi disk depth (Carlson 1977; Smith et al. 1999). Based on these indices, the trophic state of a water body can range from oligotrophic—where the water is mostly transparent and well-oxygenated—to hyper-eutrophic, characterized by an abundance of microphytes that reduce light penetration and cause anoxia in deeper waters (Harper 1992).

In Chile, the General Directorate of Water (DGA) operates a network of water quality monitoring stations that collect and publish real-time data on the physicochemical conditions of water bodies through in situ sampling and meteorological measurements (Dirección 2007). However, as these processes are associated with high economic costs (Dirección 2009) the spatial and temporal representativeness of the data can be compromised due to the limited number of sampling points and the frequency of measurements, making it challenging to inform conservation, protection, and restoration efforts.

One technique to address these limitations is the use of satellite data. Sensors such as Sentinel-2 and Landsat-8 are widely used worldwide to estimate parameter values in lake ecosystems (Bresciani et al. 2018; Cui et al. 2022; Guo et al. 2023; Hossen et al. 2022; Oyama et al. 2009; Patra et al. 2017; Sivakumar et al. 2022; Theologou et al. 2016; Ticman et al. 2018; Wang et al. 2022; Zhang et al. 2020). Other radiometric sources, such as unmanned aerial vehicles (UAVs), have been employed for parameter estimation at smaller spatial scales (Rahul et al. 2023; Wu et al. 2019).

This research focuses on Lake Llanquihue, which faces environmental challenges due to volcanic activity, urbanization, and various human activities (Secretaría 2013). The presence of the salmon industry also poses risks to the lake's health (Bohle et al. 2009). In this context, satellite data presents an opportunity to monitor the lake's water quality variability and guide conservation and prevention efforts. The objective of this study is to develop an efficient model to estimate chlorophyll-a concentrations in Lake Llanquihue by applying single or multiple regression models using Sentinel-2 satellite data. The hypothesis is that regression models based on satellite data can effectively capture the spatial variability of water quality parameters in Lake Llanquihue.

9.2 Materials and Methods

9.2.1 Study Area

Lake Llanquihue (Fig. 9.1) is located near the Andes Mountains range in the Los Lagos Region, Chile, at an elevation of 51 m above sea level (m.a.s.l.), between the provinces of Llanquihue and Osorno. Its basin covers an approximate area of 1566.8 km^2 and is administered by four municipalities (Llanquihue, Puerto Octay, Frutillar, and Puerto Varas), with their main urban centers situated along the lake's shores, along with smaller towns such as Ensenada and Las Cascadas. It is the second-largest

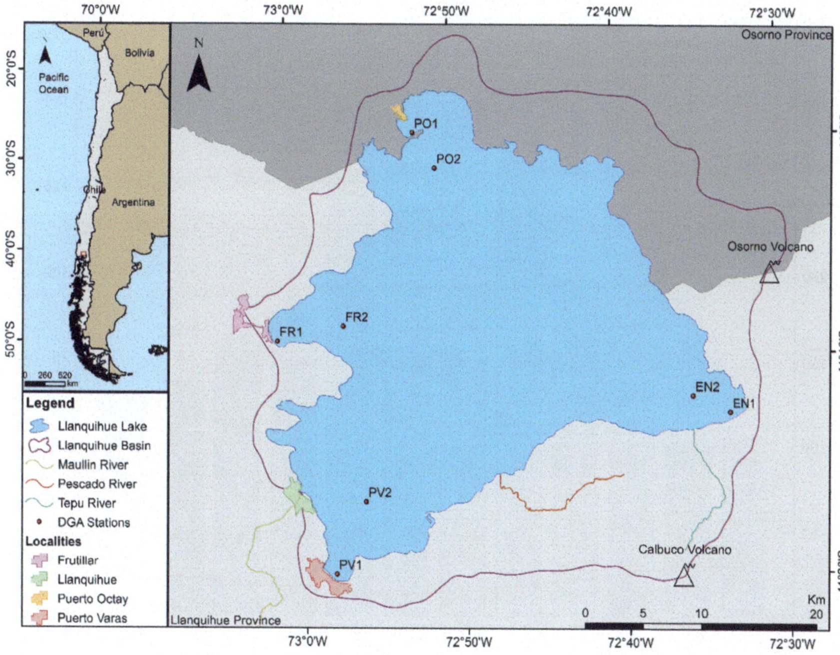

Fig. 9.1 Geographical location of Llanquihue Lake and DGA monitoring stations (red dots)

lake in Chile, after Lake General Carrera, with a surface area of approximately 870 km^2. The average depth is about 180 m, with some areas exceeding 300 m (Campos et al. 1988).

Of glacial origin, the lake has been subjected to continuous volcanic activity due to its proximity to the Osorno and Calbuco volcanoes. Past eruptions of Osorno Volcano caused part of the lake to separate, giving rise to Lake Todos Los Santos. Its main tributaries, the Pescado and Tepú rivers, have flow rates of 4 and 2 m^3/s respectively, making its hydrological regime predominantly pluvial. The lake has a single natural outflow, the Maullín River, which originates in the western sector of the lake and extends for 85 km, with an average flow rate of 72 m^3/s (Niemayer and Cereceda 1984).

The climate in the basin is temperate-rainy with Mediterranean influence (Sarricolea et al. 2017), characterized by year-round precipitation, which decreases in the summer months and reaches its maximum values in winter. According to hydrometeorological data from the DGA station "Maullín River in Llanquihue" (Alvarez-Garreton et al. 2018; Barría et al. 2021), between 1979 and 2019, the mean monthly precipitation exceeded 100 mm in all months of the year, with the lowest values recorded in February and January, at 100.5 mm and 105.8 mm, respectively. The highest values were observed in May and June, with 265.7 mm and 301.4 mm, respectively. In contrast, the mean temperature ranged between 6.8 °C in July and 15.1 °C in February (Fig. 9.2).

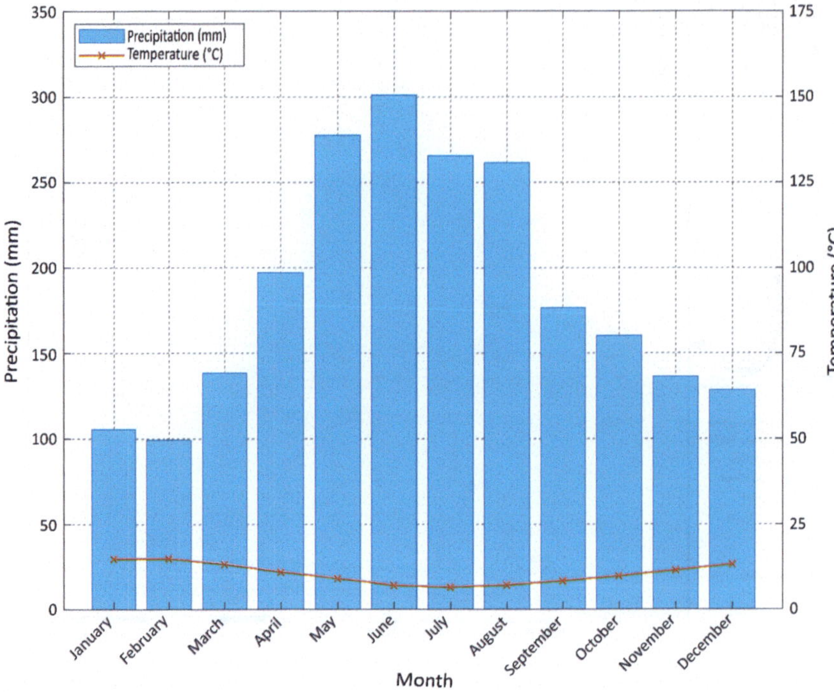

Fig. 9.2 Ombrothermic diagram for the Llanquihue Lake basin using data from the DGA station "Río Maullín en Llanquihue"

9.2.2 In-Situ Measurement

The Dirección General de Aguas (DGA) monitors water quality parameters at eight stations distributed across four sectors of the lake: Frutillar (FR1 and FR2), Puerto Octay (PO1 and PO2), Puerto Varas (PV1 and PV2), and Ensenada (EN1 and EN2) (Fig. 9.1). Chlorophyll-a data was collected through monthly monitoring campaigns conducted between October 14, 2020, and September 8, 2021, at depths of 0, 10, 20, 30, 50, 80, and 100 m. For this study, only the values obtained at the surface (0-m depth) were used.

9.2.3 Sentinel-2 Image Preprocessing

Sentinel-2 images are acquired from a pair of satellites, Sentinel-2A and Sentinel-2B, positioned at opposite sides of the globe. Together, they cover a swath of 290 km. Each satellite has a revisit time of 10 days, and combined, they enable image acquisition for the same area every 5 days. The satellites have 13 spectral bands with spatial resolutions ranging from 10 to 60 m.

Sentinel-2 images were downloaded from the Copernicus platform (https://dataspace.copernicus.eu/) and processed using SNAP, ENVI 5.6, and ArcGIS 10.5 software. A total of 17 Sentinel-2 images were selected, each within a maximum of 10 days from the sampling date (Table 9.1).

The calibration and validation periods for the model were defined as 70% and 30% of the total number of images, respectively, corresponding to 12 images for the calibration period and 5 for the validation period. The images selected for validation were chosen to represent the inter-annual variability of chlorophyll-a concentrations and include the following dates: 06/11/2020, 16/12/2020, 07/02/2021, 17/06/2021, and 11/08/2021.

Before processing, each image underwent atmospheric correction to remove the effects of atmospheric scattering and absorption of ultraviolet radiation, allowing the retrieval of reflectance values (the proportion of incident solar radiation reflected by a given surface).

Various methodologies are available for performing atmospheric correction depending on the type of imagery. In this study, Sentinel-2 images were corrected using the Sen2Cor, C2RCC, and Acolite atmospheric correction methods with the goal of identifying which method yields the most accurate chlorophyll-a estimations for Lake Llanquihue:

- **Sen2Cor** is a processor for Sentinel-2 images that performs atmospheric correction at Level 1-C to derive surface reflectance values at lower atmospheric layers.
- **C2RCC (Case 2 Regional Coast Colour)** is an atmospheric correction processor that employs a deep learning approach using neural networks trained on simulated data of water reflectance and top-of-atmosphere radiances. Its main products relate to the inherent optical properties (IOPs) of water, which are dependent solely on the absorption and scattering of its constituents (Brockmann et al. 2016). The method includes three sets of neural networks tailored to different research objectives: C2RCC-Nets (standard neural networks recommended for use in eutrophic or mesotrophic waters), C2X-NETS (specialized neural networks for water bodies with high concentrations of suspended matter and chlorophyll-a), and C2X-COMPLEX-Nets (specialized neural networks for more complex inland waters) (Soriano-González et al. 2022). In this study, we used the C2X-NETS network.
- **Acolite** is a processor developed by RBINS for atmospheric correction in coastal and inland waters, utilizing the "Dark Spectrum Adjustment" approach (Vanhellemont 2019a, b, 2020; Vanhellemont and Ruddick 2018, 2021).

Table 9.1 Sentinel-2 images used for chlorophyll-a estimation in Lake Llanquihue

ID	ID imagen	Image description	Sentinel A/B	Image date	Measurement date	Differences (days)
1	20,102,020	S2A_MSIL1C_ 20201020T143731_ N0209_R096_ T18GXV_ 20201020T181238	A	20–10–2020	14–10–2020	6
2	6,112,020	S2A_MSIL1C_ 20201106T142741_ N0209_R053_ T18GXV_ 20201106T181041	A	06–11–2020	16–11–2020	10
3	19,112,020	S2A_MSIL1C_ 20201119T143731_ N0209_R096_ T18GXV_ 20201119T181130	A	19–11–2020	16–11–2020	3
4	16,122,020	S2A_MSIL1C_ 20201216T142731_ N0209_R053_ T18GXV_ 20201216T175850	A	16–12–2020	21–12–2020	5
5	21,122,020	S2B_MSIL1C_ 20201221T142729_ N0209_R053_ T18GXV_ 20201221T175831	B	21–12–2020	21–12–2020	0
6	18,012,021	S2A_MSIL1C_ 20210118T143731_ N0500_R096_ T18GXV_ 20230527T012249	A	18–01–2021	18–01–2021	0
7	30,012,021	S2B_MSIL1C_ 20210130T142729_ N0209_R053_ T18GXV_ 20210130T180923	B	30–01–2021	08–02–2021	9
8	2,022,021	S2B_MSIL1C_ 20210202T143729_ N0209_R096_ T18GXV_ 20210202T181103	B	02–02–2021	08–02–2021	6
9	7,022,021	S2A_MSIL1C_ 20210207T143731_ N0209_R096_ T18GXV_ 20210207T181214	A	07–02–2021	08–02–2021	1

(continued)

Table 9.1 (continued)

ID	ID imagen	Image description	Sentinel A/B	Image date	Measurement date	Differences (days)
10	21,032,021	S2B_MSIL1C_ 20210321T142729_ N0500_R053_ T18GXV_ 20230605T132834	B	21–03–2021	16–03–2021	5
11	13,042,021	S2B_MSIL1C_ 20210413T143729_ N0500_R096_ T18GXV_ 20230517T033256	B	13–04–2021	14–04–2021	1
12	10,052,021	S2B_MSIL1C_ 20210510T142729_ N0500_R053_ T18GXV_ 20230207T145059	B	10–05–2021	11–05–2021	1
13	14,062,021	S2A_MSIL1C_ 20210614T142731_ N0500_R053_ T18GXV_ 20230131T203811	A	14–06–2021	15–06–2021	1
14	17,062,021	S2A_MSIL1C_ 20210617T143731_ N0500_R096_ T18GXV_ 20230201T041508	A	17–06–2021	15–06–2021	2
15	8,082,021	S2B_MSIL1C_ 20210808T142729_ N0500_R053_ T18GXV_ 20230113T154358	B	08–08–2021	09–08–2021	1
16	11,082,021	S2B_MSIL1C_ 20210811T143729_ N0500_R096_ T18GXV_ 20230211T201839	B	11–08–2021	09–08–2021	2
17	7,092,021	S2B_MSIL1C_ 20210907T142729_ N0500_R053_ T18GXV_ 20230111T021004	B	07–09–2021	07–09–2021	0

Table 9.2 Chlorophyll-a estimation algorithms utilizing Sentinel-2 satellite data

Algorithm	Formula	Atmospheric correction
B_i	$i \in [1, 11]$	Sen2Cor, C2X-COMPLEX-Nets and Acolite
B_i/B_j	$i \in [1, 11], j \in [1, 11], i < j$	Sen2Cor, C2X-COMPLEX-Nets and Acolite
NDVI	(B8 – B4)/(B8 + B4)	Sen2Cor and Acolite
GNDVI	(B8 – B3)/(B8 + B3)	Sen2Cor and Acolite
CGI	(B8/B3) – 1	Sen2Cor and Acolite

9.2.4 Algorithms for Estimating Chlorophyll-a from Satellite Data.

Several authors have identified various algorithms for estimating water quality parameters from satellite data (Chusnah et al. 2023; Cui et al. 2022; Hossen et al. 2022; Nguyen et al. 2020; Rajkumar and Sivakumar 2022). The chlorophyll-a estimation algorithms used in this study are based on individual bands, the ratios between each band and others, and specific spectral indices for Acolite and Sen2Cor, including the Normalized Difference Vegetation Index (NDVI), the Green Normalized Difference Vegetation Index (GNDVI), and the Green Coverage Index (GCI) (Table 9.2).

9.2.5 Statistic Validation

The model's efficiency is defined by its ability to simulate chlorophyll-a concentrations as closely as possible to the observed values. In this study, efficiency was assessed during the validation period using criteria and statistics commonly employed in hydrology: the root mean square error (RMSE) (Sezen et al. 2018), percent statistical bias (PBIAS), and mean absolute percent error (MAPE).

Root Mean Square Error (RMSE): RMSE is a statistical measure that quantifies the difference between the values predicted by the model and the observed values. It is calculated as the square root of the average of the squares of the residuals (Eq. 9.1). As a positive value, the optimal RMSE is 0, indicating high model efficiency in simulating flow.

$$RMSE = \sqrt{\frac{1}{n}\sum_{i=1}^{n}(\hat{y}_i - y_i)^2} \tag{9.1}$$

where

$\hat{y}_i$ = i-th simulated value (μg/L)
y_i = i-th observed value (μg/L)
n = total number of observations.

Percentage Bias Statistic (PBIAS): This statistic provides a percentage comparison between the estimated and observed values relative to the average of the observed values. A PBIAS value of zero indicates that the model projections are entirely accurate, while a non-zero PBIAS suggests the presence of bias in the predictions. Positive PBIAS values imply an overestimation of the variables, whereas negative values indicate underestimation. Thus, PBIAS helps visualize both the direction and magnitude of the bias present in the model projections (Eq. 9.2).

$$PBIAS = \frac{\sum_{i=1}^{n} (y_i - \hat{y}_i)}{\sum_{i=1}^{n} y_i} * 100 \tag{9.2}$$

where

$\hat{y}_i$ = i-th simulated value (μg/L)
y_i = i-th observed value (μg/L).

Mean Absolute Percent Error (MAPE): This statistic calculates the absolute percent difference between the estimated and observed values, then averages these percent differences. Expressed as a percentage, MAPE provides a relative measure of the model's accuracy in relation to the observed values. A lower MAPE indicates higher prediction accuracy (Eq. 9.3).

$$MAPE = \frac{\sum_{i=1}^{n} \left(\frac{|y_i - \hat{y}_i|}{y_i} \right)}{n} * 100 \tag{9.3}$$

where

$\hat{y}_i$ = i-th simulated value (μg/L)
y_i = i-th observed value (μg/L)
n = total number of observations.

9.3 Results and Discussion

9.3.1 In Situ Chlorophyll-a Trends

Chlorophyll-a concentration (μg/L) ranged from 0.05 to 0.90 μg/L, depending on the season and date. During the dry season (October to March), chlorophyll-a reaches its lowest concentrations in almost all stations, except for EN1 and PO1, with minimum values recorded at PV2 and PO2. In contrast, during the wet season (April to September), the concentration of chlorophyll-a increases, with maximum values recorded at PV1 and PO2.

From the data, it can be deduced that FR1 exhibits high seasonal variability in chlorophyll-a concentration, while EN1 and FR2 shows an inverse trend compared

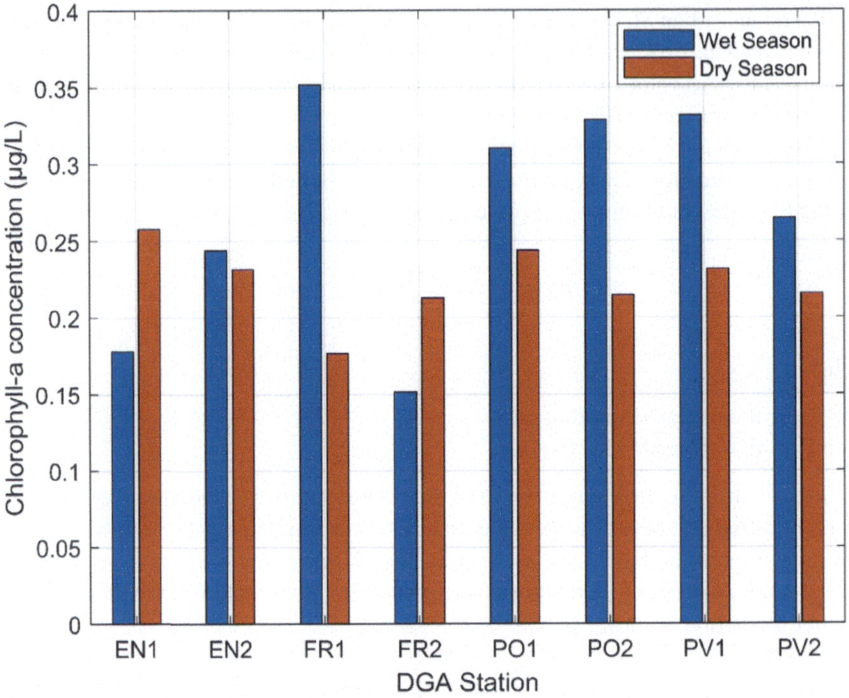

Fig. 9.3 Seasonal trends of Chlorophyll-a concentration (µg/L) at the eight DGA in situ monitoring stations for wet season (April to September) and dry season (October to March)

to the other stations, with minimum values in winter and maximum values in summer (Fig. 9.3).

Additionally, stations FR2 and FR1 have the lowest median values, meaning that half of the records correspond to minimum chlorophyll-a concentrations. The remaining stations tend to have higher median values in their coastal counterparts compared to the inland ones (Fig. 9.4).

9.3.1.1 Correlation and Algorithm Selection

The extensive surface area of Lake Llanquihue contributes to high spatial variability in chlorophyll-a concentrations and, more generally, in any other water quality parameters. Consequently, identifying algorithms for parameter estimation becomes challenging. The correlation between algorithms and stations varies depending on the atmospheric correction method used. When images are corrected using the Sen2Cor methodology, most algorithms and stations exhibit reduced correlation efficiency, with effective correlations observed only in PO1, PO2, and FR2 (Table 9.3).

Similarly, the use of the C2X-COMPLEX-Nets methodology results in low correlation efficiency across most algorithms and stations (Table 9.4).

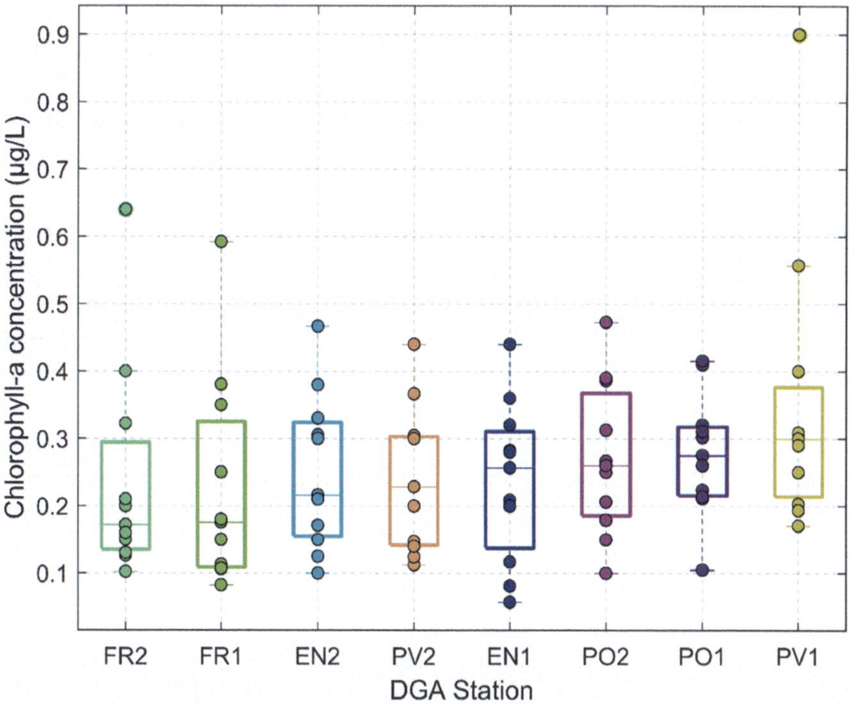

Fig. 9.4 Box-and-whisker plot for chlorophyll-a concentration (μg/L) at the eight DGA in situ monitoring stations

Table 9.3 Correlation matrix between chlorophyll-a estimation algorithms and the monitoring stations of Lake Llanquihue using the Sen2Cor atmospheric correction methodology

Band/Band combinations	PV1	PV2	PO1	PO2	FR1	FR2	EN1	EN2
B2	0.10	− 0.21	− 0.04	− 0.51	0.33	− 0.44	0.01	0.02
B3	0.10	− 0.22	0.03	− 0.54	0.33	− 0.44	0.02	0.03
B4	0.10	− 0.24	0.14	− 0.57	0.35	− 0.45	0.05	0.06
B8	0.11	− 0.25	0.16	− 0.59	0.34	− 0.47	0.06	0.07
B2/B3	− 0.01	0.38	− 0.65	0.52	− 0.10	0.56	− 0.05	0.11
B2/B4	0.05	0.39	− 0.61	0.48	− 0.08	0.61	− 0.09	0.12
B2/B8	0.05	0.41	− 0.50	0.51	− 0.06	0.63	− 0.12	0.06
B3/B8	0.07	0.42	− 0.53	0.52	− 0.04	0.52	− 0.19	− 0.10
B4/B8	− 0.02	− 0.05	− 0.30	0.08	0.01	− 0.03	− 0.33	− 0.33
NDVI	0.00	0.04	0.34	− 0.07	− 0.02	0.01	0.35	0.35
GNDVI	− 0.08	− 0.37	0.67	− 0.49	0.02	− 0.34	0.38	0.26
GCI	− 0.10	− 0.29	0.71	− 0.38	− 0.05	− 0.17	0.47	0.32

Table 9.4 Correlation matrix between chlorophyll-a estimation algorithms and the monitoring stations of Lake Llanquihue using the C2X-COMPLEX-Nets atmospheric correction methodology

Band/Band combinations	PV1	PV2	PO1	PO2	FR1	FR2	EN1	EN2
B1	0.17	0.46	0.81	0.12	0.39	– 0.26	0.36	– 0.15
B2	0.18	0.48	0.65	0.08	0.43	– 0.24	0.36	– 0.19
B3	0.12	0.52	0.07	– 0.29	0.46	0.01	0.08	– 0.21
B4	– 0.22	0.21	– 0.01	– 0.56	– 0.10	0.33	– 0.10	– 0.15
B5	– 0.22	0.04	– 0.01	– 0.54	– 0.24	0.39	– 0.08	– 0.10
B6	– 0.22	– 0.01	– 0.02	– 0.39	– 0.30	0.44	– 0.07	– 0.01
B1/B2	0.10	0.06	0.89	0.14	– 0.44	– 0.03	0.19	– 0.04
B1/B3	– 0.01	0.01	0.69	0.39	– 0.48	– 0.07	0.31	0.00
B1/B4	– 0.04	– 0.01	0.52	0.46	– 0.07	– 0.06	0.29	0.28
B1/B5	– 0.06	0.01	0.47	0.49	– 0.06	– 0.03	0.25	0.29
B1/B6	– 0.06	0.04	0.45	0.51	– 0.05	– 0.02	0.21	0.29
B2/B3	– 0.02	– 0.05	0.60	0.44	– 0.39	– 0.08	0.35	– 0.01
B2/B4	– 0.03	– 0.05	0.43	0.47	– 0.06	– 0.07	0.31	0.25
B2/B5	– 0.04	– 0.03	0.39	0.50	– 0.05	– 0.04	0.26	0.26
B2/B6	– 0.05	0.01	0.36	0.53	– 0.04	– 0.03	0.22	0.26
B3/B4	0.11	– 0.04	0.20	0.50	0.08	– 0.25	0.23	0.21
B3/B5	0.08	– 0.02	0.18	0.55	0.07	– 0.21	0.20	0.22
B3/B6	0.06	0.03	0.18	0.59	0.06	– 0.19	0.17	0.21
B4/B5	0.17	0.11	0.12	0.53	0.25	– 0.31	0.06	0.07
B4/B6	0.17	0.17	0.13	0.60	0.23	– 0.32	0.06	0.10
B5/B6	0.20	0.18	0.15	0.50	0.28	– 0.40	0.06	0.04

In contrast, the Acolite method yields the most efficient correlations for the largest number of stations, particularly between bands 4 and 5, achieving the highest correlations in stations PV2, PO1, PO2, FR1, and FR2. However, in stations PV1, EN1, and EN2, the correlations reach lower absolute values (Table 9.5).

Based on the results obtained, it is evident that the Acolite atmospheric correction methodology, combined with the B4/B5 algorithm, achieves the most efficient correlations for stations PV2, PO1, PO2, FR1, and FR2 (|correlation| > 0.5). Nevertheless, it does not yield efficient values for the remaining stations (|correlation| < 0.5).

9.3.2 *Estimation Model*

The model estimates of chlorophyll-a over the entire lake surface (Eq. 9.4) were obtained using the Acolite atmospheric correction and the B4/B5 algorithm at the stations PV2, PO1, PO2, FR1, and FR2. Stations PV1, EN1, and EN2 were not

Table 9.5 Correlation matrix between chlorophyll-a estimation algorithms and the monitoring stations of Lake Llanquihue using the Acolite atmospheric correction methodology

Band/Band combination	PV1	PV2	PO1	PO2	FR1	FR2	EN1	EN2
B1	0.26	0.41	− 0.04	0.26	− 0.43	− 0.45	0.20	0.12
B2	0.23	0.39	− 0.12	0.26	− 0.10	− 0.21	0.13	0.04
B3	0.14	0.26	− 0.59	0.17	− 0.04	− 0.01	0.07	− 0.10
B4	0.07	0.22	− 0.84	0.10	− 0.19	0.09	0.01	− 0.08
B5	0.13	0.35	− 0.70	0.19	− 0.13	0.28	− 0.01	− 0.20
B6	0.14	0.34	− 0.58	0.11	0.06	0.39	0.00	− 0.19
B7	0.17	0.22	− 0.38	0.08	0.04	0.30	0.03	− 0.09
B8	0.17	0.25	− 0.47	0.13	0.25	0.35	0.02	− 0.11
B9	0.01	0.13	− 0.44	0.00	0.07	0.28	0.32	− 0.33
B10	0.03	0.01	− 0.90	− 0.23	− 0.21	0.03	− 0.02	− 0.24
B11	0.14	0.02	− 0.80	− 0.20	− 0.03	0.10	0.01	− 0.31
B1/B2	0.13	0.18	0.15	0.01	− 0.52	− 0.71	− 0.04	0.07
B1/B3	0.25	0.02	0.52	− 0.05	− 0.30	− 0.29	0.00	0.28
B1/B4	− 0.08	− 0.29	0.66	− 0.18	− 0.14	− 0.28	− 0.10	− 0.06
B1/B5	− 0.12	− 0.38	0.51	− 0.35	− 0.19	− 0.39	0.13	0.20
B1/B6	− 0.08	− 0.40	0.59	− 0.23	− 0.37	− 0.49	0.34	0.25
B1/B7	− 0.07	− 0.21	0.42	− 0.25	− 0.34	− 0.42	0.18	0.05
B1/B8	− 0.10	− 0.32	0.50	− 0.24	− 0.54	− 0.44	0.18	0.03
B1/B9	0.03	− 0.14	0.46	− 0.23	− 0.37	− 0.42	− 0.23	0.51
B1/B10	− 0.14	− 0.04	0.67	0.54	0.13	− 0.34	− 0.10	− 0.29
B1/B11	− 0.22	0.06	0.53	0.54	− 0.26	− 0.28	− 0.16	0.13
B2/B3	0.30	− 0.06	0.71	− 0.05	− 0.07	− 0.15	0.08	0.43
B2/B4	− 0.08	− 0.33	0.75	− 0.18	− 0.06	− 0.25	− 0.05	− 0.01
B2/B5	− 0.14	− 0.42	0.57	− 0.37	− 0.11	− 0.37	0.22	0.21
B2/B6	− 0.12	− 0.44	0.57	− 0.25	− 0.25	− 0.46	0.41	0.25
B2/B7	− 0.12	− 0.26	0.40	− 0.27	− 0.20	− 0.40	0.25	0.09
B2/B8	− 0.14	− 0.37	0.49	− 0.27	− 0.49	− 0.42	0.21	0.06
B2/B9	0.01	− 0.18	0.44	− 0.23	− 0.27	− 0.40	− 0.21	0.53
B2/B10	− 0.16	− 0.06	0.70	0.55	0.22	− 0.33	− 0.04	− 0.27
B2/B11	− 0.22	0.06	0.52	0.54	− 0.18	− 0.27	− 0.16	0.13
B3/B4	− 0.13	− 0.37	0.56	− 0.17	− 0.02	− 0.28	− 0.07	− 0.11
B3/B5	− 0.21	− 0.48	0.36	− 0.41	− 0.08	− 0.44	0.21	0.13
B3/B6	− 0.21	− 0.49	0.36	− 0.24	− 0.25	− 0.46	0.36	0.12
B3/B7	− 0.21	− 0.27	0.22	− 0.26	− 0.19	− 0.42	0.17	− 0.04
B3/B8	− 0.21	− 0.39	0.23	− 0.29	− 0.53	− 0.45	0.13	− 0.09

(continued)

Table 9.5 (continued)

Band/Band combination	PV1	PV2	PO1	PO2	FR1	FR2	EN1	EN2
B3/B9	− 0.03	− 0.17	0.31	− 0.20	− 0.26	− 0.40	− 0.21	0.44
B3/B10	− 0.18	− 0.05	0.63	0.56	0.29	− 0.34	− 0.02	− 0.28
B3/B11	− 0.22	0.05	0.47	0.56	− 0.17	− 0.27	− 0.20	0.10
B4/B5	− 0.24	− 0.59	− 0.60	− 0.51	− 0.50	− 0.79	0.37	0.39
B4/B6	− 0.07	− 0.28	− 0.12	0.08	− 0.32	− 0.47	0.23	0.28
B4/B7	− 0.05	0.07	− 0.32	0.08	− 0.28	− 0.53	0.06	0.05
B4/B8	− 0.09	− 0.09	− 0.58	0.02	− 0.42	− 0.45	0.09	0.12
B4/B9	0.11	0.10	0.03	0.03	− 0.34	− 0.43	− 0.15	0.52
B4/B10	− 0.11	0.06	0.73	0.63	0.41	− 0.33	0.10	− 0.31
B4/B11	− 0.20	0.05	0.48	0.70	− 0.29	− 0.29	− 0.11	0.15
B5/B6	0.04	0.04	0.16	0.36	− 0.25	− 0.21	0.08	− 0.11
B5/B7	0.02	0.31	− 0.17	0.32	− 0.20	− 0.20	− 0.15	− 0.35
B5/B8	− 0.01	0.24	− 0.37	0.44	− 0.39	− 0.15	− 0.17	− 0.45
B5/B9	0.15	0.28	0.16	0.21	− 0.28	− 0.26	− 0.31	0.35
B5/B10	− 0.07	0.15	0.84	0.63	0.44	− 0.27	− 0.11	− 0.36
B5/B11	− 0.21	0.09	0.53	0.70	− 0.21	− 0.24	− 0.16	0.05
B6/B7	− 0.02	0.36	− 0.61	0.07	0.29	0.04	− 0.56	− 0.30
B6/B8	− 0.06	0.45	− 0.86	0.07	− 0.91	0.02	− 0.45	− 0.54
B6/B9	0.19	0.25	0.11	− 0.03	− 0.19	− 0.19	− 0.46	0.40
B6/B10	− 0.09	0.18	0.73	0.59	0.35	− 0.22	− 0.20	− 0.31
B6/B11	− 0.20	0.13	0.57	0.61	− 0.01	− 0.24	− 0.26	0.07
B7/B8	− 0.06	− 0.23	− 0.04	− 0.06	− 0.91	− 0.04	0.05	0.02
B7/B9	0.29	0.00	0.55	− 0.05	− 0.29	− 0.28	− 0.39	0.62
B7/B10	− 0.07	0.08	0.84	0.58	0.30	− 0.25	− 0.13	− 0.32
B7/B11	− 0.19	0.12	0.65	0.61	− 0.05	− 0.25	− 0.21	0.13
B8/B9	0.28	0.17	0.37	− 0.02	0.68	− 0.23	− 0.26	0.59
B8/B10	− 0.07	0.14	0.81	0.59	0.44	− 0.25	− 0.10	− 0.26
B8/B11	− 0.21	0.13	0.58	0.63	0.17	− 0.24	− 0.17	0.13
B9/B10	− 0.16	0.06	0.68	0.55	0.33	− 0.20	0.36	− 0.33
B9/B11	− 0.21	0.09	0.68	0.28	0.03	− 0.26	0.25	0.00
B10/B11	− 0.25	0.06	0.19	− 0.21	− 0.55	− 0.88	− 0.20	− 0.44
NDVI	0.12	0.06	0.57	0.01	0.38	0.42	− 0.15	− 0.11
GNDVI	0.25	0.36	− 0.16	0.21	0.49	0.45	− 0.14	0.05
GCI	0.27	0.34	− 0.12	0.17	0.46	0.45	− 0.12	0.03

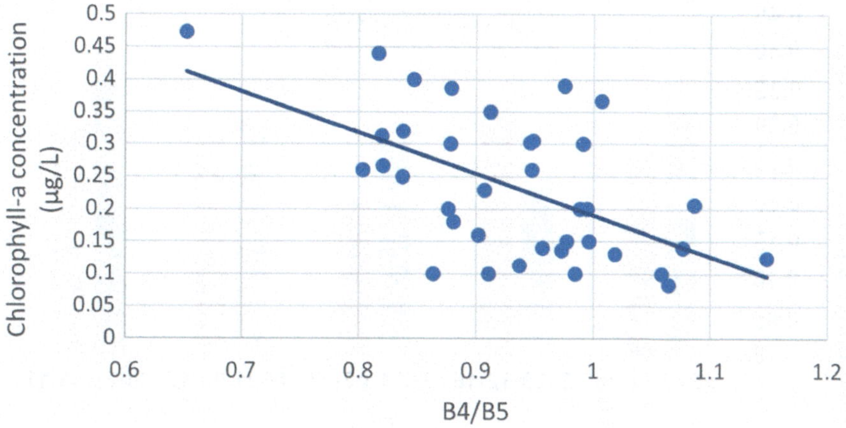

Fig. 9.5 Regression plot for the estimation of chlorophyll-a concentration (μg/L) from the B4/B5 algorithm for the total surface of Lake Llanquihue using the Acolite atmospheric correction methodology

considered in the analysis due to their low correlation. The model shows an inverse relationship between the algorithm and the variable, achieving a regression coefficient R^2 of 0.325. Thus, more information is needed to identify additional algorithms that maximize the efficiency of the regression (Fig. 9.5).

$$Ch\left(\mu\frac{g}{L}\right) = 0.8284 - 0.638 * \frac{B4}{B5} \quad (9.4)$$

In the validation period, the observed and estimated chlorophyll-a concentrations differ by a maximum of 0.1 (μg/L), where the simulated results tend to reach lower values than the observed ones, except for the day 16–12–2020 where the observed concentrations reached lower values (Fig. 9.6). The model efficiency reaches an RMSE value of 0.080 (μg/L) and BIAS and MAPE percentages of 9.6% and 25.2% respectively. Considering the magnitude of Lake Llanquihue, a BIAS and MAPE percentage variation of less than 10% reflects the potential of using remote sensors, specifically the Sentinel 2 satellite for the estimation of chlorophyll-a concentrations.

Finally, based on the results obtained above, the spatial variability of chlorophyll-a concentration in Lake Llanquihue is presented for both a summer image (07–02–2021) and a winter image (17–06–2021) (Fig. 9.7). The highest concentrations are observed near the urban centers for both images, however, in winter the concentrations are higher.

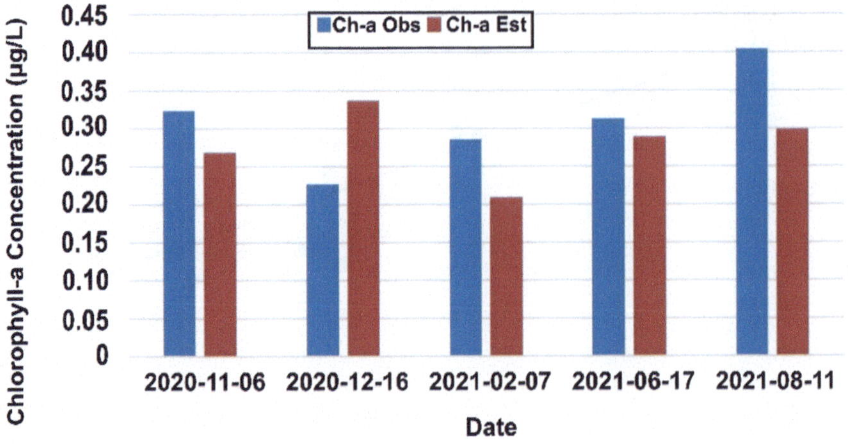

Fig. 9.6 Observed (blue) and simulated (red) Chlorophyll-a concentration for the model in validation period in Llanquihue Lake

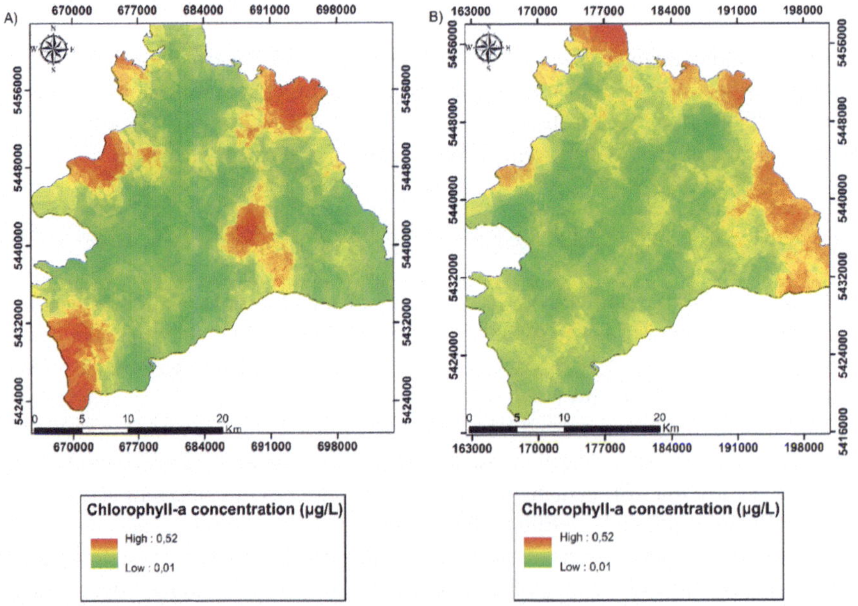

Fig. 9.7 Spatial distribution of Chlorophyll-a concentrations (µg/L) from the B4/B5 algorithm and the Acolite atmospheric correction for the total surface of Lake Llanquihue in **a** summer (07–02–2021) and **b** winter (17–06–2021)

9.4 Conclusion

Satellite information is a vital and handy tool for estimating water quality parameters. Lake Llanquihue allows estimates of chlorophyll-a concentrations (μg/L).

- The Acolite atmospheric correction method achieves more efficient Chlorophyll-a concentrations (μg/L) estimates than the Sen2Cor and C2X methods. This reflects the importance of selecting the most appropriate atmospheric correction method depending on the objectives sought, and it is not prudent to claim that one method can be the most useful for all scenarios.
- Bands 4 and band 5, corresponding to the red (665 μm length) and near-infrared (704 μm length) bands, allow the most efficient estimations for the parameter of interest in Lake Llanquihue; this combination of bands is based on the absorption of chlorophyll-a and water and allows identifying the type, health, and structure of the vegetation.
- The rest of the algorithms vary in their estimation efficiency. However, this variability is due to the extensive and heterogeneous surface that covers the water body. However, other algorithms may be practical depending on the lake and the homogeneity of its surface area. However, the model's efficiency is reduced due to the low spatial representativeness of the monitoring stations and the number of satellite images used, so it is necessary to increase the number of measurements and satellite images to optimize the model and minimize the values of the efficiency statistics.

The model is beneficial as a first approximation of this methodology applied to a large lake such as Lake Llanquihue. They demonstrated the high potential of satellite information as a tool for the characterization of the trophic state of a water body from the estimation of parameters such as Chlorophyll-a. This methodology can be extended to more parameters such as turbidity, suspended solids, and water transparency, allowing a comprehensive view of the surface water quality of the water body under study.

Acknowledgements ANID/Fondecyt Iniciación/11250177 and ANID/Fondecyt Regular/1221091.

Author Contributions Conceptualization: [Lien Rodríguez-López, Neftalí Flores Betansson], …; Methodology: [Neftalí Flores Betansson; Lien Rodríguez-López, Santiago Yépez], …; Formal analysis and investigation: [Neftalí Flores Betansson, Lien Rodríguez-López], …; Writing—original draft preparation: [Lien Rodríguez-López, Neftalí Flores Betansson]; Writing—review and editing: [Lien Rodríguez-López; Neftalí Flores Betansson]; Funding acquisition: [Lien Rodríguez-López]; Resources: [Lien Rodríguez-López and Neftalí Flores Betansson], …; Supervision: [Lien Rodríguez-López and Santiago Yépez].

Funding ANID/Fondecyt Iniciación/11250177 and ANID/Fondecyt Regular/1221091.

Data Availability To access the data contact the corresponding author.

Conflict of Interest The authors declare that they have no financial interests or conflicts.

References

Alvarez-Garreton C, Mendoza PA, Boisier JP, Addor N, Galleguillos M, Zambrano-Bigiarini M, Garreaud R et al (2018) The CAMELS-CL dataset: catchment attributes and meteorology for large sample studies–Chile dataset. Hydrol Earth Syst Sci 22(11):5817–5846

Barría P, Sandoval IB, Guzman C, Chadwick C, Alvarez-Garreton C, Díaz-Vasconcellos R, Fuster R et al (2021) Water allocation under climate change: a diagnosis of the Chilean system. Elem Sci Anth 9(1):00131

Bohle H, Tapia E, Martínez A, Rozas M, Figueroa A, Bustos P (2009) Francisella philomiragia, bacteria asociada con altas mortalidades en salmones del Atlántico (Salmo salar) cultivados en balsas-jaulas en el lago Llanquihue. Archivos De Medicina Veterinaria 41(3):237–244

Bresciani M, Cazzaniga I, Austoni M, Sforzi T, Buzzi F, Morabito G, Giardino C (2018) Mapping phytoplankton blooms in deep subalpine lakes from Sentinel-2A and Landsat-8. Hydrobiologia 824:197–214

Brockmann C, Doerffer R, Peters M, Kerstin S, Embacher S, Ruescas A (2016) Evolution of the C2RCC neural network for Sentinel 2 and 3 for the retrieval of ocean colour products in normal and extreme optically complex waters. In: Living planet symposium, vol 740

Campos H, Steffen W, Aguero G, Parra O, Zúñiga L (1988) Limnological study of Lake Llanquihue (Chile): morphometry, physics, chemistry, plankton and primary productivity. Archiv Für Hydrobiologie. Supplement Band. Monographische Beiträge 81(1):37–67

Carlson RE (1977) A trophic state index for lakes 1. Limnol Oceanogr 22(2):361–369

Chusnah WN, Chu H-J, Tatas, & Jaelani, L. M. (2023) Machine-learning-estimation of high-spatiotemporal-resolution chlorophyll-a concentration using multi-satellite imagery. Sustain Environ Res 33(1):11

Cui Y, Yan Z, Wang J, Hao S, Liu Y (2022) Deep learning–based remote sensing estimation of water transparency in shallow lakes by combining Landsat 8 and Sentinel 2 images. Environ Sci Pollut Res 29(3):4401–4413

Dirección General de Aguas (2007) Manual de normas y procedimientos del departamento de conservación y protección de recursos hídricos. Departamento de Conservación y Protección de Recursos Hídricos. Chile

Dirección General de Aguas (2009) Evaluación de costos de laboratorio ambiental y de las redes de calidad de aguas de la DGA. Asesores Consultores de Empresas Ltda. Chile, Optimiza

Glibert P, Seitzinger S, Heil C, Burkholder J, Parrow M, Codispoti L, Kelly V (2005) The role of eutrophication in the global proliferation of harmful algal blooms. Oceanography 18:198–209

Guo H, Zhu X, Huang JJ, Zhang Z, Tian S, Chen Y (2023) An enhanced deep learning approach to assessing inland lake water quality and its response to climate and anthropogenic factors. J Hydrol 620:129466

Harper DM (1992) Eutrophication of freshwaters: principles, problems and restoration. Chapman and Hall, London

Hossen H, Mahmod WE, Negm A, Nakamura T (2022) Assessing water quality parameters in Burullus lake using sentinel-2 satellite images. Water Resour 49(2):321–331

Nguyen H-Q, Ha N-T, Pham T-L (2020) Inland harmful cyanobacterial bloom prediction in the eutrophic Tri An Reservoir using satellite band ratio and machine learning approaches. Environ Sci Pollut Res 27:9135–9151

Niemayer H, Cereceda P (1984) Hidrografía. Geografía de Chile. Instituto Geográfico Militar (IGM), Santiago, 8, 309

Oyama Y, Matsushita B, Fukushima T, Matsushige K, Imai A (2009) Application of spectral decomposition algorithm for mapping water quality in a turbid lake (Lake Kasumigaura, Japan) from Landsat TM data. ISPRS J Photogramm Remote Sens 64(1):73–85

Patra PP, Dubey SK, Trivedi RK, Sahu SK, Rout SK (2017) Estimation of chlorophyll-a concentration and trophic states in Nalban Lake of East Kolkata Wetland, India from Landsat 8 OLI data. Spat Inf Res 25:75–87

Rahul TS, Brema J, Wessley GJJ (2023) Evaluation of surface water quality of Ukkadam lake in Coimbatore using UAV and Sentinel-2 multispectral data. Int J Environ Sci Technol 20(3):3205–3220

Rajkumar SVPB, Sivakumar R (2022) Spatio-temporal assessment and monitoring of chlorophyll-a in lake water through geo-informatics algorithm approach (GAA). Arab J Geosci 15(14):1252

Sarricolea P, Herrera-Ossandon M, Meseguer-Ruiz Ó (2017) Climatic regionalisation of continental Chile. J Maps 13(2):66–73

Secretaría Regional Ministerial de Vivienda y Urbanismo (SEREMI) (2013) Informe Ambiental: Adecuación estudios previos actualización plan regulador intercomunal Ribera Lago Llanquihue e Hinterland Puerto Montt. Región de Los Lagos, Chile

Sezen C, Bezak N, Šraj M (2018) Hydrological modelling of the karst Ljubljanica River catchment using lumped conceptual model. Acta Hydrotechnica 31(55):87–100

Sivakumar R, Prasanth BRSV, Ramaraj M (2022) An empirical approach for deriving specific inland water quality parameters from high spatio-spectral resolution image. Wetlands Ecol Manage 30(2):405–422

Smith VH, Tilman GD, Nekola JC (1999) Eutrophication: impacts of excess nutrient inputs on freshwater, marine, and terrestrial ecosystems. Environ Pollut 100(1–3):179–196

Soriano-González J, Urrego EP, Sòria-Perpinyà X, Angelats E, Alcaraz C, Delegido J, Moreno J et al (2022) Towards the combination of C2RCC processors for improving water quality retrieval in inland and coastal areas. Remote Sensing 14(5):1124

Theologou I, Kagalou I, Papadopoulou MP, Karantzalos K (2016) Multitemporal mapping of chlorophyll–α in Lake Karla from high resolution multispectral satellite data. Environ Processes 3:681–691

Ticman KDV, Medina JM, Gubatanga EV, Jalbuena RL, Santos JAS, Sta Ana RRC, Blanco AC (2018) Assessment of Landsat 8-based indices for water quality parameter estimation in Laguna de Bay, Philippines In: 39th Asian Conference on Remote Sensing (ACRS 2018), Kuala Lumpur, Malaysia

Vanhellemont Q (2019) Adaptation of the dark spectrum fitting atmospheric correction for aquatic applications of the Landsat and Sentinel-2 archives. Remote Sens Environ 225:175–192

Vanhellemont Q (2019) Daily metre-scale mapping of water turbidity using CubeSat imagery. Opt Express 27(20):A1372–A1399

Vanhellemont Q (2020) Sensitivity analysis of the dark spectrum fitting atmospheric correction for metre-and decametre-scale satellite imagery using autonomous hyperspectral radiometry. Opt Express 28(20):29948–29965

Vanhellemont Q, Ruddick K (2018) Atmospheric correction of metre-scale optical satellite data for inland and coastal water applications. Remote Sens Environ 216:586–597

Vanhellemont Q, Ruddick K (2021) Atmospheric correction of Sentinel-3/OLCI data for mapping of suspended particulate matter and chlorophyll-a concentration in Belgian turbid coastal waters. Remote Sens Environ 256:112284

Wang D, Tang B-H, Fu Z, Huang L, Li M, Chen G, Pan X (2022) Estimation of Chlorophyll-A concentration with remotely sensed data for the nine Plateau lakes in Yunnan province. Remote Sensing 14(19):4950

Wu D, Li R, Zhang F, Liu J (2019) A review on drone-based harmful algae blooms monitoring. Environ Monit Assess 191:211

Zhang T, Huang M, Wang Z (2020) Estimation of chlorophyll-a Concentration of lakes based on SVM algorithm and Landsat 8 OLI images. Environ Sci Pollut Res 27:14977–14990

Chapter 10
Geospatial and Hydrogeochemical Insights for Monitoring Water Quality and Salinity in Coastal Regions of Southern Karnataka, India

Vijay Suryawanshi, H. Ramesh, and T. Nasar

Abstract Coastal areas face significant challenges due to the depletion of groundwater and seawater intrusion into freshwater aquifers. Additionally, insufficient monitoring of freshwater quality is a major concern for consumers. In Karnataka's Dakshina Kannada district, groundwater is crucial for meeting the needs of the community, industry, and agriculture. This study investigates the impact of excessive use, human activities, and agricultural chemicals on groundwater quality, with a focus on the hydrogeochemistry of the Natravathi and Gurapura catchments. The study analyzed 32 groundwater samples collected seasonally from 2021 to 2022 for 18 physiochemical parameters. The Water Quality Index (WQI) was determined using factors such as pH, Dissolved Solids, Oxidatio Reduction Potentisl, Electrical Conductivity, Total Hardness, Total Dissolved Solids, Calcium, Chlorides, Potassium, and Sodium. WQI scores ranged from 0 to 52 post-monsoon and 0 to 42 pre-monsoon. An ArcGIS-based spatial distribution map was created to show temporal changes in groundwater quality. Post monsoon measurements showed significant cations ranging from 4.25 to 64.54 mg/l, calcium from 40 to 520 mg/l, chloride from 40 to 200 mg/l, and potassium from − 8.05 to 15.44 mg/l. Pre-monsoon measurements indicated sodium levels from 28 to 208 mg/l, calcium from 240 to 840 mg/l, chloride from 19.99 to 159.9 mg/l, and potassium from 0 to 61.79 mg/l. WQI results for the post-monsoon season showed 36% of sampling sites as excellent, 52% good, 8% poor, and 4% very poor, while pre-monsoon results indicated 13% excellent, 46% good, and 42% poor. The research reveals higher toxin concentrations in drinking water during pre monsoon period compared to post monsoon, with increased salinity in freshwater aquifers making the water unsuitable for consumption.

Keywords WQI · Hydrogeochemistry · Multivariate statistics · GIS · Western ghats

---

V. Suryawanshi (✉) · H. Ramesh · T. Nasar
Department of Water Resources and Ocean Engineering, National Institute of Technology Karnataka, Surathkal, Mangalore, India
e-mail: suryawanshi.vijay73@gmail.com

V. Saritha et al. (eds.), *Remote Sensing for Environmental Monitoring*,
https://doi.org/10.1007/978-981-96-5546-5_10

10.1 Introduction

For thousands of years, groundwater has been a crucial source of drinking water. However, global groundwater reserves are now approaching their limits, with some regions in the world already overexploited. The potential for further groundwater development worldwide is minimal. As a result, tapping into these resources may not be economically viable and could harm the environment (Liu et al. 2013). With 97% of the freshwater on Earth, groundwater is a vital resource for both household consumption and industry. But over-extraction and pollution create serious hazards, which emphasizes the need of sustainable development methods (Odukoya 2015). Groundwater, which is derived from precipitation, is a constantly evolving and profound component of the Earth's natural water cycle. Globally, it is a significant and invaluable sustainable asset that supports human survival and economic advancement (Sivakumar et al., 2023; Tiwari et al., 2017).

Groundwater deterioration and quality, such as all physicochemical parameters, heavy metals, harmful substances, and all significant ions, is a global concern (Singh and Noori 2022) It is found in aquifers, which are geologic formations with permeable structures capable of storing and conveying water at rates sufficient to supply significant quantities to well (Karunanidhi et al. 2021). Many anthropogenic factors, particularly industry, excessive fertilizers, population growth, and agricultural pesticides, rapidly pollute groundwater resources (Arslan & Çolak, 2023; Hinge et al., 2022). The Water Quality Index (WQI) generally combines various parameters like pH, total dissolved solids, and dissolved oxygen to produce a single value representing water quality. In river basin assessments, for instance, a WQI score under 50 might suggest clean water, whereas scores exceeding 100 could indicate contamination (Brown et al. 1970). Water's chemical and physical properties fluctuate temporally and spatially due to natural events, human interference, and saltwater intrusion. In coastal regions, the combination of seawater intrusion and agricultural and industrial activities significantly impacts groundwater quality changes (Uddin et al. 2021). The Mangalore delta has experienced increased groundwater salinity attributed to seawater intrusion and intensive farming. These issues underscore the importance of developing reliable WQI models specifically for coastal ecosystems. (Pradesh et al. 2011) emphasized the importance of WQI in water resource management and monitoring by observing shifts in water quality patterns. Similarly, (Kanagaraj et al. 2018) stressed the influence of human and natural factors, such as agricultural drainage and biogeochemical processes, in intensifying groundwater salinity. The water quality index score will be determined through the analysis of various physicochemical parameters. These include Total Dissolved Solids, pH, Oxidation–Reduction Potential, Dissolved Oxygen, Electrical Conductivity, Salinity, and Temperature. Additionally, the assessment will encompass Magnesium, Calcium, Total Hardness, Chlorides, Potassium, Sodium, Sulphate, Carbonates, and Bicarbonates (Gunes 2023). Although certain regions receive substantial precipitation from July to September, they still encounter shortages of potable water. In contrast, other areas suffer from excessive rainfall, resulting in floods (Sitharam et al. 2017). For example, during the monsoon season, surplus water in flood-prone regions often goes unused due to insufficient infrastructure. On the other hand, dry areas experience severe shortages in the non-rainy months because of uneven distribution and increasing water demands from urban growth and agricultural activities (Parthasarathy et al. 2019).

Coastal reservoirs present a promising approach to store flood waters for use throughout the year, addressing both flooding issues and water scarcity. The Marina Barrage in Singapore exemplifies the efficacy of coastal reservoirs in water resource management and prevention of seawater intrusion (Yang et al. 2013). In areas with plentiful water resources, contamination from agricultural runoff, industrial waste, and untreated sewage makes water unsafe for consumption, further exacerbating water scarcity (Sitharam et al. 2019). The pollution of the Yamuna River serves as an example, significantly restricting its use for drinking and irrigation purposes. (Kolathayar et al. 2018) emphasized the difficulties in managing seasonal fluctuations in water supply and their effects on drinking and irrigation needs. Likewise, (Yang 2013) highlighted the dual concerns of water scarcity and quality, highlighting the necessity for innovative solutions such as coastal reservoirs, which (Yang and Lin 2012) suggested as an effective method for storing flood waters near river mouths.

10.2 Materials and Methods

10.2.1 Description of the Study Area

The Natravathi River (Fig. 10.1), a significant westward-flowing waterway in Karnataka State's Dakshina Kannada district, spanning latitude 12° 29′ 11″ to 13° 11′ 11″ N and longitudes 74° 49′ 08″ to 75° 47′ 53″ E (Suryawanshi et al. 2024b). This river serves multiple purposes throughout the year, including irrigation, supplying drinking water to over 0.6 million people, supporting petrochemical industries in Mangalore city, and fulfilling religious needs in locations such as Dharmasthala and Kukke Subramanya (Babar and Ramesh 2015). The Natravathi River begins its journey in the Western Ghats, specifically in the Bangrabalige valley of Chikkamagaluru district, at an elevation of 1000 m above MSL. This river system encompasses a catchment area of 3657 km^2 (Suryawanshi et al. 2024a). In the west, it flows for about 125 km before meeting the Arabian Sea at Mangalore. The geology of the area is made up of lateritic earth covered in gneiss (CGWB 2012).

On average, the catchment experiences 4030 mm of annual rainfall, with temperatures fluctuating between 16° and 42° C. The area's geological composition consists of an early Precambrian tonalitic gneisses foundation, which has been infiltrated by granites, granulite's, and dolerite dykes (Shankar and Manjunatha 1994). The older gneissic rocks have been encroached upon by granulite's and granites (Ravindra and Venkat Reddy 2011). Although the Gurpur River originates at a higher elevation, the Natravathi River exhibits a steeper overall elevation profile in relation to its total length (Kumar et al. 2010). The Natravathi River's catchment is monitored by a river

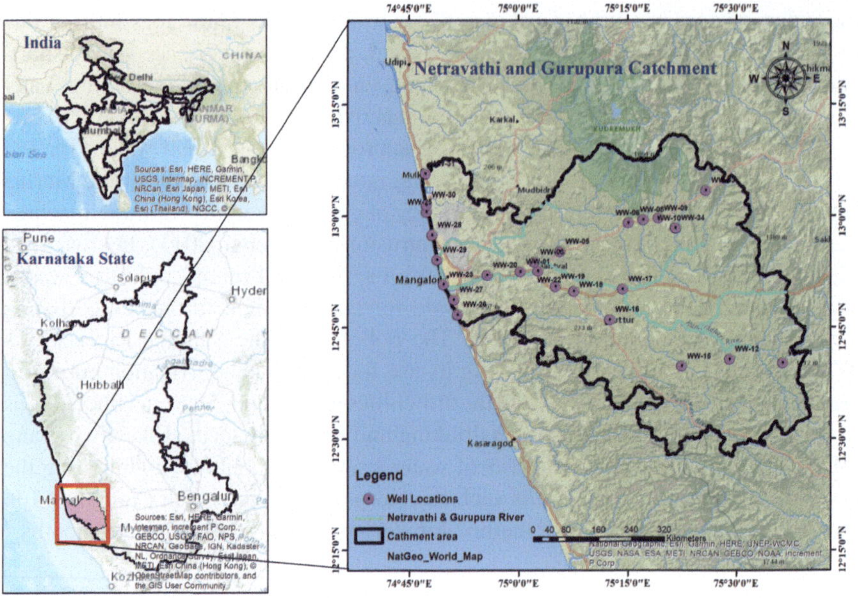

Fig. 10.1 Location map of the study area with monitoring well

gauging station at Bantwal, while the Gurpur River's catchment is observed by two gauging stations located at Addoor and Polali. These monitoring facilities are operated by the Central Water Commission of the Government of India and the Public Works Department of the Government of Karnataka (Sharannya et al. 2020).

10.2.2 Groundwater Sample Monitoring

Researchers monitored 32 groundwater samples from the Natravathi and Gurpur catchment regions. Laboratory analysis was conducted on various physical and chemical parameters, including Mg, Cl, TH, chlorides Cl, K, Na, SO_4, CO_3, and HCO_3. The findings were then evaluated against WHO 2004 and BIS 2012 standards. The Water Quality Index (WQI), developed by Horton (1965) in the United States, incorporates ten key water quality indicators such as dissolved oxygen (DO), pH, coliforms, specific conductance, alkalinity, and chloride. This index has gained widespread acceptance in the scientific community (Tyagi et al. 2013; Uddin et al. 2021). The WQI is computed by considering multiple parameters and assessing their overall significance based on the water's intended use. Examining the hydro geochemistry of a groundwater source is essential for determining its quality. This analysis can be employed to investigate the hydrogeochemical processes responsible for variations in groundwater chemistry across time and space (Curry and Stiff 2021).

10.2.3 Quantitative Assessment of Water Quality Using WQI

The Water Quality Index assesses the combined impact of key physical and chemical factors on water quality. Developed by the National Sanitation Foundation in 1970, the WQI employs a mathematical equation to deliver a standardized and easily comprehensible measure of water quality (Brown et al. 1973). This index builds on Horton's previous work by allocating weights to each parameter based on its importance to overall water quality (Brown et al. 1970), conducted an extensive analysis of the Horton index in 1970, enhancing it to create an improved WQI model (Eti et al. 2024). This enhancement allowed the index to better adapt to various water quality behaviors identified through physicochemical studies. The WQI methodology involves evaluating essential water quality parameters against the BIS 2012 standards, ensuring compliance with national quality criteria. Utilizing formulas (10.1, 10.2 and 10.3 and 10.4), Researchers computed the relative weight (Wi) and quality rating scale (qi) for each parameter. This method facilitates a systematic assessment of water quality and highlights areas requiring enhancement.

Calculation of Unit Weight (Wn) factors for each parameter:

$$Wn = \frac{K}{Sn} \tag{10.1}$$

where

$$K = \left[\frac{1}{\frac{1}{S1} + \frac{1}{S2} + \frac{1}{S3} + \ldots + \frac{1}{Sn}}\right] = \frac{1}{\sum \frac{1}{Sn}} \tag{10.2}$$

Sn = Standard Desirable Value of the nth Parameter.

On summation of all selected parameter unit weight factors, Ws = 1 (Unity).

Calculate the sub-Index (Qn) Value by using the formula

$$Qn = \frac{[(Vn - Vo)]}{[(Sn - Vo)]} * 100 \tag{10.3}$$

where

Vn = Mean Concentration of the nth Parameter

Sn = Standard desirable value of the nth Parameter.

VO = Actual Value of the parameter in pure water (generally VO = 0, for most parameter except pH)

Combining step-1 and Step-2, WQI is calculated as fallow

$$Overall\ WQI = \left[\frac{\sum Wn * Qn}{\sum Wn}\right] \tag{10.4}$$

The process of evaluating groundwater quality by integrating spatial and non-spatial data is depicted in the flowchart presented in Fig. 10.2. Specifically, the Spatial Data component involves digitizing the study area using ArcGIS, creating a study location map, and developing spatial distribution maps of different water parameters. On the other hand, the Non-Spatial Data component involves collecting well locations using GPS and conducting groundwater quality tests on site, with a focus on physical, chemical, and biological parameters. The results of these tests are then used for Water Quality Index Assessment, which is compared to BIS and WHO standards to evaluate groundwater quality. Ultimately, this integrated approach leads to the creation of a comprehensive Water Quality Index Map that visualizes groundwater quality in the study area.

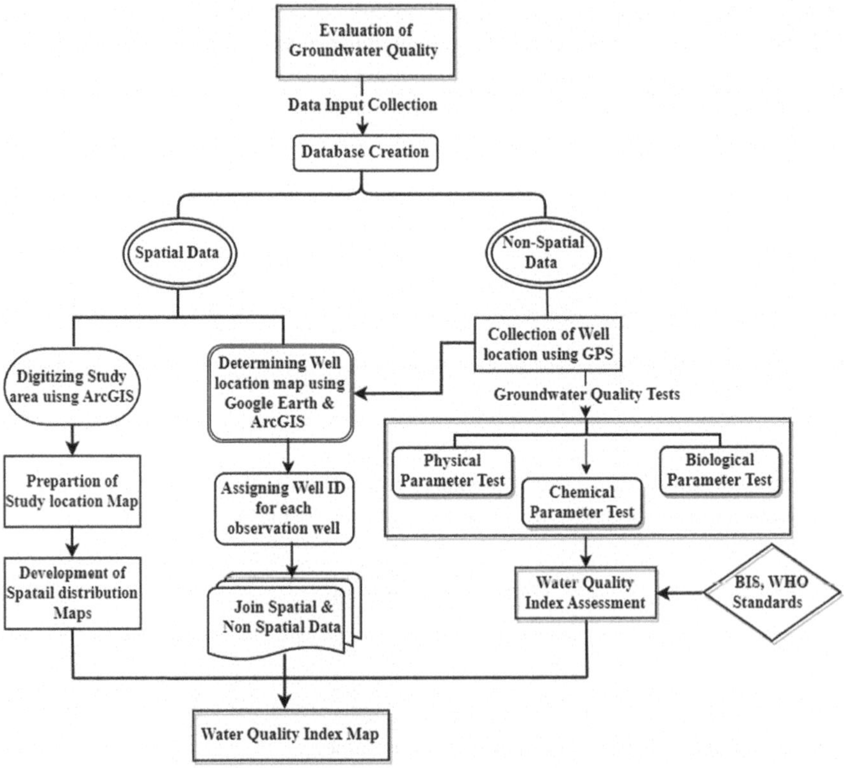

Fig. 10.2 Flow chart of the methodology

10.3 Analysis and Interpretation of Results

This study aims to assess the physicochemical parameters of groundwater to evaluate its suitability for drinking and other purposes in the study area as per APHA (2012). Groundwater quality standards vary depending on its intended use, with drinking water requiring the most stringent criteria. The World Health Organization (2017) and BIS Standards (IS:10500–2012) outline comprehensive criteria for assessing drinking water quality, including parameters such as pH, TDS, and microbial contamination. Figures 10.3 and 10.4, groundwater levels in the study area range from 0.4 to 12.2 m, with an increase observed during the post-monsoon season, indicating significant recharge. The pH levels range from 6 to 8.5, with a peak value of 7.81 in the coastal belt and a low of 5.5 upstream. Anthropogenic activities, such as agricultural runoff, industrial discharge, and over-extraction, have led to localized changes in groundwater quality, particularly affecting shallow aquifers with increased nitrate and chloride levels. An area with a high oxidation–reduction potential score, such as a lake, river, or well, has an abundance of oxygen and can effectively eliminate pollutants and decomposing organisms from the surrounding environment. There is a direct relationship between atmospheric pressure and the amounts of dissolved oxygen in water. The dissolved oxygen concentrations in the research area range from 1.3 to 4.6 mg/l. The World Health Organization states that water should have a minimum concentration of 4 mg/l of dissolved oxygen to sustain a healthy aquatic environment.

In this study area, the total dissolved solids value varies from 29 to 590 mg/l, suggesting that most groundwater samples fall within the permissible limit. High TDS concentrations in groundwater samples are caused by salt leaching from the soil, and domestic sewage may seep into the groundwater. The main contributors to hard water in a solution are calcium and, to a lesser extent, magnesium. Figure 10.4i displays Total Hardness between 80 and 720 during the post-monsoon season, while Fig. 10.3i displays Total Hardness as $CaCO_3$ ranges between 40 and 360 $CaCO_3$.

10.3.1 Multivariate Statistical Analysis

Multivariate statistical analysis offers a convenient and effective approach for examining large environmental datasets, particularly by reducing dimensionality and addressing skewness (Ghaemi and Noshadi 2022; Sharma et al. 2020). Techniques such as Principal Component Analysis (PCA) and Cluster Analysis (CA) were employed to identify dominant features impacting groundwater quality and to group sampling locations based on water quality characteristics (Nath et al. 2021; Mishra and Lal 2023). These methods enhance understanding by revealing patterns and correlations that are not immediately evident. The analysis utilized WHO (2004) and BIS (2012) standards, as they provide comprehensive guidelines for safe drinking

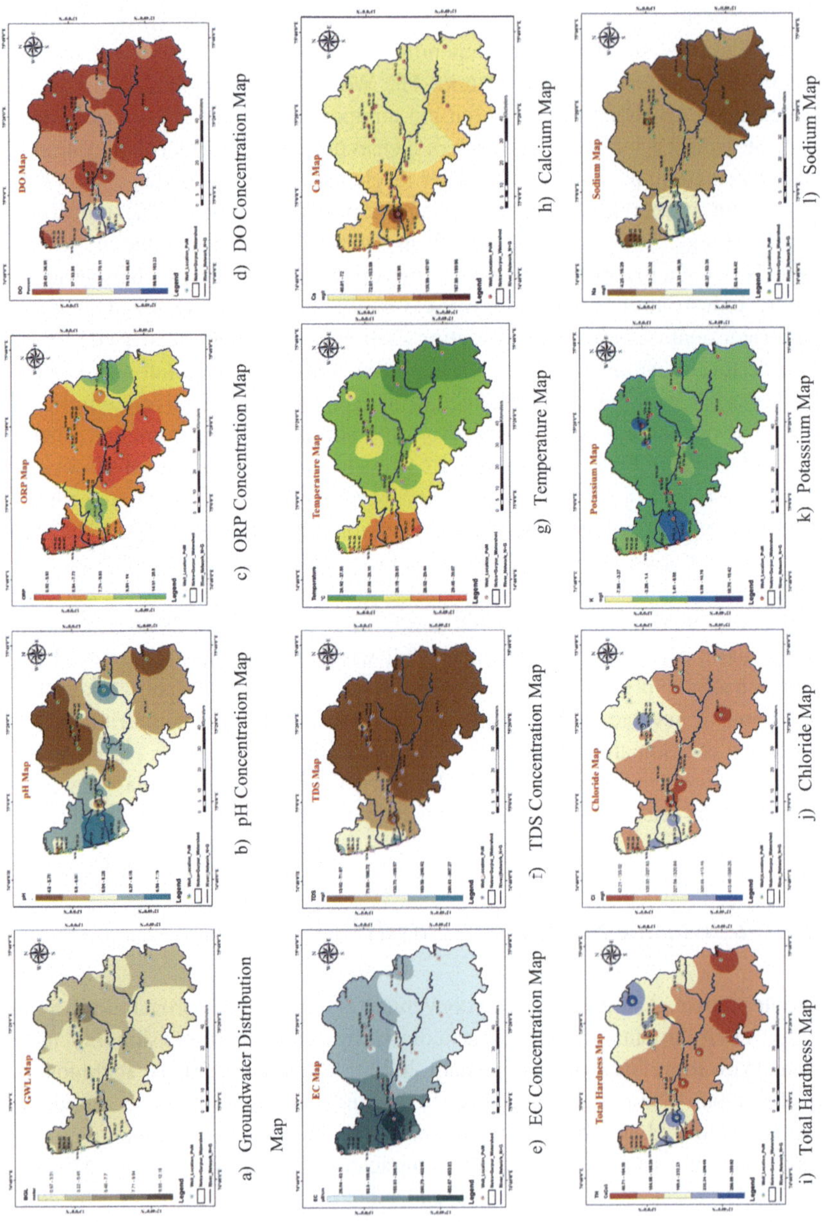

Fig. 10.3 Post-monsoon 2021: spatial hydrogeochemical patterns in the catchment

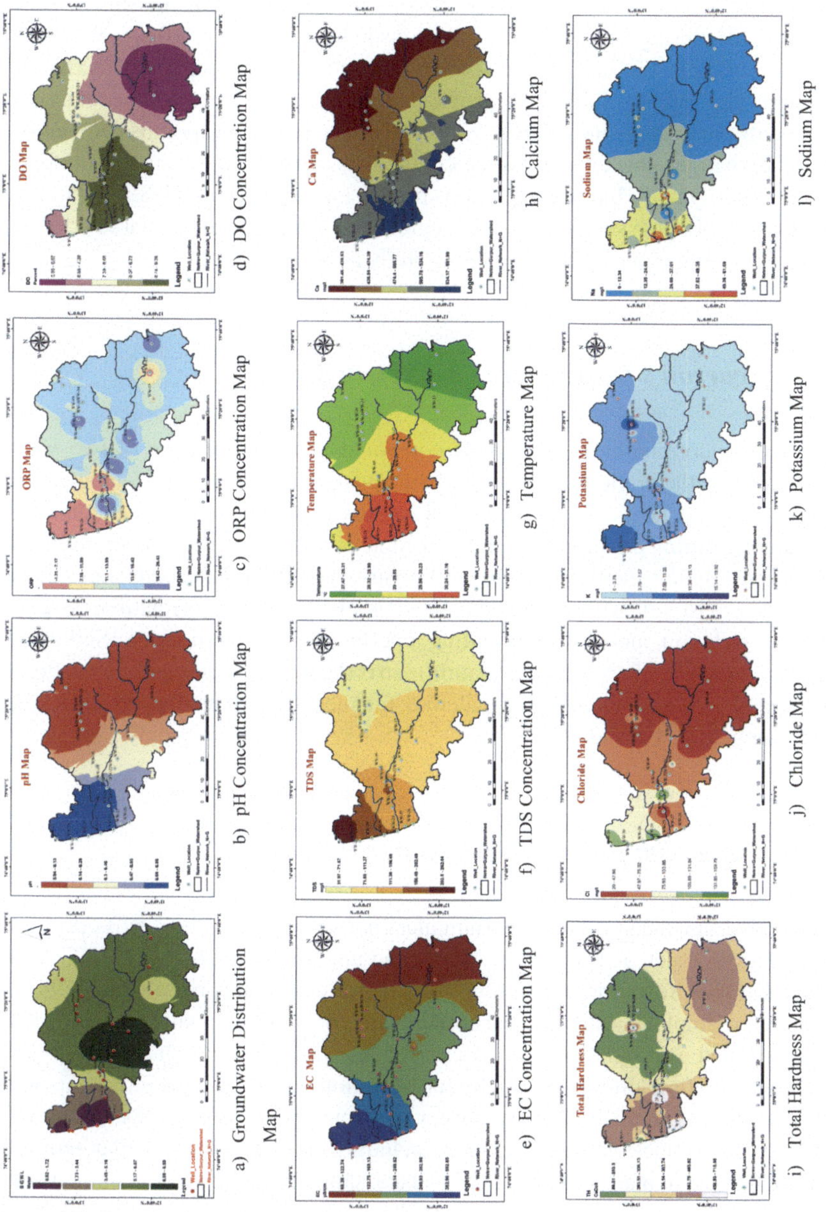

Fig. 10.4 Pre-monsoon 2022: spatial hydrogeochemical patterns in the catchment

water quality, addressing key chemical and physical parameters critical for public health.

Box and whisker plots revealed significant seasonal variability in hydrogeochemical parameters (Hema and Subramani 2013), with notable trends in salinity, TDS levels, and pH during the pre- and post-monsoon seasons (Fig. 10.5a, b). These variations highlight the influence of recharge events and anthropogenic activities on groundwater quality. Table 10.1 provides descriptive statistics for the analyzed water quality parameters, including mean, standard deviation, and permissible limits. Table 10.2a, b highlights the correlation coefficients, identifying strong associations between Water Quality Index (WQI) and specific metrics such as TDS and salinity levels. Figures 10.5a, b show the coastal region's hydrogeochemical characteristics, seasonal distribution, and variability in the study area.

10.3.2 Pearson's Correlation Coefficient Matrix

The Pearson's correlation coefficient is computed as the covariance of two variables divided by the product of their standard deviations. To assess the relationships between hydrogeochemical parameters and identify potential groundwater contamination sources during pre-monsoon and post-monsoon seasons, researchers employ a Pearson's correlation coefficient matrix (Bhuiyan et al. 2016). The coefficient values range from -1 to $+1$, with -1 indicating a complete negative correlation, 0 signifying no correlation, and $+1$ representing a perfect positive correlation (Shil et al. 2019). Analysis of the Pearson's correlation matrix demonstrates notable connections between various elements, including calcium, magnesium, sulphate, potassium, EC, and TH, with TDS, Cl, and pH.

10.3.3 Piper Plot for Hydrogeochemical Analysis

The Piper diagram incorporates two triangular charts that display the relative proportions of cations (such as Ca^{2+}, Na^{+}, Mg^{2+}) and anions (like $SO_4{}^{2-}$, Cl^{-}, $HCO_3{}^{-}$) in water samples, featuring a common baseline for easy comparison (Piper 1944). This diagram employs a diamond-shaped layout for multiple analyses, with each sample depicted as a circle whose size correlates with the Total Dissolved Solids (TDS) concentration, offering a visual representation of both chemical makeup and salinity (Hwang et al. 2017). By examining a sample's position on a Piper diagram, researchers can deduce the water's source, including the effects of particular geological formations or human activities. For instance, samples dominated by calcium and sulfate may indicate gypsum dissolution, while a predominance of sodium-chloride could suggest seawater intrusion. This technique enables scientists to track changes in water chemistry along a flow path, aiding in the identification of natural and human-induced influences (Sayyed et al. 2013).

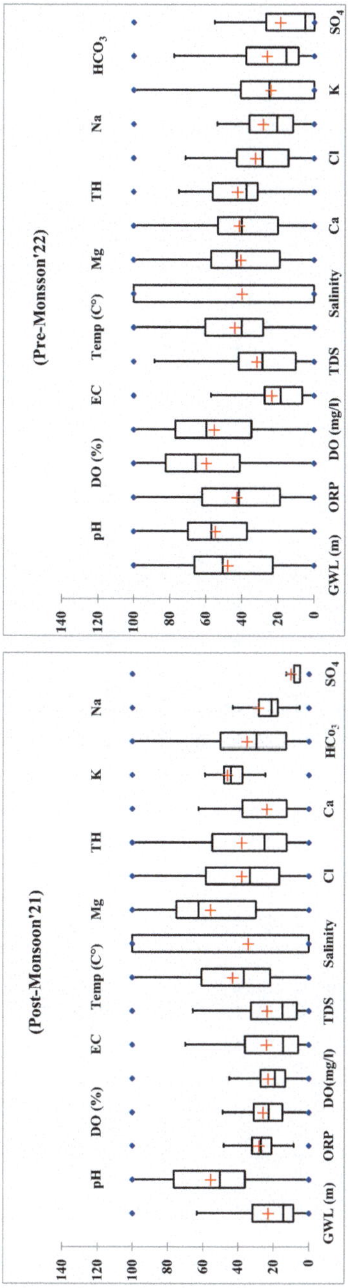

Fig. 10.5 **a, b** Box and whisker plot representation of key water quality parameters

Table 10.1 Groundwater quality assessment: statistical summary for post-monsoon 2021 and pre-monsoon 2022

Sl. No.	Parameter	Unit	Post-monsoon' 21					Pre-monsoon' 22					WHO (2004)		BIS Std. (2012)	
			Min	Max	Mean	Median	St. Dev.	Min	Max	Mean	Median	St. Dev.	acceptable limit	Permissible limit	Acceptable limit	Permissible limit
1	GWL	meter	0.8	12.3	3.45	2.45	2.55	0.5	9.7	5	5.33	2.76			–	–
2	pH	pH-unit	4.8	7.2	6.14	6.005	0.6	4.88	7.55	6.39	6.43	0.64	7–8.5	9.2	6.5–8.5	–
3	ORP	millivolts	0.9	20.9	6.55	6.29	3.9	– 1.3	26.9	10.5	10.2	7.91	– 100	300	0	300
4	DO	%	20.2	103.6	41.75	39.05	16.89	3.7	11	8.07	8.4	2.16	–	–	80	120
5	DO	mg/l	1.57	8.84	3.25	2.955	1.42	0.34	0.86	0.63		0.15	1	5	6.5	8
6	EC	mS/cm	26	684	184.25	120.5	171.2	44	954	263.12	216	212.91	–	200	300	900
7	TDS	mg/l	13	308	82.72	56.5	72.85	22	316	118.19	108	79.87	1500	500	500	2000
8	Temp	°C	26.92	30.09	28.29	28.08	0.87	27.06	32.55	29.52	29.28	1.29	10	20	10	20
9	Salinity	ppm	0	1	0.34	0	0.48	0	1	0.33	0	0.48	0.5	35	0.5	30
10	Mg	$CaCO_3$	– 240	80	– 61.25	– 40	97.58	– 480	360	– 110.3	– 80	192.36	50	150	30	100
11	Ca	$CaCO_3$	40	520	223.13	200	122.88	240	840	493.85	480	165.48	75	200	75	200
12	TH	$CaCO_3$	40	360	161.88	120	100.75	80	720	353.85	340	139.8	100	500	200	600
13	Cl	mg/l	40	200	78.13	60	34.59	19.99	159.9	64.59	59.96	38.47	200	600	75	1000
14	K	mg/l	– 8.05	15.44	2.8	2.31	4.86	0	61.79	17.41	13.64	16.96	–	12	12	–
16	HCO_3	mg/l	48	352	154.72	138	84.25	0	0	0	0	0	500	–	125	–
17	Na	mg/l	4.25	64.54	21.56	16.965	13.82	28	208	78.09	60	46.1	–	200	45	–
18	SO_4	mg/l	0.048	89.7	9.37	4.65	15.9	0	0.172	0.03	0.01	0.05	200	400	200	400

Table 10.2 **a** Water quality parameter correlations—post-monsoon 2021 **b** Water quality parameter correlations—pre-monsoon 22

	Mg	sl	Alkalinity	pH	EC	TDS	Ca	TH	Temp	K	Na	DO %	DO mg/l	Cl	SO_4
(a)															
Mg	0.24														
Salinity	0.59	− 0.03													
Alkalinity	− 0.1	0.51	0.21												
pH	− 0.21	− 0.21	0.39	0.41											
EC	− 0.1	0.24	0.4	0.62	0.57										
TDS	− 0.26	− 0.31	0.57	0.42	0.48	0.79									
Ca	− 0.12	− 0.6	0.07	0.43	0.13	0.43	0.31								
TH	0.09	0.23	0.05	0.13	− 0.04	0.29	0.08	0.64							
Temp	− 0.15	− 0.17	− 0.01	0.16	0.16	0.32	0.27	0.24	0.13						
K	0.22	− 0.31	− 0.08	0.33	0.05	0.47	0.35	0.63	0.46	0.22					
Na	0.23	− 0.27	− 0.08	0.29	0.23	0.58	0.39	0.38	0.2	0.3	0.72				
DO %	0.06	− 0.13	− 0.07	0.13	0.29	0.34	0.05	0.19	0.11	0.39	0.33	0.57			
DO mg/	0.03	− 0.15	− 0.09	0.09	0.27	0.31	0.04	0.19	0.09	0.35	0.33	0.57	0.99		
Cl	− 0.14	− 0.08	0.19	0.49	0.34	0.67	0.32	0.24	0.21	0.29	0.29	0.35	0.44	0.41	
SO_4	0.34	0.11	− 0.17	0.38	0.19	0.52	− 0.06	0.21	0.37	0.19	0.34	0.53	0.48	0.43	0.59
(b)															
Mg	0.24														
Salinity	− 0.03	− 0.39													
Alkalinity	− 0.41	0.21	− 0.1												
pH	− 0.21	0.51	0.41	0.21											

(continued)

Table 10.2 (continued)

	Mg	sl	Alkalinity	pH	EC	TDS	Ca	TH	Temp	K	Na	DO %	DO mg/l	Cl	SO_4
EC	– 0.24	0.4	0.62	0.57	– 0.1										
TDS	– 0.31	0.57	0.42	0.48	0.79	0.26									
Ca	– 0.6	0.07	0.43	0.13	0.43	0.31	0.12								
TH	0.52	0.05	0.13	0.04	0.29	0.08	0.64	0.09							
Temp	– 0.17	– 0.01	0.16	0.16	0.32	0.27	0.24	0.13	– 0.15						
K	– 0.31	– 0.08	0.33	0.05	0.47	0.35	0.63	0.46	0.22	0.22					
Na	– 0.27	– 0.08	0.29	0.23	0.58	0.39	0.38	0.2	0.3	0.72	0.23				
DO %	– 0.13	– 0.07	0.13	0.29	0.34	0.05	0.19	0.11	0.39	0.33	0.57	0.06			
DO mg/l	– 0.15	– 0.09	0.09	0.27	0.31	0.04	0.19	0.09	0.35	0.33	0.57	0.99	0.03		
Cl	– 0.08	0.19	0.49	0.34	0.67	0.32	0.24	0.21	0.29	0.29	0.35	0.44	0.41	0.14	
SO_4	0.11	– 0.17	0.38	0.19	0.52	0.06	0.21	0.37	0.19	0.34	0.53	0.48	0.43	0.59	0.34

Figure 10.6a, b demonstrate the practical application of this method, showing various water samples plotted within the triangular charts, with their relative positions reflecting the underlying geochemical processes. By contrasting the positions of these samples, it becomes possible to determine the contributions of both lithogenic and anthropogenic sources to the water's composition.

10.3.4 Groundwater Quality Assessment

ArcGIS software was employed to create Water Quality Index (WQI) maps for the 2021 and 2022 pre- and post-monsoon periods, illustrating the physicochemical fluctuations in groundwater quality. The WQI serves as a comprehensive metric that combines various water quality factors, including pH, hardness, and pollutant levels, to offer an overall water quality evaluation. By developing WQI maps for different seasons, this study examines temporal and spatial changes in groundwater quality. The physicochemical variation in groundwater, as depicted in Table 10.3, encompasses shifts in water hardness, a crucial indicator for water quality assessment. Hardness measurements were taken from 18 hydrogeochemical samples and utilized to evaluate groundwater quality across different parts of the catchment area. The resulting WQI maps provide a thorough overview of groundwater quality during both pre- and post-monsoon seasons. Seasonal water quality differences played a significant role in this study. Pre-monsoon periods typically exhibit higher contaminant concentrations due to reduced precipitation, while post-monsoon periods may experience some dilution but also an influx of surface runoff that introduces new pollutants. Table 10.4 showcases Horton's indexing system, which was utilized to categorize groundwater quality into groups such as 'excellent,' 'good,' 'fair,' and 'poor,' based on hardness concentration and other relevant parameters. This classification system, applied to both pre- and post-monsoon data, offers valuable insights into areas experiencing water quality improvement or deterioration, aiding in the prioritization of regions for continued monitoring and intervention.

Figure 10.7a presents water quality data from the post-monsoon season of 2021. The highest Water Quality Index (WQI) recorded is slightly above 100, and the lowest is around 40. Most bars in the figure are above the 60 mark on the vertical scale, indicating moderate to high water quality for many wells. This figure highlights variations in water quality across different wells. Figure 10.7b, a bar graph, shows WQI values ranging from 40 to 120 for wells in the post-monsoon season, reflecting diverse water quality among the sampled regions. Higher WQI values denote poorer water quality, while lower values denote better quality.

The water quality classification for the Natravathi and Gurpur Catchments is divided into four categories: Poor (represented by a brownish shade), Fair (represented by a greenish shade), Good (represented by a blue shade), and Excellent (represented by the darkest blue shade). Figure 10.8a illustrates the water quality status across different well locations in the Catchments after the monsoon season. This classification provides a clear visualization of water quality, with each color

Fig. 10.6 a, b Visualizing hydrochemical facies in groundwater using piper diagram

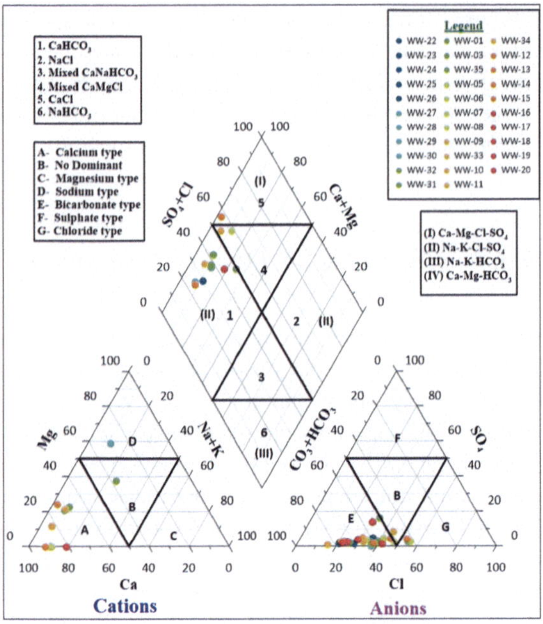

(a) Pre-Monsoon

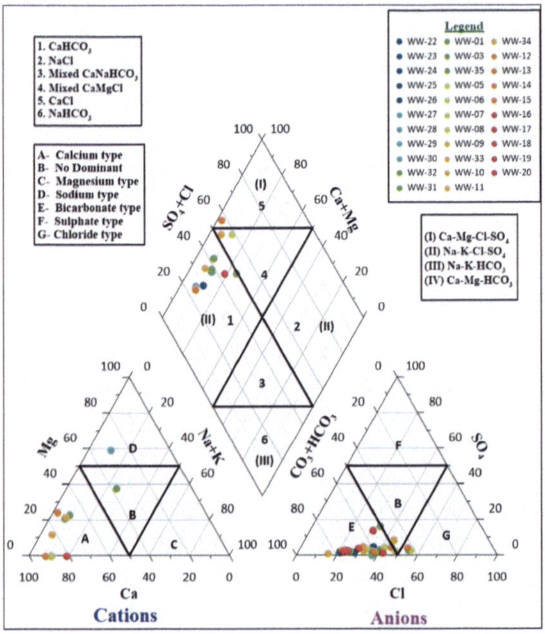

(b) Post-Monsoon

Table 10.3 Categorizing well water samples based on water hardness

Hardness as $CaCO_3$ (mg/l)	Water quality status	% of water Post-monsoon'21	% of water Pre-monsoon'22
0–75	Soft	0	12.5
75–150	Moderately hard	3.85	40.625
150–300	Hard	38.46	28.125
> 300	Very hard	57.69	18.75

Table 10.4 Classification of water quality using standard WQI values (Brown et al. 1970)

WQI value	Water quality	% water sample Post-monsoon'21	% water sample Pre-monsoon'22
90–100	Excellent water quality	36	13
71–90	Good water quality	52	46
51–70	Poor water quality	8	42
31–50	Bad water quality	4	0
0–30	Unsuitable for drinking water	0	0

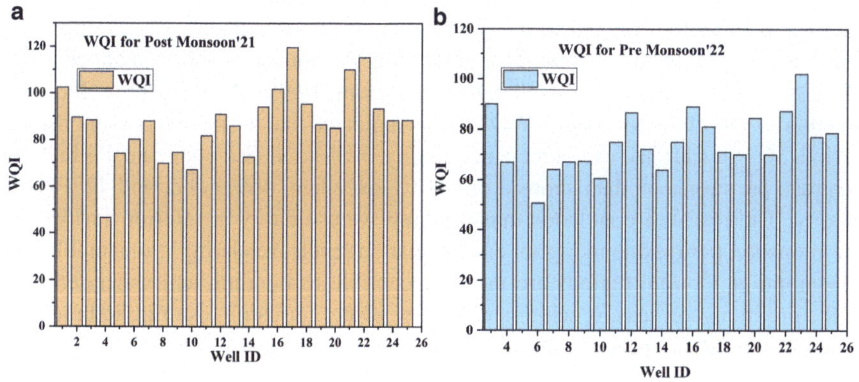

Fig. 10.7 WQI versus well location

corresponding to a specific water quality category. In contrast, Fig. 10.8b presents the Water Quality Index (WQI) map for the pre-monsoon season. This map uses circles of varying sizes and colors to represent water quality at different well locations. Dark green circles indicate poor water quality, orange circles suggest good quality, and light pink circles denote excellent water quality. By comparing the pre- and post-monsoon maps, this visual representation highlights the temporal variations in groundwater quality across the region. These maps are essential for environmental management, public health monitoring, and resource planning. For instance, areas identified with poor water quality can be prioritized for water treatment or further investigation to assess health risks. Similarly, this information can be used in resource planning to identify regions requiring improved water management strategies.

10.4 Conclusion

This study utilized geostatistical and GIS methods to evaluate and map groundwater quality in the Dakshina Kannada district. The study analyzed field observations and laboratory data to determine the spatial distribution of groundwater quality indicators, providing insights into aquifer conditions and water table fluctuations affecting fresh and saltwater dynamics. The investigation combined freshwater data from upstream areas with saline groundwater information from the coastal region. The Water Quality Index (WQI) was employed to simplify complex water quality data into a single value by integrating multiple parameters. Findings revealed that 13% of pre-monsoon well samples exhibited good water quality, while 36% of post-monsoon samples were classified as poor. The Piper diagram, which simplifies intricate hydrologic data, showed that ($SO_4 + Cl$) was dominant during the pre-monsoon period, indicating high groundwater alkalinity and acidity. Bicarbonate ($CO_3 + HCO_3$) was the primary anion, while calcium (Ca) and magnesium (Mg) were the main cations. Post-monsoon cation concentrations ranged from 4.25 to 64.54 mg/l for calcium, 40–520 mg/l for chloride, and − 8.05 to 15.44 mg/l for potassium. Pre-monsoon values spanned from 28–208 mg/l for sodium, 240–840 mg/l for calcium, 19.99–159.9 mg/l for chloride, and 0–61.79 mg/l for potassium. WQI results indicated that during the post-monsoon period, 36% of sites had excellent water quality, 52% good, 8% poor, and 4% bad. In contrast, pre-monsoon data showed only 13% excellent, 46% good, and 42% poor quality. The research concludes that drinking water quality is inferior during the pre-monsoon period due to higher contamination levels, rendering it unsuitable for consumption. Coastal regions are more susceptible to flooding and water scarcity in the pre-monsoon season, with increasing seawater intrusion resulting in brackish groundwater quality near the shore.

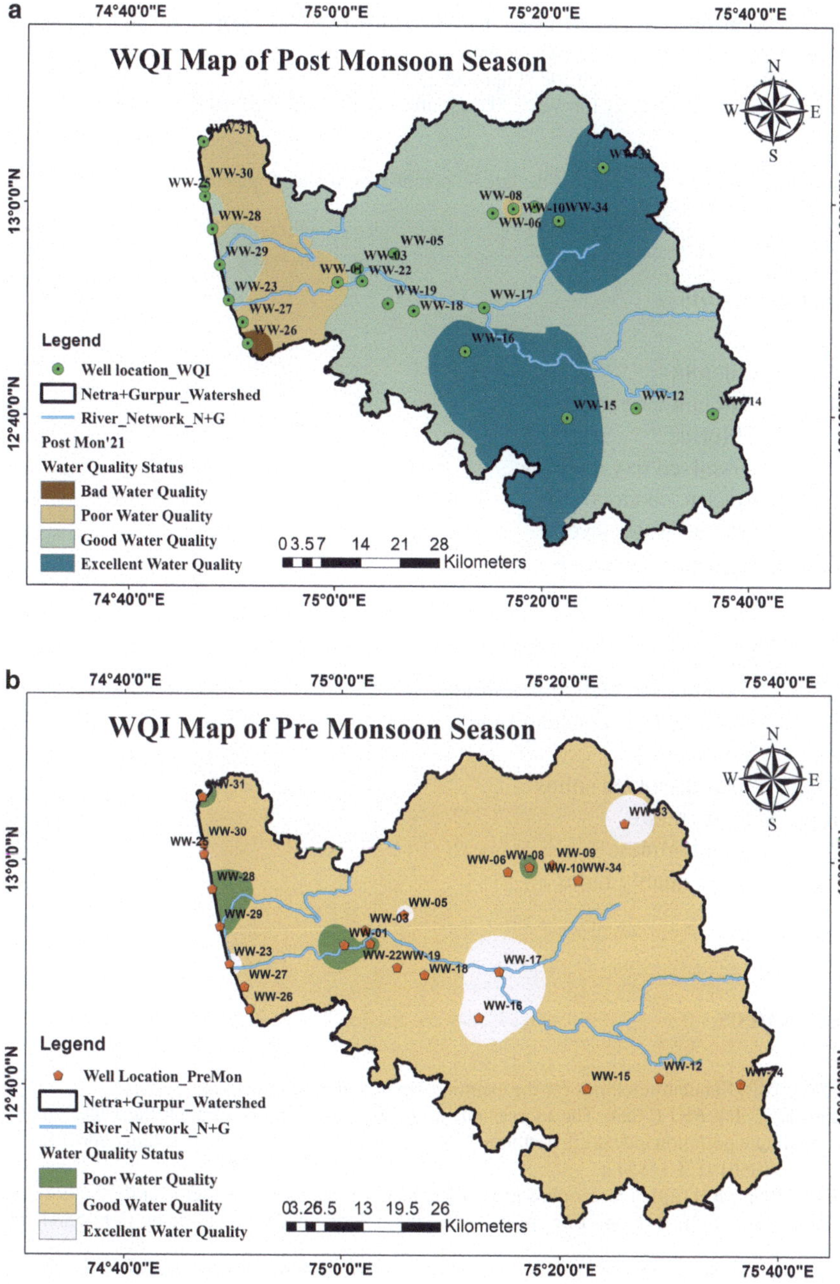

Fig. 10.8 Spatial distribution of water quality index classification map

Acknowledgements The authors thank the DST-SERB Govt. of INDIA (Ref No: IMP/2018/001298/WR) for providing 207 financial support to carry out this project work. This study is part of an ongoing R&D project titled 208 "Impounding of River flood waters along Dakshina Kannada Coast: A Sustainable Strategy for water 209 resource development IMPRINT-II Project".

Competing Interests The authors declare no potential conflicts of interest.

Abbreviations

Ca	Calcium
$CaCO_3$	Calcium carbonate
Cl	Chloride
DO	Dissolved oxygen
EC	Electric conductivity
GWL	Ground water level
HCo_3	Hydrogencarbonate
K	Potassium
Mg	Magnesium
Na	Sodium
ORP	Oxidation reduction potential
pH	Hydrogen ion concentration
SO_4	Sulphate
TDS	Total dissolved solids
Temp	Temperature
TH	Total hardness
WQI	Water quality index

References

APHA (2012) Standard methods for the examination of water and wastewater. 1496

Arslan H, Çolak MG (2023) The assessment of groundwater quality through the water quality and nitrate pollution indexes in northern Türkiye. Environ Monit Assess 195. https://doi.org/10.1007/s10661-023-11854-x

Babar S, Ramesh H (2015) Streamflow response to land use-land cover change over the Nethravathi River Basin, India. J Hydrol Eng 20:05015002. https://doi.org/10.1061/(asce)he.1943-5584.0001177

Bhuiyan MAH, Bodrud-Doza M, Islam ARMT, Rakib MA, Rahman MS, Ramanathan AL (2016) Assessment of groundwater quality of Lakshimpur district of Bangladesh using water quality indices, geostatistical methods, and multivariate analysis. Environ Earth Sci 75:1–23. https://doi.org/10.1007/s12665-016-5823-y

Brown RM, McClelland NI, Deininger RA, Tozer RG (1970) A-water-quality-index-do-we-dare-BROWN-R-M-1970. Water Sew Work 10:339–343

Brown RM, McClelland NI, Deininger RA, O'Connor MF (1973) A Water quality index—crashing the psychological barrier. Adv Water Pollut Res:787–797. https://doi.org/10.1016/b978-0-08-017005-3.50067-0

CGWB (2012) District, groundwater information booklet, Dakshina Kannada, Karnataka, 283

Curry BB, Stiff BJ (2021) Hydrochemistry ostracoda as proxies for quaternary climate change, 1–9

Eti FA, Easin I, Aronna MM (2024) Assessment of water quality index (Wqi) in different locations of Bangladesh using groundwater quality parameters, pp 0–11

Ghaemi Z, Noshadi M (2022) Surface water quality analysis using multivariate statistical techniques: a case study of Fars Province rivers, Iran. Environ Monit Assess 194. https://doi.org/10.1007/s10661-022-09811-1

Gunes G (2023) Evaluation of groundwater quality with microbiological and physicochemical parameters in Bartın, Turkey. Environ Monit Assess 195. https://doi.org/10.1007/s10661-023-11323-5

Hema S, Subramani T (2013) Study of physico-chemical characteristics of surface water using regression analysis of cauvery river and its tributaries in Tamilnadu, India. Asian J Chem 25:3199–3203. https://doi.org/10.14233/ajchem.2013.13590

Hinge G, Bharali B, Baruah A, Sharma A (2022) Integrated groundwater quality analysis using Water Quality Index, GIS and multivariate technique: a case study of Guwahati City. Environ Earth Sci 81:1–15. https://doi.org/10.1007/s12665-022-10544-0

Hwang JY, Park S, Kim H, Kim M, Jo H, Kim J, Lee G, Shin I, Kim T (2017) Hydrochemistry for the assessment of groundwater quality in Korea, 1–29. https://doi.org/10.4236/jacen.2017.61001

Kanagaraj G, Elango L, Sridhar SGD, Gowrisankar G (2018) Hydrogeochemical processes and influence of seawater intrusion in coastal aquifers south of Chennai, Tamil Nadu, India. Environ Sci Pollut Res 25:8989–9011. https://doi.org/10.1007/s11356-017-0910-5

Karunanidhi D, Aravinthasamy P, Deepali M, Subramani T, Bellows BC, Li P (2021) Groundwater quality evolution based on geochemical modeling and aptness testing for ingestion using entropy water quality and total hazard indexes in an urban-industrial area (Tiruppur) of Southern India. Environ Sci Pollut Res 28:18523–18538. https://doi.org/10.1007/s11356-020-10724-0

Kolathayar S, Sitharam TG, Yang S (2018) Overview. 1–15. https://doi.org/10.2166/ws.2018.140

Kumar A, Jayappa KS, Deepika B (2010) Application of remote sensing and geographic information system in change detection of the Netravati and Gurpur river channels, Karnataka, India. Geocarto Int 25:397–425. https://doi.org/10.1080/10106049.2010.496004

Liu J, Yang S, Jiang C (2013) Coastal reservoirs strategy for water resource development—a review of future trend. J Water Resour Prot 5:336–342. https://doi.org/10.4236/jwarp.2013.53a034

Mishra A, Lal B (2023) Assessment of groundwater quality in Ranchi district, Jharkhand, India, using water evaluation indices and multivariate statistics. Environ Monit Assess 195. https://doi.org/10.1007/s10661-023-11101-3

Nath AV, Selvam S, Reghunath R, Jesuraja K (2021) Groundwater quality assessment based on groundwater pollution index using Geographic Information System at Thettiyar watershed, Thiruvananthapuram district, Kerala, India. Arab J Geosci 14. https://doi.org/10.1007/s12517-021-06820-1

Odukoya AM (2015) Geochemical and quality assessment of groundwater in some Nigerian basement complex. Int J Environ Sci Technol 12:3643–3656. https://doi.org/10.1007/s13762-015-0789-y

Parthasarathy CR, Sitharam TG, Kolathayar S (2019) Geotechnical considerations for the concept of coastal reservoir at Mangaluru to impound the flood waters of Netravati River. Mar Georesources Geotechnol 37:236–244. https://doi.org/10.1080/1064119X.2018.1430194

Piper AM (1944) A graphic procedure in the geochemical interpretation of water-analyses. Eos, Trans Am Geophys Union 25:914–928

Pradesh A, Kumar GNP, Head F, Pradesh A, Prasad AS, Pradesh A, Hemalatha T, Pradesh A (2011) Generation of groundwater quality index map—a case study. 1:9–21

Ravindra BM, Venkat Reddy D (2011) Neotectonic evolution of Coastal Rivers of Mangalore, Karavali Karnataka, India. Int J Earth Sci Eng 4:561–574

Sayyed MRG, Wagh GS, Supekar A (2013) Assessment of impact on the groundwater quality due to urbanization by hydrogeochemical facies analysis in SE part of Pune city, India. Proc Int Acad Ecol Environ Sci 3:148–159

Shankar R, Manjunatha BR (1994) Elemental composition and particulate metal fluxes from Netravati and Gurpur Rivers to the coastal Arabian Sea. J Geol Soc India 43:255–265

Sharannya TM, Al-Ansari N, Barma SD, Mahesha A (2020) Evaluation of satellite precipitation products in simulating streamflow in a humid tropical catchment of India using a semi-distributed hydrological model. Water (Switzerland) 12. https://doi.org/10.3390/w12092400

Sharma A, Ganguly R, Kumar Gupta A (2020) Impact assessment of leachate pollution potential on groundwater: an indexing method. J Environ Eng 146:1–16. https://doi.org/10.1061/(asce)ee.1943-7870.0001647

Shil S, Singh UK, Mehta P (2019) Water quality assessment of a tropical river using water quality index (WQI), multivariate statistical techniques and GIS. Appl Water Sci 9:1–21. https://doi.org/10.1007/s13201-019-1045-2

Singh SK, Noori AR (2022) Groundwater quality assessment and modeling utilizing water quality index and GIS in Kabul Basin, Afghanistan. Environ Monit Assess 194. https://doi.org/10.1007/s10661-022-10340-0

Sitharam TG, Kolathayar S, Samuel M, Rao RS (2017) Feasibility study on formation of fresh water reservoir and impounding the surface runoff for urban water survival in a coastal brackish water region of Kollam, India. J Sustain Urban Plan Prog 2:34–37. https://doi.org/10.26789/jsupp.2017.02.004

Sitharam TG, Kolathayar S, Yang S, Krishnan A (2019) Concept of a geotechnical solution to address the issues of sea water intrusion in Ashtamudi Lake, Kerala. Springer International Publishing

Sivakumar V, Chidambaram SM, Velusamy S, Rathinavel R, Shanmugasundaram DK, Sundararaj P, Shanmugamoorthy M, Thangavel R, Balu K (2023) An integrated approach for an impact assessment of the tank water and groundwater quality in Coimbatore region of South India: implication from anthropogenic activities. Environ Monit Assess 195. https://doi.org/10.1007/s10661-022-10598-4

Suryawanshi V, Ramesh H, Nasar T (2024a) Integrated ecological river health assessment of netravathi basin based on physiochemical and hydrochemical analysis. In: Sivakumar Babu GL, Mulangi RH, Kolathayar S (eds) Technologies for sustainable transportation infrastructures. SIIOC 2023. Lecture Notes in Civil Engineering, vol 529. Springer, Singapore, pp 917–929

Suryawanshi V, Honnasiddaiah R, Thuvanismail N (2024b) Large-scale flood forecasting in coastal reservoir with hydrological modeling. Arab J Geosci. https://doi.org/10.1007/s12517-024-12109-w

Tiwari AK, Singh AK, Singh AK, Singh MP (2017) Hydrogeochemical analysis and evaluation of surface water quality of Pratapgarh district, Uttar Pradesh, India. Appl Water Sci 7:1609–1623. https://doi.org/10.1007/s13201-015-0313-z

Tyagi S, Sharma B, Singh P, Dobhal R (2013) Water quality assessment in terms of water quality index. 1:34–38. https://doi.org/10.12691/ajwr-1-3-3

Uddin MG, Nash S, Olbert AI (2021) A review of water quality index models and their use for assessing surface water quality. Ecol Indic 122:107218. https://doi.org/10.1016/j.ecolind.2020.107218

Yang S (2013) Coastal reservoir-the trend of water supply in new era

Yang S-Q, Lin P (2012) Coastal reservoir by soft-dam and its possible applications. Recent Patents Eng 5:45–56. https://doi.org/10.2174/1872212111105010045

Yang S, Liu J, Lin P, Jiang C (2013) Coastal reservoir strategy and its applications. Water Resour Planning Dev Manag:95–115. https://doi.org/10.5772/52315

Chapter 11
GIS-Based Study on Groundwater Depletion in NCR Regions of Uttar Pradesh, India

Ayush Tyagi, Raj Singh, and Shalu Kumar

Abstract Groundwater is one of the most important resources that is gifted by nature to mankind, but as the world moves towards urbanization and industrialization, this valuable resource is being ignored, and people tend to misuse groundwater while irrigating and industrial requirements. In this study the problem of groundwater depletion in NCR regions of the Uttar Pradesh (UP) state is discussed. In this study, both primary and secondary data is used. The study is divided into two main objectives; first is to calculate the water demand of the districts in order to find which sector is consuming most of the water, whether it be a groundwater and surface water and for that Central Public Health and Environmental Engineering Organization (CPHEEO) manual guidelines are used and geospatial tools were used to, and second one is to find out the vulnerable zones and suggest recommendations. To achive the second objective, trend analysis of the 5 years Pre and Post monsoon groundwater data is used. By integrating geospatial data and analytical techniques, the research contributes significantly to the field, offering a comprehensive framework for addressing water scarcity challenges in the NCR regions of Uttar Pradesh and potentially serving as a model for other areas grappling with similar issues.

Keywords Groundwater · Depletion of water · Contamination · Water demand

A. Tyagi (✉) · R. Singh
Dr. K. N. Modi University, Newai, Rajasthan, India
e-mail: tyagiayush348@gmail.com

R. Singh
GITAM Deemed to be University, Visakhapatnam, India

S. Kumar
Dev Bhoomi Uttarakhand University, Naugaon, India

V. Saritha et al. (eds.), *Remote Sensing for Environmental Monitoring*,
https://doi.org/10.1007/978-981-96-5546-5_11

11.1 Introduction

Water is a most precious resource for the survival of each organism on earth, without water no one on the earth can sustain their life. The water can be available on the surface as well as in groundwater. However, groundwater is one of the most precious gifts by God to humanity; it is found below the surface of the earth, under the cracks and spaces of the soil, sand, and rock, and can be called subsurface water. These water resources are being exploited extensively in many parts of the world, with a massive increase in extraction in the past few decades due to the availability of new and cheaper drilling and pumping technologies (Laghari et al. 2012; Scanlon et al. 2020; Turner et al. 2019). Most groundwater comes from rainfall, when the soil reaches saturation level, the water percolates downwards and gets deposited into aquifers (Gavrilescu 2021; Salama et al. 1999). The usage of groundwater has gradually increased because of the increase in water demand and the shortage of surface water during the growth of population and rapid industrialization (Craswell 2021; Han and Currell 2022; Singh et al. 2024). As the world is moving toward development it is exploring new techniques of exploiting this resource. Not only in India, this is one of the most common problems worldwide in some of the most developed countries like China, Turkey, and parts of the United States of America is facing this problem (Dangar et al. 2021).

In India, mostly groundwater is one of the most exploited resources. The people tend to rely too much on the resource for both their personal and business needs (Sidhu et al. 2020; Swain et al. 2022). India is an agriculturally rich country, the large portion of staple crops like wheat and rice in the country can be grown with the help of groundwater (Priyan 2021). Major industrial and urban activities depend upon the groundwater in the country and most of the parts of the country like Tamil Nadu, Rajasthan, Punjab, and Uttar Pradesh even were facing a crisis of water (Raghavan et al. 2021). In a country like India, water demand can be dependent on various other factors like changes in temporal, temperature geomorphology of an area, and many more.

The NCR regions of UP are rich in natural resources and people's abilities. Over the years, there has been a significant rise in population and groundwater usage, straining aquifers (Saha et al. 2022). These areas, known for agriculture and industries, are now facing challenges in meeting domestic water demand due to increased household needs (Saxena et al. 2021).

The region, especially districts like Meerut and Bulandsahar, is agriculturally and industrially prosperous. Water-intensive crops such as sugarcane and rice thrive here. Additionally, major industrial clusters contribute to the economic vibrancy of the area (Salim et al. 2024). The NCR geographic area relies heavily on groundwater to meet its agricultural, domestic, and some industrial needs (Das et al. 2021). While the presence of the Upper Ganga canal is beneficial, there is a concerning trend of continued misuse of these vital resources.

Moreover, groundwater contamination is one of the serious problems that is rising nowadays all over the world (Papazotos 2021). As the world moves towards globalization, the increased use of technology in their daily life the more they contaminate and deplete this vital resource (Li et al. 2017; Raza et al. 2017; Xie et al. 2023). Groundwater once contaminated it is nearly impossible to restore it to its original quality (Akhtar et al. 2021). However, there is official law on a central level for groundwater use but several rules and policies were imposed by different states in their region and their implementation is also dependent upon the different levels.

Therefore, the current study calculates the water demand of the districts of NCR regions of UP and identifies vulnerable locations of groundwater depletion using Geospatial Tools, and suggests recommendations. Employing advanced Geospatial Tools, the study not only quantifies water demand but also uniquely identifies specific vulnerable locations susceptible to groundwater depletion. This innovative approach provides a nuanced understanding of the spatial dynamics of water resources, enabling targeted and effective recommendations for sustainable management.

11.2 Material and Methods

11.2.1 Study Area

The NCR regions of UP has an estimated area of 11,275 sq km which accounts 31.19% of area (Fig. 11.1). It plays a major role for industrial; infrastructure and agriculture point of view and it lies on a western plain which is surrounded by river Ganga, Yamuna and Hindon form the sides. Because of this region comes in the NCR regions, the vast development has been seen across the regions in terms of agriculture, industrial, and development point of view. The justification for picking this as a subject of study is because groundwater contamination is an intense issue worldwide and the horticulture and modern activities in both the metropolitan country spaces of the NCR regions of UP have been expanded by and large inside the range of 10 years and it influence groundwater table gravely. The map below represents the exact geographical location of the study area.

11.2.2 Dataset and Experiment Design

11.2.2.1 Calculation of the Water Demand of the District

For the calculation of the demand of particular sector like industrial, agricultural, livestock, hospitals, hotels, schools and colleges, data was collected by using both primary and secondary sources and different assumptions were made which are based on the pilot survey of the cities. Data for the populations of 1991, 2001 and 2011

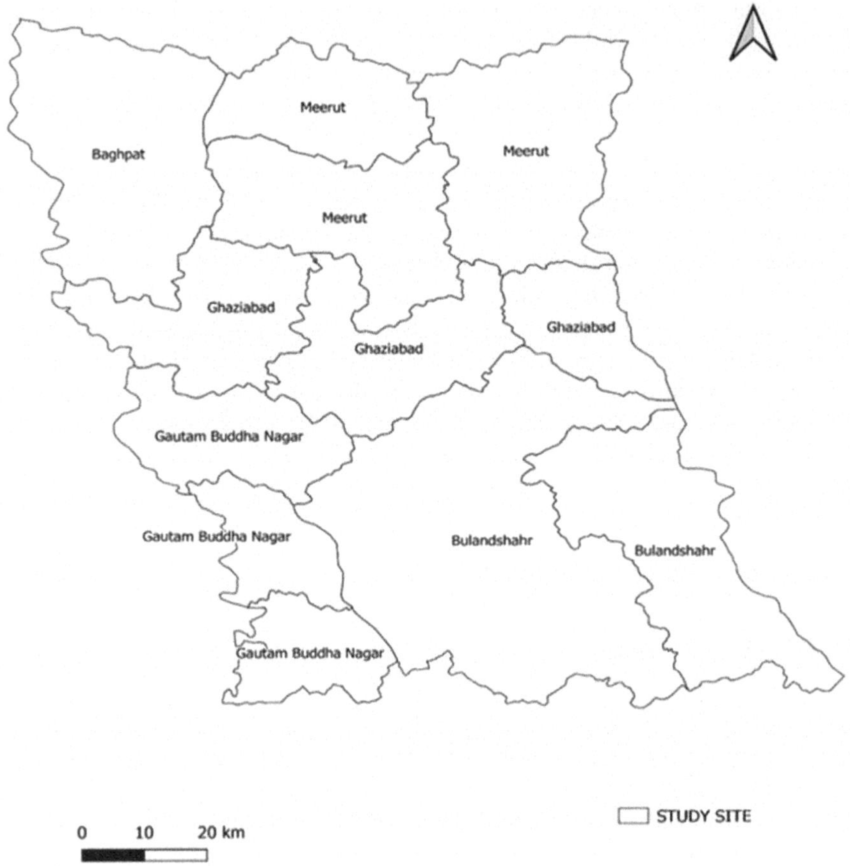

Fig. 11.1 Location of study sites

was taken from the census of India website and methodologies will be discussed for each parameter separately below.

11.2.2.2 Agriculture

For agriculture the crop wise land statistics were taken from the district agriculture board Meerut and data for each crop were also taken from the department and for the calculation the below stated formula has used (Fig. 11.2).

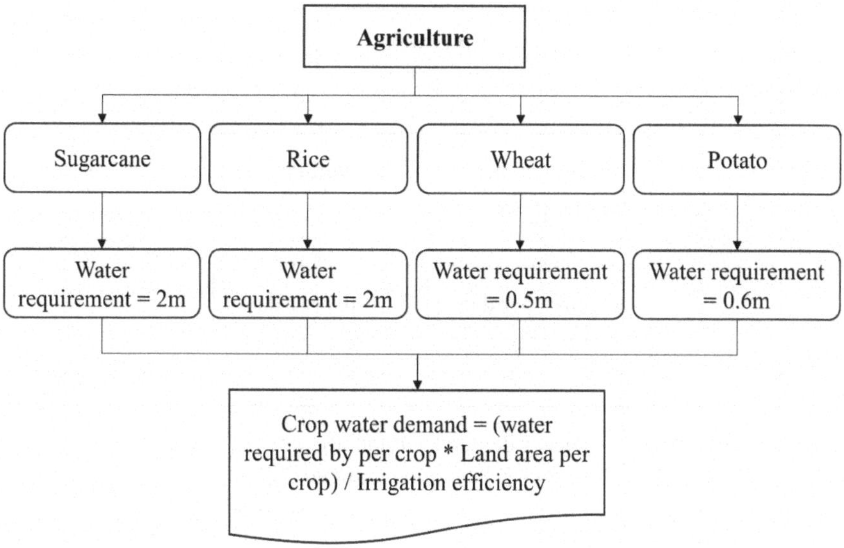

Fig. 11.2 Data collection and calculation

11.2.2.3 Livestock, School, College, Industries, Hotels and Hospitals

The data was taken from the secondary sources and calculation standards were taken from the CPHEEO manual (https://cpheeo.gov.in/). The calculations of water demand for livestock, school, college, industries, hotels and hospital are mentioned in Figs. 11.3, 11.4, 11.5, 11.6, 11.7 and 11.8 respectively.

11.2.3 Process of Identifying Vulnerable Locations of Groundwater Depletion

The groundwater table data is taken from the district groundwater board, and for performing geospatial analysis QGIS software is used and shapefile were extracted using DIVA-GIS website and categorized maps was prepared for the same. The linear trend analysis is performed by using MS-Excel software and for the suggestion of the recommendation some of the content analysis like literature review, peer reviewed journals will be adapted.

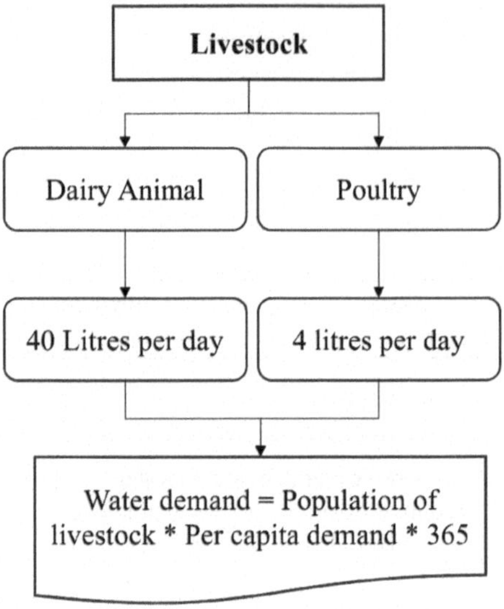

Fig. 11.3 Water demand for livestock

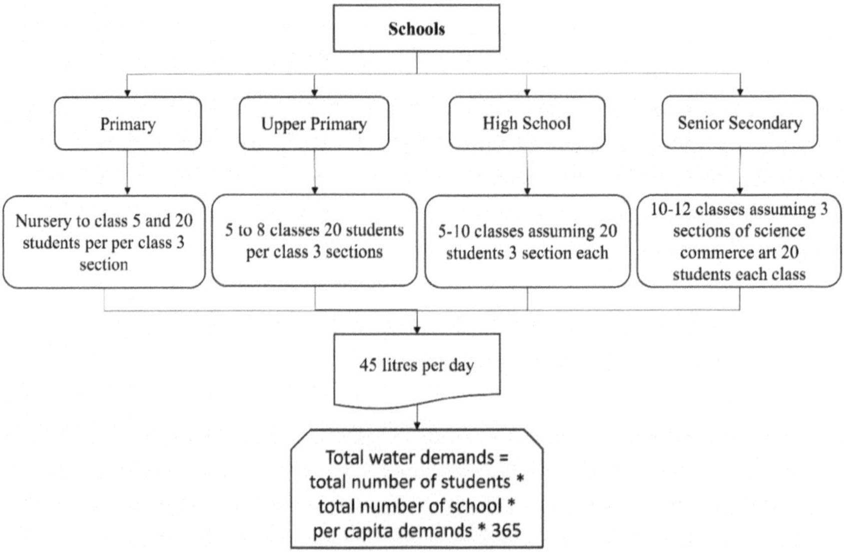

Fig. 11.4 Water demand for school

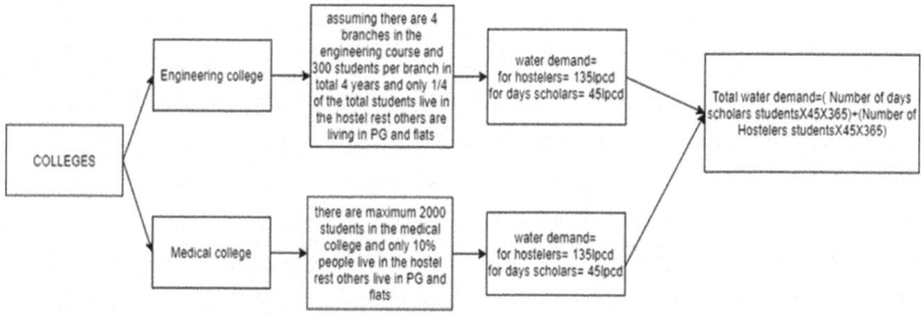

Fig. 11.5 Water demand for college

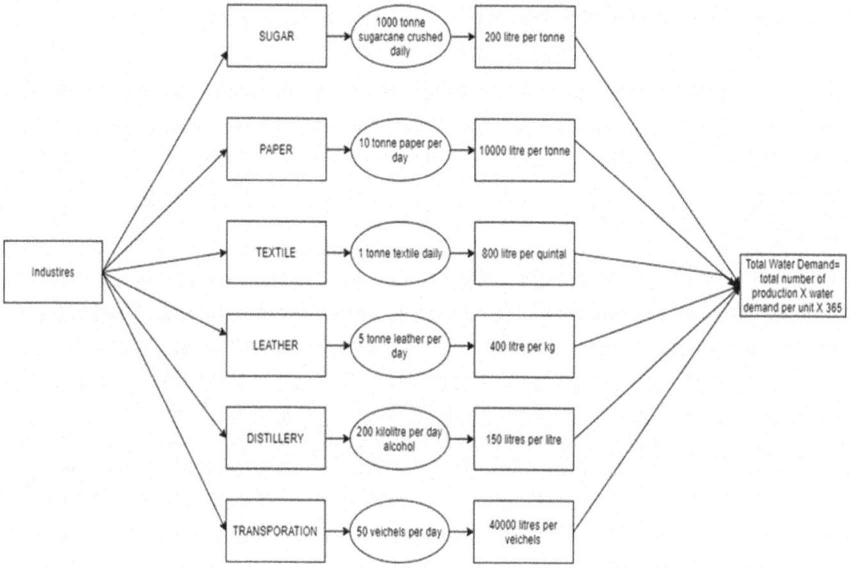

Fig. 11.6 Water demand for industries

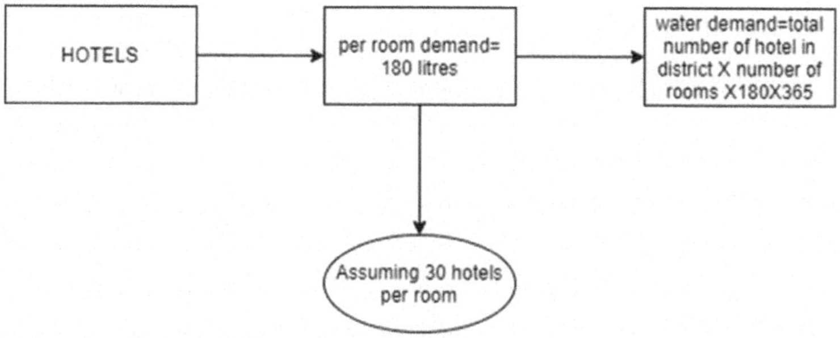

Fig. 11.7 Water demand for hotels

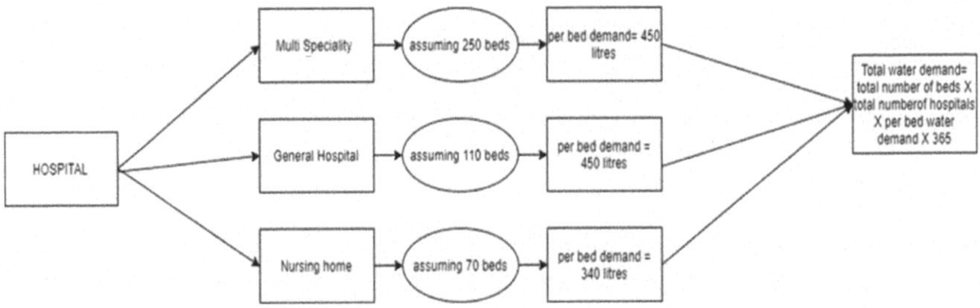

Fig. 11.8 Water demand for hospital

11.3 Results and Discussion

All 5 districts of this region were very different in population, water consumption, economic, industrial, and agricultural activities. To calculate the demand for water in each district of NCR regions of UP, first, the forecasting of the population was done by using the incremental increase method, and their water demand also be calculated based on the population forecasted (Table 11.1).

So, the district Meerut has a forecasted population of 66,62,679 and has water forecasted water demand of 328.3 MCM/YR which is comparatively higher than Baghpat, Gautam Buddha Nagar because Meerut is experiencing a drastic change in the development, people from the NCR regions were migrating towards the city in search of a better and comparatively cheaper lifestyle than NCR regions and also increase in connectivity from the nearby NCR regions attracted builders and industrialists to set up their industries in the outskirts of the city, which subsequently increases the populations of the city. Baghpat district has a forecasted population of 31,97,321 and forecasted water demand is 157.54 MCM/YR the city has the lowest water demand because it has a comparatively smaller land area and the city has fewer employment opportunities as compared to nearby districts like Ghaziabad, Meerut so there is less population in the city. Bulandshahar has the highest forecasted population of 84,58,651 and the highest water demand of 416.8 MCM/YR it has the highest forecasted water demand because the area of the district is comparatively bigger and there are mostly village clusters located around the city and the villagers tend to use water carelessly. The forecasted population of Gazhiabad+Hapur is 58,15,232

Table 11.1 Population with water demand

S. No.	Name of district	Population	Forecasted water demand (MCM/YR)
1	Meerut	66,62,679	328.3
2	Baghpat	31,97,321	157.54
3	Bulandshahar	84,58,651	416.8
4	Gazhiabad+Hapur	58,15,232	296.5
5	Gautam Buddha Nagar	35,83,428	176.5

and has an estimated water demand of 296.5 MCM/YR this city is very close to the NCR region and is facing the extreme development and a lot of corporates and builders were investing in the development of the city and also the location of major industries like Dabur were located around the city which is increasing the population of the city and in case of Hapur the city is associated with the Ghaziabad and people working in the nearby NCR region were settled around this city in search of cheaper lifestyle as compared to New Delhi, Ghaziabad, Meerut, Gautam Buddha Nagar. The Forecasted population of Gautam Buddha Nagar is 35,83,428 and the forecasted water demand of the district is 176.5 MCM/YR the city is the main attraction of builders and industrialists because most of the landholders of the city were ready to sell their land to them in greed for more money and better lifestyle construction of skyscrapers and big industrial clusters were attracting people from nearby areas to live in the city.

11.3.1 Water Demand Estimation

The water demand is estimated based on the following parameters.

11.3.1.1 Agriculture

These districts were important from the agriculture point of view as well as the presence of the Upper Ganga canal, Yamuna River, Hindon river, sufficient rainfall and have plenty amount of groundwater, majorly 4 types of crops were grown in this the districts namely Sugarcane, Rice, Wheat, Potato and other horticulture crops mustards, pulses, and some fruits like mangoes, watermelon, guava etc. In kharif crops there are various qualities of rice were grown like 79,98,113 (numbers assigned by the authority according to quality) and in this region generally Has a yield of near around 2000–2500 ha/ton the rice crop generally takes near around 90–120 days from sown to grown and being harvested. This produced rice is generally being sold at the rate of Rs 2100 per quintal, this is the minimal price which is being set by the government but the farmers sold at different prices according to the quality of rice so in greed of producing good quality they tend to use more groundwater and being responsible for groundwater depletion. The sugarcane is also one of the most water incentive monocrops which is practiced in this region at a large scale because of the 26 presence of large sugar mill in this area this area is known for sugar cultivation, large lands of farmers and many subsidies by the government and sugarcane is soled at the minimum price of Rs 350 per quintal these are the major factors which promoted the growth of sugarcane in the region and put stress on the groundwater table of the region specially district of Meerut and nearby region witnessed large growth of sugarcane in the region and also sugar industries were also encouraging farmers, no doubt this crop is helping a farmers lot in improving financial condition but it

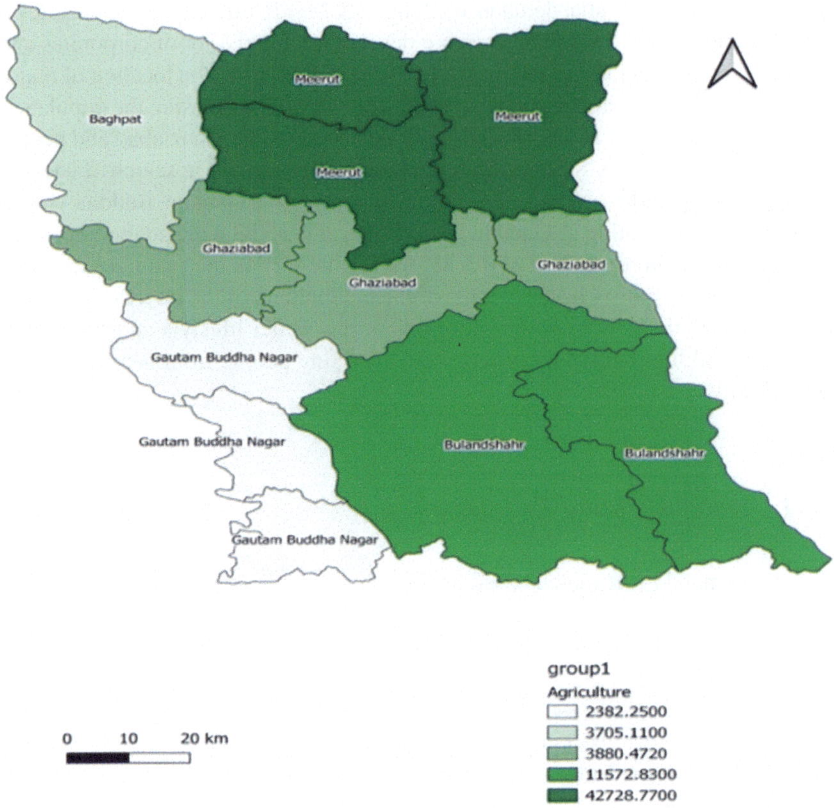

Fig. 11.9 Agriculture water demand

is consuming groundwater lot of groundwater. The water demand of agriculture is shown in Fig. 11.9.

11.3.1.2 Livestock

Apart from the rapid industrialization and urbanization this city plays an important role in terms of an agricultural point of view despite having modern farming techniques they are also dependent upon cows and buffalos not only from the farming point of view but also from the dairy products business which will be proved to be extra income for them and also a business of poultry proved to be the great income saver for them, however, the water they consume is not very much but it's an important parameter in the calculation of water demand more deeply, the data of them is provided in Fig. 11.10.

In the Meerut district the combined water demand of both cows and buffaloes is 9.75 MCM/YR the nearby villagers of the district were relatively dependent on

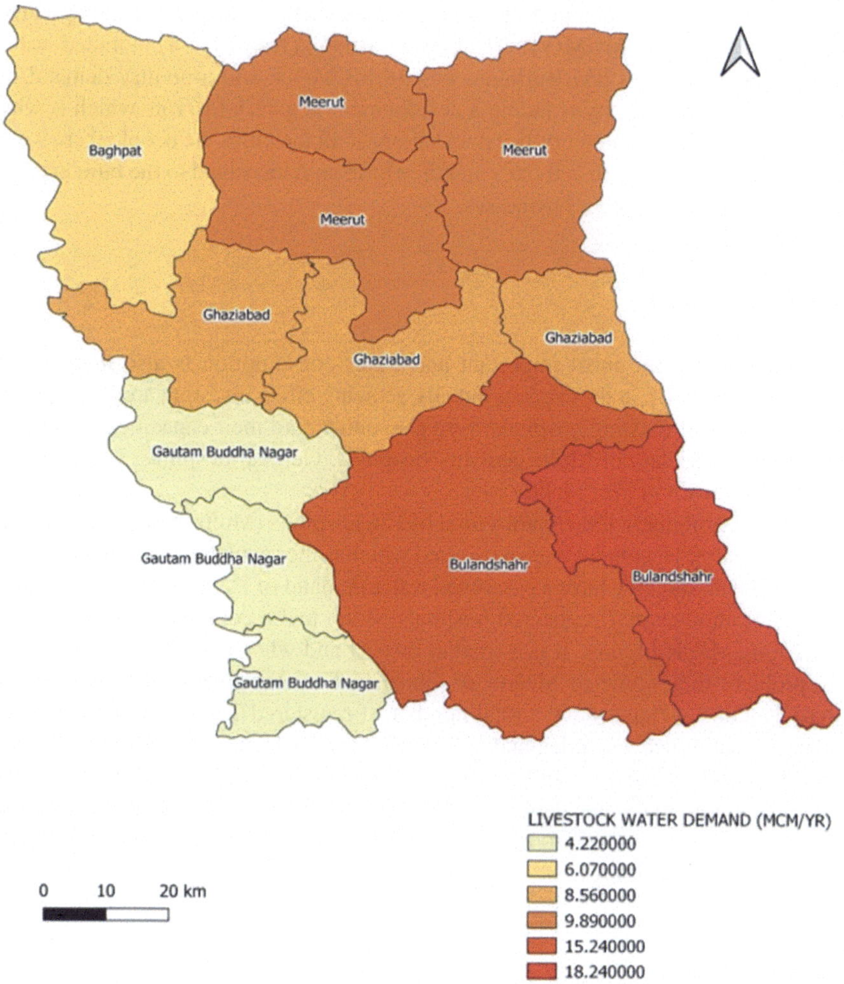

Fig. 11.10 Livestock water demand

them for agriculture and business activities, the farmers mostly prefer to use animals instead of using tractors and other techniques in midst of saving money, the poultry demand of Meerut district is 0.30 MCM/YR. Baghpat district has a combined water demand for both cows and buffaloes is 5.95 MCM/YR and poultry demand is 0.30 MCM/YR the area is comparatively smaller than other districts so it has less number of livestock and poultry than other districts. District Bulandshahar has the highest livestock water demand of 17 MCM/YR because this district is rich in agriculture point of view most people of the villages in this district were in the agriculture and dairy farming businesses, and the poultry water demand is 0.22 MCM/YR which is comparatively higher than other districts. The Ghaziabad + Hapur district has a combined water demand for both cows and Buffaloes is 8.815 MCM/YR and a poultry water demand of 0.19 MCM/YR. The Gautam Buddha Nagar has a combined water demand for both cows and Buffaloes is 4.18 MCM/YR and a poultry demand of 0.02 MCM/YR as the city is facing a development at a very fast rate which is why they have the lowest number of livestock amongst all the cities the people there were less indulge in agriculture activities and they have sold their land to the builders and shifted themselves to other businesses.

11.3.1.3 Hospital

The hospitals are the most important and water consumption is also high in the hospitals, especially in these economically growing cities investors are investing in the hospitals and existing hospitals were also enhancing their capacities of beds in the hospitals the data of Multispeciality hospitals, General hospitals, and Nursing homes were given in Fig. 11.11.

In the Meerut district, there are a total of 176 hospitals (Multi Speciality, Nursing Homes, General Hospitals) were registered which is the secondhighest after Gautam Buddha Nagar and they have a combined water demand of 2.29 MCM/YR. Baghpat has a total number of 7 registered hospitals which had a combined water demand of 0.13 MCM/YR because it is a smaller district and whenever people suffer from any problem they come to Meerut or Ghaziabad for the treatment. The districts Ghaziabad and +Hapur have a total number of registered 248 registered hospitals and have a combined water demand of 3.14 MCM/YR which is the highest among all the districts. The district Bulandshahar has a total number of 105 registered hospitals and has a combined water demand of 2.5 MCM/YR, the district Gautam Buddha Nagar has a total number of 181 registered hospitals and has a combined water demand of 2.5 MCM/YR.

11.3.1.4 Hotels

The hotels in these districts were not as much high demand because there were no such tourists' attractions in this area and the hotels also consume less amount of water as compared to the other sectors the data on the hotels and their water consumption were given in Fig. 11.12.

The water consumed by 48 hotels in the Meerut district is 0.09 MCM/YR which is the second-highest after Gazhiabad+Hapur which has 64 hotels and consumes 0.12 MCM/YR. Baghpat has the lowest number of hotels 1 which is consuming the lowest amount of water 0.0019 MCM/YR as compared to these districts because of the more village background area and no tourist attraction spots in the city. Gautam Buddha Nagar has 23 hotels that consume 0.04 MCM/YR of water as the city is facing development more new hotel projects were coming up in the city and the number of hotels were increasing significantly.

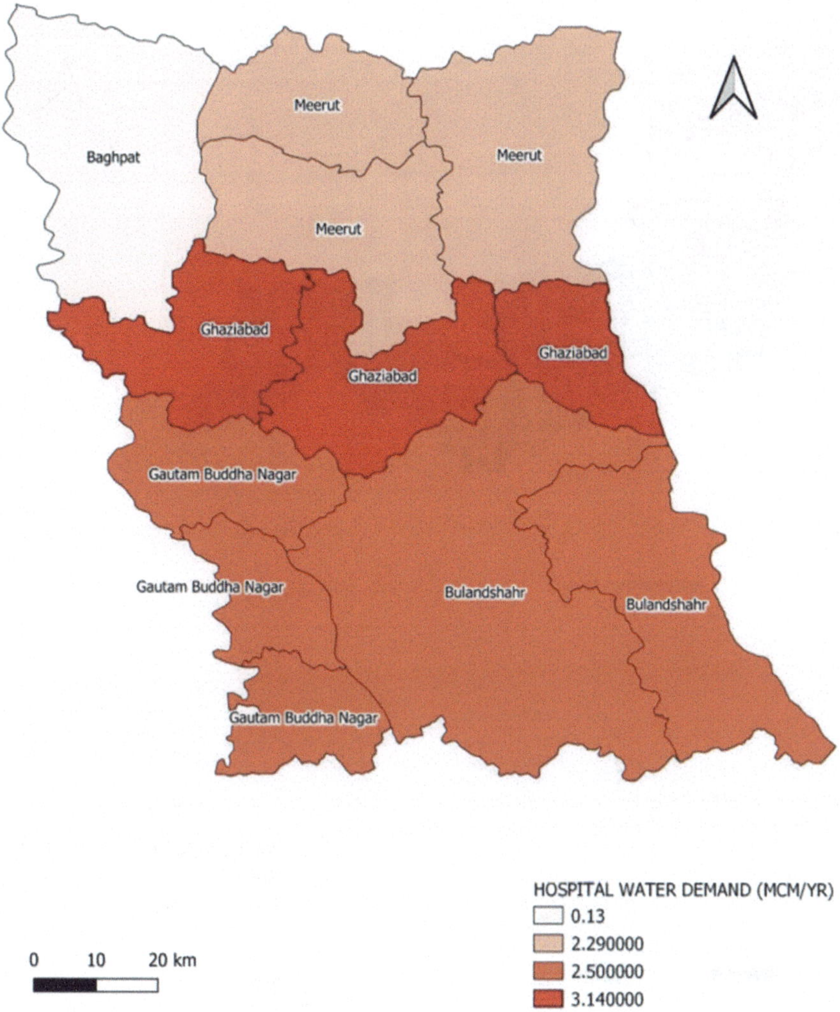

Fig. 11.11 Hospital water demand

11.3.1.5 Industries

This is one of the most water-consuming sectors in these cities, there were majorly 5 types of industries that were located in these cities which consumes more water can it be ground or surface water. Some industries in these areas were involved in the theft of ground and surface water, as well as installation of unnotified submersible pumps in the production unit to withdraw groundwater for their production and cleaning of that area and unusual withdrawal of nearby river or ponds water apart from the theft they were also responsible for the contamination of both ground as

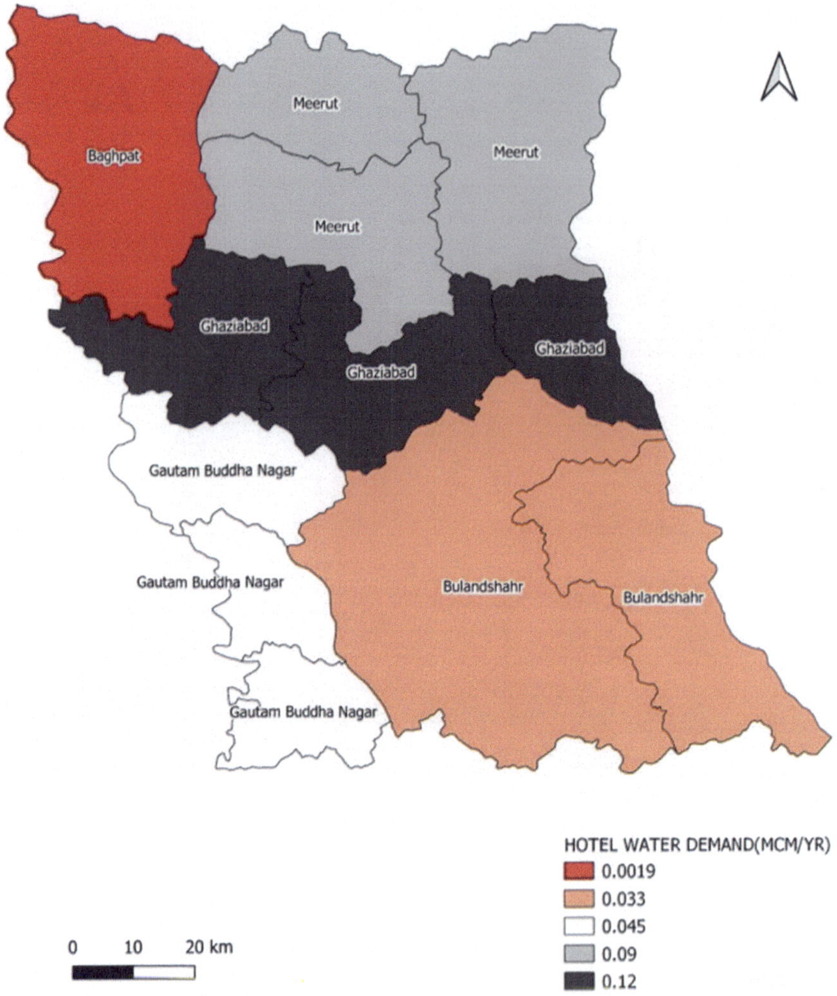

Fig. 11.12 Hotel water demand

well as surface water discharge of their waste directly into the river and surface, contaminates the water, unfunctional ETP Plants contribute more to the problem, apart from this inattentiveness of officials toward this serious issues contributes more to the problem and made the situation worse, the water consumption by different clusters of industries and with the type of industries in each district were given in Fig. 11.13.

Different industries have different quantities of production and consumption of water is also not same in different cities like in case of Sugar industries the city Meerut have 5 sugar industries where on an average 1000 tonnes of sugarcane crushed daily so they have a combine water demand of 3.65 MCM/YR and it has 1 textile industry

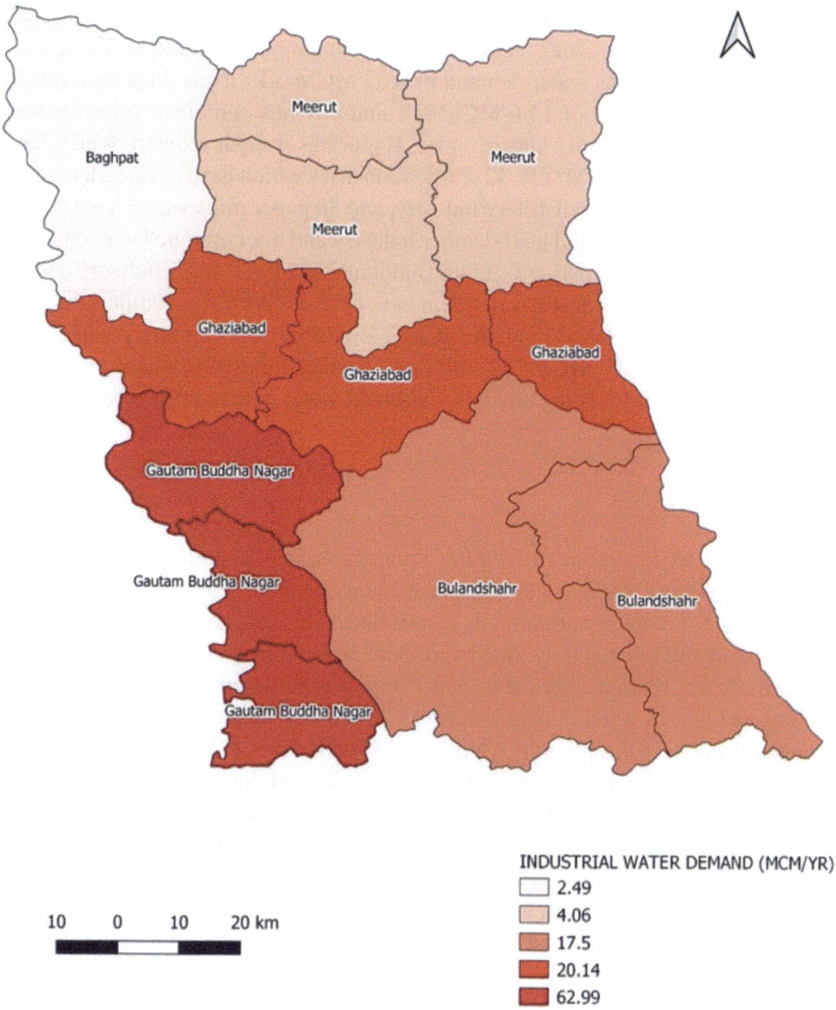

Fig. 11.13 Industrial water demand

in which 1-tonne textile made daily and have combined water demand of 0.292 MCM/YR, 5 Distillery units in which 200 kilolitre of alcohol made daily and have combine water demand of 0.05 MCM/YR and 2 paper industry in which 10 tonnes of paper made daily which having the combined water demand of 0.073 MCM/YR the city has no leather industry and it has total water demand of 4.069 MCM/YR. The assumptions of production were considered the same for all the districts. Ina Baghpat the city has 3 Sugar industries and has a water demand of 2.19 MCM/YR 1 distillery unit has a water demand of 0.0109 MCM/YR it has no paper and leather industry and it has a combined water demand of 2.49 MCM/YR. In Bulandsahar there are 3 sugar mills which have a water demand of 2.19 MCM/YR, and has 2 textile industry which has a water demand of 0.5 MCM/YR, the city has the highest number of distillery industries among all the cities and has a water demand of 0.12 MCM/YR it has 1 paper industry which has water demand of 0.03 MCM/YR, it has 2 leather industry which has water demand of 14.6 MCM/YR and city has combined water demand of 17.53 MCM/YR. The city Ghaziabad + Hapur has 9 sugar industry which has a water demand of 6.57 MCM/YR, 42 textile industries which have a water demand of 12.2 MCM/YR city has no distillery industry, and 36 paper mills which have a water demand of 1.3 MCM/YR and has 0 leather industry and has combined water demand of 20.14 MCM/YR. The district Gautam Buddha Nagar has 0 sugar industry and has 33 textile industry which has a water demand of 9.6 MCM/YR 6 distillery industry which have a water demand of 0.06 MCM/YR and 60 paper industry which have a water demand of 2.19 MCM/YR and has 7 leather industries which have water demand of 51.1 MCM/YR the city has combined water demand 62.99 MCM/YR.

11.3.1.6 School

These cities in NCR regions of UP play a major role in education point of view the education of nearby villages people and residents of the city rely on these schools event there are mostly private schools comes under this category but there are some are controlled and managed by the government authorities mentioned Fig. 11.14.

In the Meerut district, there are 1392 Primary schools which are the highest in numbers and have a water demand of 9.60 MCM/YR. There are 577 upper primary schools in number in the city which have a water demand of 1.70 MCM/YR and have 126 high schools, and 189 senior secondary schools which have a water demand of 0.012 MCM/YR and 2.79 MCM/YR, and have a combined water demand of 14.12 MCM/YR. The district Baghpat has 371 primary schools which as water demand of 2.56 MCM/YR, and have 185 upper primary schools which have a water demand of 0.54 MCM/YR, the district has 37 high schools and 71 senior secondary schools which have a water demand of 0.47 MCM/YR and 1.04 MCM/YR and have total water demand of 4.63 MCM/YR. The district of Ghaziabad has 684 primary schools which have a water demand of 4.72 MCM/YR, and has 694 upper primary schools which have a water demand of 2.05 MCM/YR, and has 103 high schools and 86 senior secondary schools which have a water demand of 1.32 MCM/YR and 1.27 MCM/YR respectively and have total water demand of 4.63 MCM/YR. The district Bulandshahar has 1226 primary schools which have a water demand of 8.46 MCM/YR, the district has 475 upper primary schools which have a water demand of 1.40 MCM/YR and the district has 160 high schools and 153 senior secondary schools which have water demand of 2.05 MCM/YR and 2.26 MCM/YR and has total water demand of 14.18 MCM/YR. the district Meerut and Bulandshahar have same almost the same water demand it is because Bulandshar has a comparatively bigger area and Meerut city has more schools than Bulandshahar. The district Gautam Buddha Nagar has 514 primary schools having a water demand of 3.55 MCM/YR, the district has 194 upper primary schools which have a water demand of 0.57 MCM/YR, there are

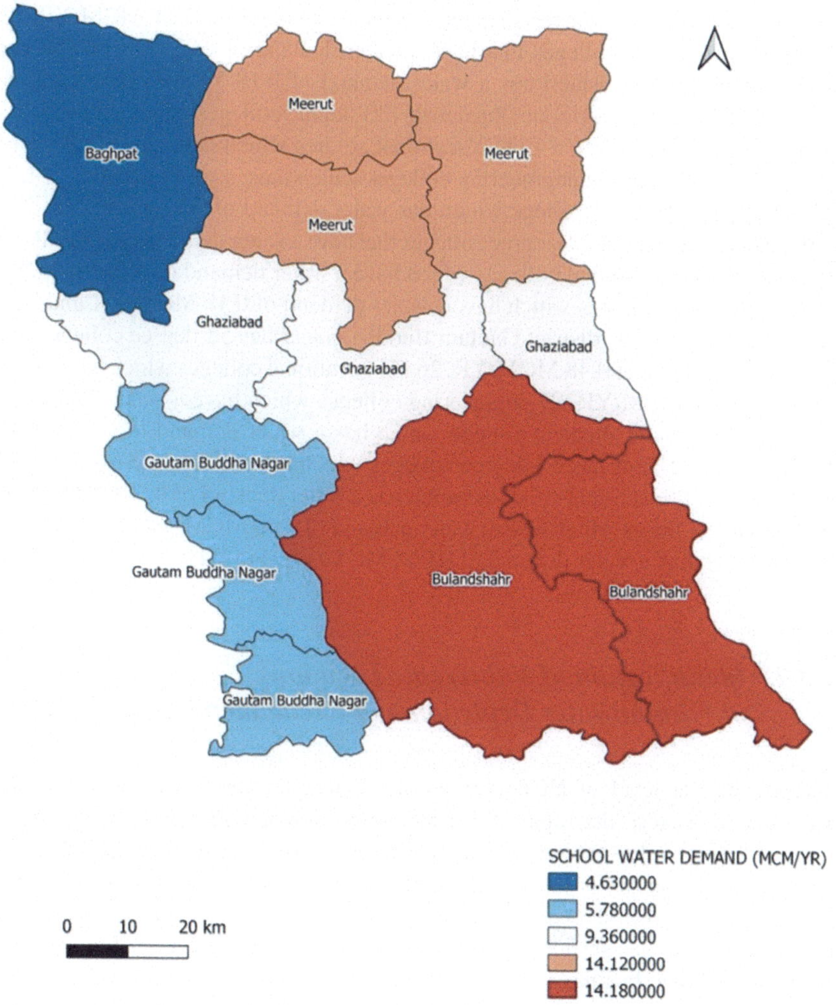

Fig. 11.14 School water demand

76 high schools and 46 senior secondary schools which have a water demand of 0.97 MCM/YR and 0.46 MCM/YR and has total water demand of 5.78 MCM/YR.

11.3.1.7 Colleges

In the Meerut district, 171-degree colleges have a water demand of 1.24 MCM/YR and 21 ITI Vocational studies colleges have a water demand of 0.12 44 MCM/YR. The district has 33 engineering colleges that have a water demand of 1.46 MCM/YR and 2 medical colleges which has a water demand of 0.06 MCM/YR. The Baghpat district has 49-degree colleges that have a water demand of 0.24 MCM/YR, 11 ITI Vocational studies colleges that have a water demand of 0.021 MCM/YR, and 4 engineering colleges which has a water demand of 0.18 MCM/YR and has no medical colleges. The district of Ghaziabad 133-degree colleges which have a water demand of 0.99 MCM/YR, 34 ITI Vocational studies which have a water demand of 0.32 MCM/YR, and 39 engineering colleges which have a water demand of 1.73 MCM/YR and 1 medical college which has water demand of 0.03 MCM/YR. The district Bulandshahar has 59-degree colleges that have a water demand of 0.51 MCM/YR, it has 36 ITI vocational colleges which have a water demand of 0.34 MCM/YR and 4 engineering colleges which have a water demand of 0.18 MCM/YR and have 0 medical colleges. The district Gautam Buddha Nagar has 58-degree colleges that have a water demand of 0.48 MCM/YR, 26 ITI vocational colleges which has a water demand of 0.17 MCM/YR, 27 engineering colleges which have a water demand of 1.20 MCM/YR and 1 medical college which has a water demand of 0.03 MCM/YR. The district Hapur has 20-degree colleges which have a water demand of 0.12 MCM/YR, 3 ITI Vocational colleges which has a water demand of 0.009 MCM/YR, 9 engineering colleges which have a water demand of 0.40 MCM/YR and 2 medical colleges which has a water demand of 0.068 MCM/YR, Fig. 11.15.

11.3.2 Identification of Vulnerable Locations of Groundwater Depletion and Recommendation

The groundwater level of NCR regions of UP faced a sever depletion from the past 5 years, because the rapid industrialization, urbanization, and change in the lifestyle of the people are continuously putting pressure on the existing aquifers and depleting the groundwater level of the regions, most of the activities of the people of these regions were dependent on the groundwater and the nearby rivers. The lifestyle of people has shifted more toward western culture because of which they have increased their businesses which resulted in the more demand for water so they see groundwater as a great option of use because there were no such regulations and norms for extracting groundwater, agriculture was the main reason, growing of water incentive crops such as sugarcane, rice has made the situation worse, despite this region receives an average amount of rainfall near around 900 mm which is not enough for satisfying the water demand of the fields because farmers of these districts have comparatively large farms and fields in which they grow rice and sugarcane and different crops and sell them to the industries and local vendors. This region has loamy soil which is perfect for the growth of these crops. The groundwater level of Pre and Post monsoon was different in different regions. The map which is shown below shows the fluctuations in the groundwater level in pre and post-monsoon in the past 5 years which region has experienced (Fig. 11.16).

The groundwater table map of the region states that the groundwater level in the Meerut district in pre-monsoon shows a growth of 0–0.5 m and in the post-monsoon,

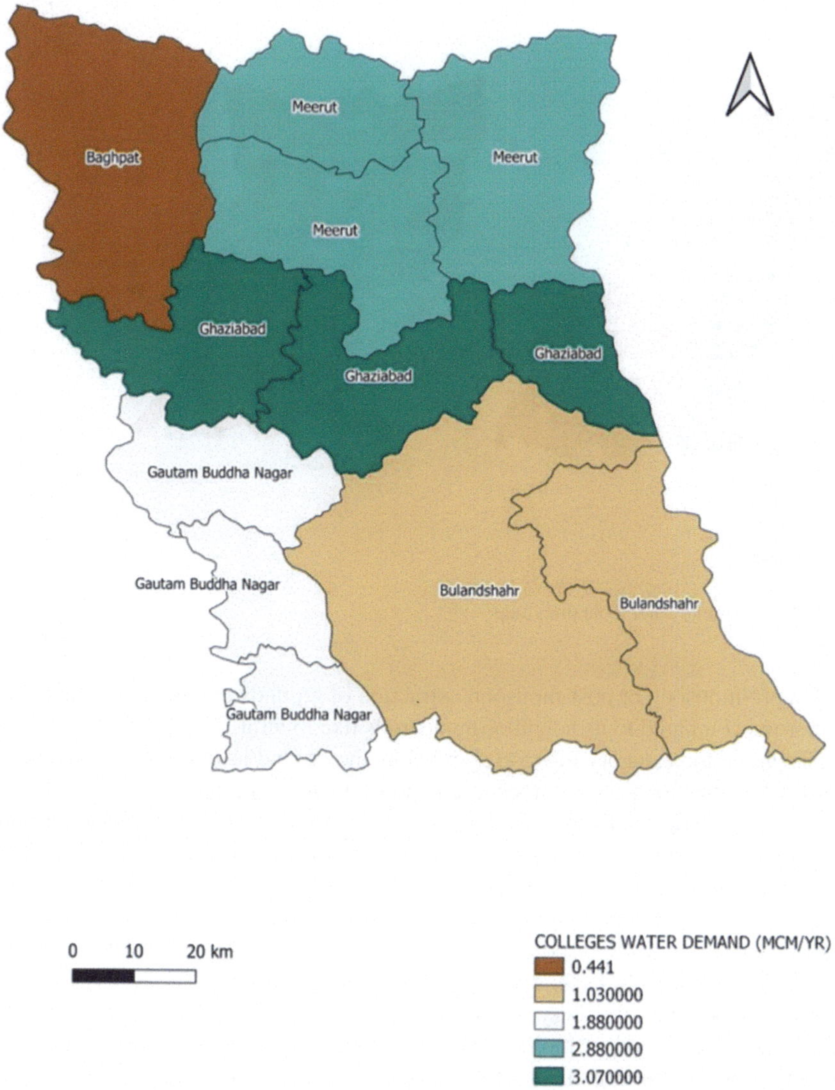

Fig. 11.15 College water demand

it shows a decline of more than 1.5 m below the ground level and cultivation of rice along with the sugarcane is possibly the reason for the depletion because during post-monsoon season cultivation of rice takes place in the region and some requirement of water for the rice is taken from the rainfall but still it needs a lot of more water for proper cultivation. The groundwater level drop of the Baghpat region remains more than 1.5 m below the ground level because there is a mostly rural area and people were majorly dependent on the groundwater so it is playing in the danger category whether

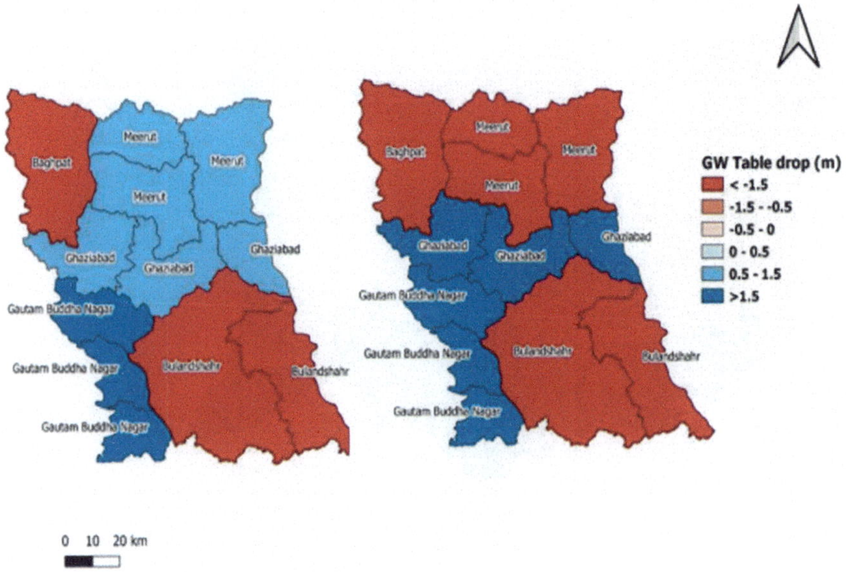

Fig. 11.16 Groundwater table drop map

it be a pre-monsoon or post-monsoon extraction of groundwater in the Baghpat city is more as compared to its neighboring districts like Meerut. The district Ghaziabad comes under the category of 0.5–1.5 m below the ground level in the pre-monsoon season but after the post, the recharge of more than 1.5 m has been seen because there is no such rice cultivation taking place in the city, and if places the local farmers took water from the small canal named Ramganga which is flowing nearby area. The district Bulandshahar always comes under the overexploited category when it comes to the groundwater depletion and the main reason is the more cultivation area and village area because more population has deviated towards the dairy business for which they are mainly dependent on groundwater mostly rich farmers apart from farming indulge into the dairy businesses which supply milk-based products to the NCR regions in greed of more money and poor and landless farmers worked as an employee of rich farmers apart from relocating themselves to a new place. The district Gautam Buddha Nagar is the same in terms of groundwater usage there is no such drastic change in the extraction and it has seen a recharge of more than 1.5 m in the past 5 years. it is because people of the district have sold their land to builders and developers in greed for more money and started indulging themselves in other businesses the district is very near to the National capital of India New Delhi so that's why the land of the district is in the eyes of builders and developers.

The residents, 47 however, do not pay much attention to the agriculture side to make money, the development schemes of New Delhi have benefited the Gautam Buddha Nagar to a great extent like the construction of highways, and high-rise buildings in the city have shifted the people towards urbanization and industrialization. The current scenario of this region simply states that the groundwater is an easily available resource that is not being taken care of well, even after people know that once groundwater is depleted it will take many years to recharge the aquifers again. There are no such strict policies of the government related to extraction and usage of groundwater, in some of the regions of these districts groundwater is not only depleted but contaminated also the presence of heavy metals such as Nickel, cs, Fe, Pb, etc. were found in the groundwater table these districts. However, Uttar Pradesh Government has shown some activeness regarding groundwater depletion and tried to introduce some new rules like the mandate of rainwater harvesting systems in houses and complexes which has an area of more than 150 sqm, the mandate of installation of STP and ETP in industrial and development projects, restrictions on the installation of submersibles pumps without the permission of the government, etc. but the implementation of these laws was not checked regularly. The farmers secretly installed submersibles pumps in their fields and continue extractions, and also withdrawal of river water like the upper Ganga canal is a problem and because of the carelessness of the administration, the groundwater depletion is slowly becoming one of the serious problems not only in these districts but also on the state level. Local people should understand the importance of the groundwater and the government should make citizens aware of the seriousness of the problem and future risks, regular audits of the farms, and industries should be done by the government officials, and imposing fines and penalties should be done by the government against the violators.

The trend analysis of the following districts also reveals some of the unexpected results related to groundwater, the linear trend analysis was performed from the groundwater data that was taken from the groundwater board and the water table.

11.3.2.1 Meerut Seasonal Groundwater Trend

The seasonal groundwater level trend of Meerut city reveals that in the pre monsoon 2015 the groundwater level was 11.4 mbgl in 2015 and in 2020 it was 12.24 mbgl (Fig. 11.17). The post monsoon groundwater level in 2015 was 12.54 mbgl and huge variation of 14.99 mbgl was seen in 2020. Because of the cultivation of rice and sugarcane was the main reason for the fluctuation in groundwater table of the city.

11.3.2.2 Bulandsahar Seasonal Groundwater Trend

The seasonal groundwater level of Bulandsahar district reveals that the in pre monsoon 2015 the groundwater level was 12.17 mbgl in 2015 and 14.36 mbgl in 2020 (Fig. 11.18). The post monsoon groundwater level was 12.67 mbgl and the water level in 2020 has seen change of 14.8 mbgl in 2020. The city has compared bigger area as compared to the other districts and the district has more influence of village so lot of people were engaged into primary sectors of employment. The fluctuation of groundwater shows that the residents were much more dependent upon

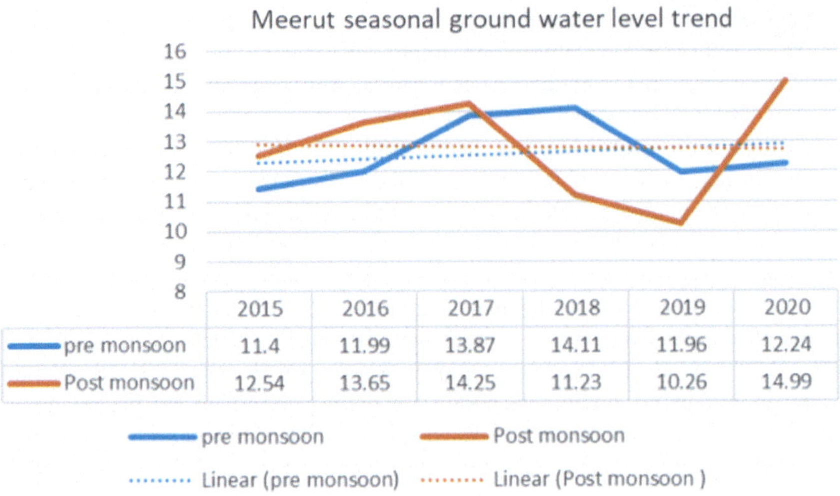

Fig. 11.17 Groundwater trend (Meerut)

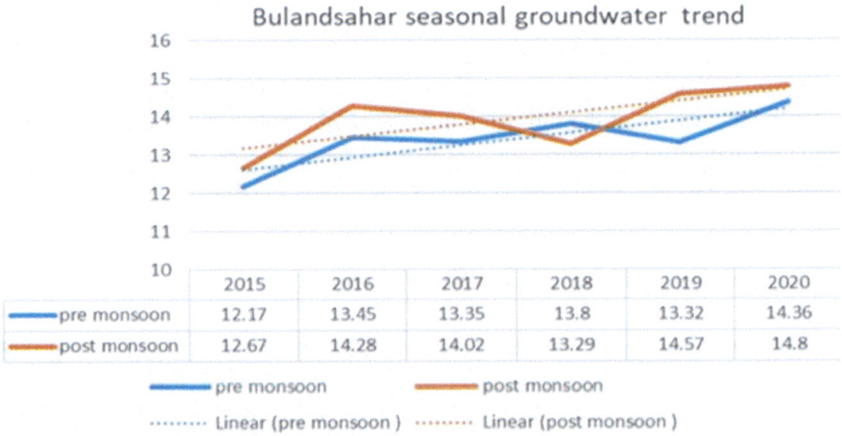

Fig. 11.18 Groundwater trend (Bulandsahar)

groundwater and have installed submersibles pumps in their houses and using this water for their household as well as business purposes and some industries of the district were involved in the theft of groundwater.

11.3.2.3 Ghaziabad + Hapur Seasonal Groundwater Trend

The seasonal groundwater trend of Gazhiabad + Hapur in pre monsoon reveals that the in 2015 the groundwater level was 10.28 mbgl and in 2020 the water level

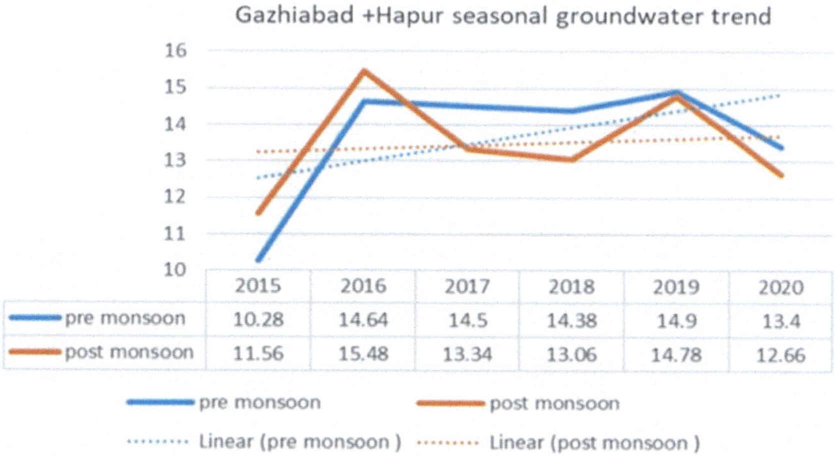

Fig. 11.19 Groundwater trend (Gazhiabad + Hapur)

was 13.4 mgbl (Fig. 11.19). The ground water level in post monsoon on 2015 was 11.56 mbgl and 2020 in 12.66 mbgl. The post monsoon ground water trend of district in 2015 was 11.56 mbgl and in 2020 was 12.66 mbgl. The fluctuation in groundwater table and possibly reasons where the city was facing development both in terms of infrastructure and in agriculture point of view the Ghaziabad was facing advancement in development as it was close to Delhi and Hapur region is facing advancement in growing of water incentive crops.

11.3.2.4 Gautam Buddha Nagar Seasonal Groundwater Trend

The district Gautam Buddha Nagar show the groundwater level in pre monsoon 2015 was 10.07 mbgl and in 2020 it was 12.55 mbgl and in post monsoon 2015 it was seen at 10.53 mbgl and in 2020 it was seen 12.87 mbgl (Fig. 11.20). The groundwater level has shown less fluctuation in past years because the city has face advancements in development and the agriculture land has been sold to the builders and developers and villagers and nearby people indulged in other businesses. The condition of groundwater in the city is slightly better than other districts because farming was not so much promoted in this region.

11.3.2.5 Baghpat Seasonal Groundwater Trend

The groundwater level in pre monsoon 2015 is 13.6 mbgl and 13.99 mbgl in 2020. The ground water level in post monsoon 2015 is 12.58 mbgl and 15.22 mbgl in 2020 (Fig. 11.21).

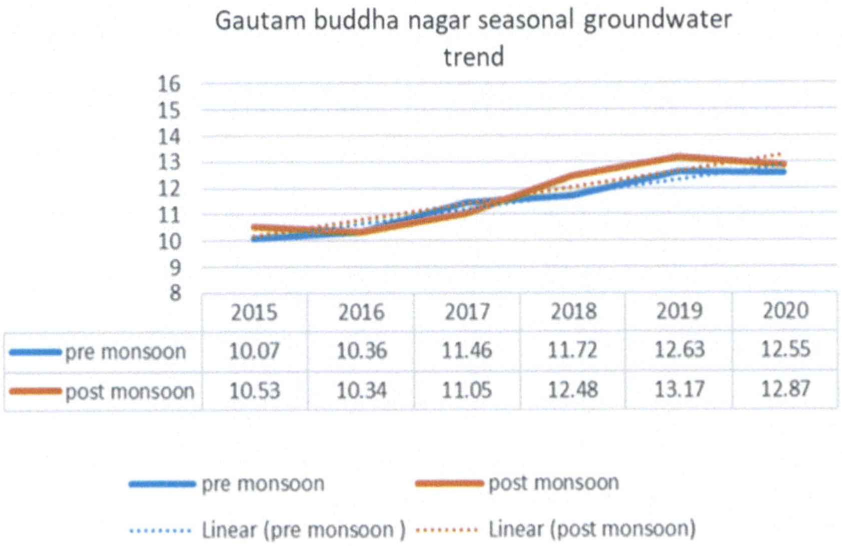

Fig. 11.20 Groundwater trend (Gautam Buddha Nagar)

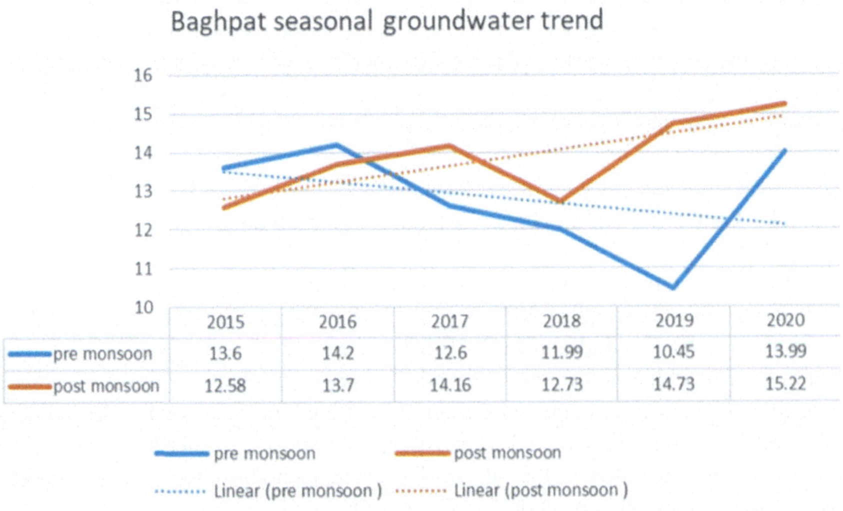

Fig. 11.21 Groundwater trend (Baghpat)

11.4 Conclusion

From the first objective, it is estimated that the water demand of all the 5 districts were very different in terms of water consumption and it can be concluded that the population of these districts has increased at an exponential rate, residents were using this resource carelessly and causing depletion. Apart from this, agriculture consumes most of the water whether it be a ground or surface and continuously putting pressure on the aquifers and production of water incentive crops. Subsidies by the government is the main reason for the depletion, Gautam Buddha Nagar consumes most of the water in industrial sector and Baghpat consumes least. More than 56MCM/YR of water can be consumed by school sector and 8.05 MCM/YR consumed by colleges some of them uses ground water and some uses surface water. Hotels in these districts consumes least amount of water because there are no as such tourist attraction places in these districts. The hospital in the district consumes 10.56 MCM/YR of water some use groundwater and some use surface water for their operation. From the second objective it can be concluded that the districts Baghpat and Bulandsahar is worst in terms of ground water level and they extract most of the groundwater. The condition of Meerut has shown a decline and cultivation of sugarcane and rice is more in the district, on the other hand the condition of Ghaziabad and Hapur has shown an improvement in groundwater table in past 5 years. The condition of Gautam Buddha Nagar seems stable in past 5 years and because of rapid urbanization and industrialization in the district and the trend analysis of groundwater table of each district reveals the constant fluctuation in the groundwater table despite of being having upper ganga canal. Yamuna river and river Hindon is available in these districts. These districts continuously extract excess groundwater and government is equally responsible for this because of announcing too much policies on farming and urbanization motivate residents towards development without thinking of the resources.

Competing Interests The authors declare no potential conflicts of interest.

References

Akhtar N, Syakir Ishak MI, Bhawani SA, Umar K (2021) Various natural and anthropogenic factors responsible for water quality degradation: a review. Water 13(19):2660

Craswell E (2021) Fertilizers and nitrate pollution of surface and ground water: an increasingly pervasive global problem. SN Appl Sci 3(4):518

Dangar S, Asoka A, Mishra V (2021) Causes and implications of groundwater depletion in India: a review. J Hydrol 596:126103

Das IC, Kumar KV, Rajasekhar D, Bhattacharya A, Reddy PR, Subramanian SK, Srivastav SK et al (2021) IRS-1C satellite data utilization for groundwater prospects mapping for the entire country under national rural drinking water program (NRDWP): a state of the art initiative. J Indian Soc Remote Sens 49:111–120

Gavrilescu M (2021) Water, soil, and plants interactions in a threatened environment. Water 13(19):2746

Han D, Currell MJ (2022) Review of drivers and threats to coastal groundwater quality in China. Sci Total Environ 806(Pt 4):150913

Laghari AN, Vanham D, Rauch W (2012) The Indus basin in the framework of current and future water resources management. Hydrol Earth Syst Sci 16(4):1063–1083

Li P, Tian R, Xue C, Wu J (2017) Progress, opportunities, and key fields for groundwater quality research under the impacts of human activities in China with a special focus on western China. Environ Sci Pollut Res Int 24(15):13224–13234

Papazotos P (2021) Potentially toxic elements in groundwater: a hotspot research topic in environmental science and pollution research. Environ Sci Pollut Res Int 28(35):47825–47837

Priyan K (2021) Issues and challenges of groundwater and surface water management in semi-arid regions. In: Pande CB, Moharir KN (eds) Groundwater resources development and planning in the semi-arid region. Springer, Cham, pp 1–17

Raghavan R, Britz R, Dahanukar N (2021) Poor groundwater governance threatens ancient subterranean fishes. Trends Ecol Evol 36(10):875–878

Raza M, Hussain F, Lee J-Y, Shakoor MB, Kwon KD (2017) Groundwater status in Pakistan: a review of contamination, health risks, and potential needs. Crit Rev Environ Sci Technol 47(18):1713–1762

Saha D, Sikka AK, Goklani R (2022) Artificial recharge endeavours in India: a review. Water Secur 16:100121

Salama RB, Otto CJ, Fitzpatrick RW (1999) Contributions of groundwater conditions to soil and water salinization. Hydrogeol J 7:46–64

Salim MZ, Choudhari N, Kafy A-A, Nath H, Alsulamy S, Rahaman ZA, Al-Ramadan B et al (2024) A comprehensive review of navigating urbanization induced climate change complexities for sustainable groundwater resource management in the Indian subcontinent. Groundwater Sustain Dev 25:101115

Saxena R, Rath R, Gupta S, Sood N (2021) A review on ecological degradation, its causes and sustainable development in Delhi, India. J Appl Nat Sci 13(4):1294–1304

Scanlon BR, Reedy RC, Xu P, Engle M, Nicot JP, Yoxtheimer D, Ikonnikova S (2020) Can we beneficially reuse produced water from oil and gas extraction in the US? Sci Total Environ 717:137085

Sidhu BS, Kandlikar M, Ramankutty N (2020) Power tariffs for groundwater irrigation in India: a comparative analysis of the environmental, equity, and economic tradeoffs. World Dev 128:104836

Singh R, Saritha V, Pande CB (2024) Monitoring of wetland turbidity using multi-temporal Landsat-8 and Landsat-9 satellite imagery in the Bisalpur wetland, Rajasthan, India. Environ Res 241:117638

Swain S, Taloor AK, Dhal L, Sahoo S, Al-Ansari N (2022) Impact of climate change on groundwater hydrology: a comprehensive review and current status of the Indian hydrogeology. Appl Water Sci 12(6):120

Turner SWD, Hejazi M, Yonkofski C, Kim SH, Kyle P (2019) Influence of groundwater extraction costs and resource depletion limits on simulated global nonrenewable water withdrawals over the twenty-first century. Earth's Future 7(2):123–135

Xie X, Shi J, Pi K, Deng Y, Yan B, Tong L, Ma L et al (2023) Groundwater quality and public health. Ann Rev Environ Resour 48:395–418

Chapter 12
Evaluating Soil Erosion in Dehradun Using the RUSLE Model: Challenges, Impacts, and Policy Strategies for Effective Soil Conservation

Himanshu Sahu, Jyoti Nagarkoti, Purnendu Sardar, Pooja Purohit, Arun Pratap Mishra, Mriganka Shekhar Sarkar, and Ali R. Alruzuq

Abstract Soil erosion poses a significant environmental challenge in Dehradun due to its complex, steep terrain, heavy rainfall, and rapid urbanization. This study employed a Google Earth Engine-based approach to assess the average soil loss annually using the Revised Universal Soil Loss Equation (RUSLE). The analysis used various factors, including rainfall-runoff erosivity (R), soil erodibility (K), slope length (L), slope steepness (S), cropping management (C), and supporting conservation practices (P), to evaluate regional soil erosion potential. Our findings indicated

H. Sahu · P. Purohit
Department of Animal Ecology and Conservation Biology, Wildlife Institute of India, Dehradun, India

H. Sahu · J. Nagarkoti · P. Purohit
Bhomya Foundation, Monal Enclave, Dehradun, India

J. Nagarkoti
Soil Conservation Division, Uttarakhand Forest Department, Ranikhet, India

P. Sardar
Department of Physical Geography and Ecosystem Science, Centre for Advanced Middle Eastern Studies (CMES), Lund University, Lund, Sweden

A. P. Mishra (✉)
Department of Forestry and Remote Sensing, Earthtree Enviro Private Limited, Shillong, Meghalaya, India
e-mail: arunpratap7371@gmail.com

M. S. Sarkar
G.B. Pant National Institute of Himalayan Environment, North-East Regional Centre, Itanagar, Arunachal Pradesh, India

A. R. Alruzuq
Department of Geography and Geographic Information Systems, Imam University, Riyadh, Kingdom of Saudi Arabia

H. Sahu
Academy of Scientific and Innovative Research (AcSIR), Ghaziabad, Uttar Pradesh, India

V. Saritha et al. (eds.), *Remote Sensing for Environmental Monitoring*,
https://doi.org/10.1007/978-981-96-5546-5_12

substantial soil loss in most of the study area (78%) with an average soil erosion rate of 488.58 t/ha/year annually. The distribution of soil erosion strongly correlated with areas of steeper slopes. Our study emphasized the critical roles of factors such as topography, land use and rainfall erosivity in influencing erosion patterns in Dehradun. To mitigate such risks, we recommend a holistic approach towards soil management by using sustainable development goals (SDGs) as a road map and implementing targeted conservation measures, increasing vegetation cover with strict policies on deforestation and community engagement in land conservation. These strategies are essential for preserving Dehradun's natural resources, promoting long-term environmental sustainability, and supporting the SDG goals.

Keywords Afforestation · Dehradun · Hydrological impact · Soil erosion · Sustainability

12.1 Introduction

Soil erosion is a natural phenomenon that results in the detachment and displacement of soil particles from the soil surface (Govers et al. 1990; Flanagan 2002) through natural factors such as wind, water, and gravity (Fayas et al. 2019). It involves a diverse host of factors with different combinations, variations, and interactions that affect the soil cover, as well as interactions that affect the soil cover and are further alleviated by human activities (Gao et aL 2018; Römkens et al. 2002). Soil erosion depletes soil fertility, negatively impacts the environment and poses significant challenges to agricultural productivity and water quality of the region (Prasannakumar et al. 2012), which becomes a significant environmental concern (Elsen et al. 2003; Singha et al. 2006). Overgrazing, inadequate technology, poor agricultural practices, steep topography, gully development, rainwater run-off and nutrient loss due to sedimentation are key factors exacerbating erosion (Zerihun et al. 2018). The occurrence and intensity of rainwater run off are influenced by land use and land cover (Hovius 1998; Karvonen et al. 1999; Chen et al. 2001). Globally, from the World's terrestrial ecosystems soil erosion leads to at least 75 billion tonnes of soil loss (Pimentel and Kounang 1998) among which India's average annual soil loss is about 5.3 billion tonnes (Dogra 2011), which leads to decreased productivity, representing a significant degradation hazard all over India (Gupta 2007; Mahapatra et al. 2018).

In the Indian Himalayan Region (IHR), studies in the river basin have shown that the rate of soil loss is extremely high, ranging from 20 to 9p t/ha/yr. Indian Himalayan region, characterized by its young mountains, steep and dissected topography, shallow soil depth, predominantly coarse-textured soils with limited water-holding capacity, ongoing agricultural practices on slopes and active tectonic processes, is highly susceptible to erosion (Sati et al. 2011; Shukla et al. 2020; Rawat et al. 2017). Soil erosion in these catchment areas, followed by sediment deposition in waterways, significantly impacts reservoir storage capacity and deteriorates downstream water quality (Walsh et al. 2005; Yereseme et al. 2022). Additionally, Uttarakhand, situated within this region, is highly susceptible to natural disasters like cloudbursts, flash floods, earthquakes and landslides due to its proximity to geographically active zones and irregular climatic patterns (Kansal and Singh 2022). These natural events result in soil displacement, adversely affecting the agricultural productivity and the health of natural ecosystems (Behera et al. 2020; Jamal and Sen 2024; Negese et al. 2021). It leads to reduced soil fertility and diminished crop yields, posing a threat to regional and national food security and environmental sustainability (Dapin and Ella 2023). There is a growing recognition of the strong connections between soil degradation and issues such as food insecurity, biodiversity loss, diminished environmental security, societal instability, poverty and conflict (Keesstra et al. 2016). Further, poor land use practices, deforestation, burning and clearing of forests cause a loss of soil material of about 4.1 tonne per ha per year due to rolling towards foothills in steep slopes (Mahapatra et al. 2018).

Presently, water-induced soil erosion is identified as one of the most critical environmental challenges worldwide (Jacobson 2011; Devatha et al. 2015), especially in Uttarakhand, facing severe challenges of water-induced soil erosion due to extreme precipitation therefore enhanced storm run-off volumes (Shushter et al. 2015; Mahapatra et al. 2018). Rainfall and soil erosion are closely interconnected through the combined effects of raindrop impact, which detaches soil particles, and surface run-off, which transports them (Mkhonta 2000). Expansion of the impervious surfaces disrupts the natural hydrological cycle, leading to increased run-off, reduced groundwater recharge, and heightened flood risks (Mandal et al. 2024; Arnold and Gibbons 1996; Misra 2011; Rodak et al. 2020). Impervious surfaces combined with removing forest cover and land degrading affect the water hydrology and soil topography (Watson et al. 1981; Reinelt et al. 2023; Walsh 2000; Rhodes et al. 2001; Faulkner 2004; NEERI 2016). Extreme rainfall contributes majorly to soil loss (Okorafor et al. 2017), which is susceptible to rainfall erosivity is characterized in terms of volume, duration, and intensity of the rainfall and the physical properties of the soil (Costea 2012; Oduro-Afriyie 1996). Stormwater run-off carries soil sediments, which often cause surface clogging due to sedimentation (Xiong et al. 2023, Gogate and Rawal 2012), which causes failure of traditional drainage systems (Tsihrintzis and Hamid 1997; Urbonas 2000; McGrane 2016 Booth and Jackson 1997; Jones and Macdonald 2007).

The capital city of the state of Uttarakhand, Dehradun, is popular for its diverse and undulated topography (Pathak et al. 2024). Dehradun is in the Doon Valley, bordered by the Himalayas to the north and the Shivalik Hills to the south, with average annual precipitation exceeding 2000 mm (Indian Meteorological Department 2020). The combination of steep slopes and heavy rainfall results in high run-off volumes, increasing the risk of flash floods and waterlogging, particularly in low-lying areas (Dhruva 1987; Jain and Kumar 2012; Singh et al. 2022; Pandey and Vishwakarma 2019), leading to soil displacement. Thus, assessing soil erosion in hilly regions like Dehradun is crucial for implementing effective soil management and mitigation measures (Bansal et al. 2015). Hence, policies must be reconstructed using historical and predicted data to mitigate water erosion, including gully erosion and other environmental challenges such as droughts and flood risks. With the likelihood of increased extreme weather events and associated economic losses in the future, such measures are vital (Winsemius et al. 2016; Ghosh and Mukherjee 2022).

The soil erosion prediction primarily relies on models developed to measure soil loss under natural run-off or simulated rainfall conditions across diverse soil topographies and their management practices. Popular models include the Water Erosion Prediction Project (WEPP) (Flanagan and Nearing 1995), the Universal Soil Loss Equation (USLE) (Wischmeier and Smith 1978), later refined as the RUSLE (Renard et al. 1997), which operates on similar principles as USLE but incorporates several improvements in the calculation of various factors (Kumar and Kushwaha 2013). Built on extensive experimental and monitored data, these models are robust tools for estimating soil erosion rates based on rainfall, topography, soil characteristics, and management (Römkens et al. 2002). RUSLE for water-based erosion estimates long-term average annual soil loss by combining multiple factors: rainfall erosivity (R), soil erodibility (K), slop length (L), terrain steepness (S), Vegetation cover (C), and conservation practices (P) (Kumar and Kushwaha 2013). Numerous studies in Asian countries such as Nepal, India, Sri Lanka, Thailand, China and Myanmar (Saha et al. 2018; Chang, 2010; Dissanayake et al. 2019; Gayen et al. 2020; Koirala et al. 2019; Chen et al. 2024 Sourn et al. 2022) have undertaken erosion studies at national level employing this approach (Gilani et al. 2021; Gyeltshen et al. 2022; Uddin et al. 2018). Similar studies across Indian Himalayan Region used remote sensing and field-based data with the RUSLE or variants to estimate soil erosion across the Indian Himalayan Region (Kumar et al. 2014; Chanyal 2020; Gupta and Kumar 2017; George et al. 2021). These studies have assessed soil loss rates across diverse land-use systems under varying scenarios, including sub-watersheds in the Shivalik Hills, hilly watersheds in the mid-Himalayan region of Uttarakhand, and the broader Himalayan ecosystem of Uttarakhand (Kumar and Kushwaha 2013; Kalambukattu and Kumar 2017; Mahapatra et al. 2018; Olokeogun and Kumar, 2020). Such studies suggested crucial measures for soil and water management can manage soil loss, improve hydrological balance, and enhance agricultural and forestry productivity, ultimately reducing the rate of reservoir siltation (Naryana 1987). A similar study across Nepal's Jhimruk watershed used the RUSLE model to calculate the erosion status within the watershed (Pandey and Gurung 2022), many such studies across Nepal Himalayas cover different land types with varying elevations and climatic conditions provide suggestions to plan mitigation strategies (Shrestha 1997; Uddin et al. 2016; Gardner et al. 2000). While these studies demonstrate the utility of the RUSLE model in assessing spatial erosion patterns in the IHR, more research is needed to understand the complex interactions between climate change, land use, and soil properties that drive erosion processes in this ecologically sensitive and economically important region. Climate change is expected to significantly impact soils' physical, chemical and biological properties in many areas due to rising temperatures and shifting precipitation patterns, influencing microbial communities and their activity rates. More frequent extreme precipitation rates are expected to accelerate erosion rates (Eekhout et al. 2018).

Regional-scale studies, which enable large-scale assessments of soil erosion, play a critical role in supporting stakeholders and policymakers in developing regulatory policies that promote the sustainable and efficient use of land resources (Wei et al. 2012). Achieving land restoration and management of soil health can be done through SDGs (Lal et al. 2021). The SDGs provide a framework with specific goals, targets and indicators for sustainable land management (Bouma et al. 2019). Enhancing soil productivity and rehabilitating degraded soils are recognized as critical focus areas for sustainable land management to support the achievement of key sustainable development goals (Lal et al. 2021). In their study, Lal et al. 2021 demonstrated the SDGs such as SDG 1 (No Poverty), SDG 2 (Zero Hunger), SDG 3 (Good Health and Well Being), SDG 6 (Clean Water and Sanitation), SDG 14 (Life Below Water), SDG (Climate Action), and SDG 15 (Sustainable Use of Terrestrial Ecosystems) have direct connections to soil conservation efforts. They also contribute indirectly to achieving other SDGs (Bouma and Montanarella 2016; Lal et al. 2021; Vanino et al. 2023).

The use of RUSLE and spatial modeling in Uttarakhand is attempted by George et al. (2021) to predict the soil erosion rates and their spatial distribution in the state. Mahapatra et al. (2018) assessed soil erosion in Uttarakhand using 10 km × 10 km grid data to address the various factors responsible for erosion and interpolating data, which neglected high intra-grid spatial variability concerning the various factors for erosion in these terrains. Despite several river basin studies employing Google Earth Engine (GEE) and other geospatial techniques to predict land degradation rates in Uttarakhand (Krishna et al. 2016; Kumar et al. 2024a b; Kumar et al. 2022; Raj et al. 2023), not many attempts have been made for the erosional rate of soil loss in Dehradun. Accurate soil erosion monitoring results are essential to analyze the patterns and trends of soil erosion (Senanayake et al. 2024; Garg and Anand, 2022).

Moreover, several studies focus on soil erosion and soil management. Still, many policymakers and researchers remain uncertain about effectively translating these findings into actionable strategies for promoting sustainable land management to achieve the SDGs (Senanayake et al. 2024). Therefore, in this study, we attempted to predict the annual Soil loss in the Dehradun region of Uttarakhand using the Google Earth Engine (GEE) through RUSLE modeling to suggest measures for soil management and effective water run-off policies in Dehradun to achieve SDGs goals in hilly terrain such as Dehradun.

Our research specifically addresses the unique challenges faced by Dehradun. Unlike previous studies that primarily focus on larger Himalayan basins or broad regional scales, we concentrate on Dehradun's distinct vulnerabilities stem from steep topography, high rainfall, and rapid urbanization. Additionally, we incorporated the Normalized Difference Vegetation Index (NDVI) into the RUSLE calculations, allowing us to account for seasonal differences in green cover. This enhancement provides a more dynamic and accurate assessment of the cropping management factor (C) in the RUSLE equation. Moreover, our study connects soil erosion modeling with actionable policy recommendations that align with the Sustainable Development Goals (SDGs). This approach bridges the gap between geospatial analysis and practical conservation strategies. Finally, we present high-resolution insights into erosion dynamics. By utilizing finer spatial resolution datasets and incorporating detailed land-use classifications, our study more effectively captures intra-grid variability than broader-scale studies, allowing for more precise intervention planning.

12.2 Study Area

Dehradun district is located between latitudes 29° 95′ N and 30° 99′ N and longitudes 77° 57′ E and 78° 31′ E. The Ganges River bounds it to the East and the Yamuna River to the West. The Himalayan Range flanks the north part of the district, while the Shivalik Range lies to its south. As one of the most populous districts in Uttarakhand, Dehradun spans an area of 3088 km^2. According to the 2011 Census, it had a population of 1.6 million and a population density of 549 people per km^2 (Census of India 2011). By 2023, the population is estimated to have grown to 2.25 million. The district's elevation varies between 269 and 3062 m.

Dehradun district is administratively organized into seven tehsils and six development blocks, encompassing 767 villages. During winter, particularly in January, the minimum temperature can drop to as low as 3.6 °C, while the maximum temperature typically reaches 19.3 °C. In contrast, the summer months, especially June, experience significantly higher temperatures, with minimums around 29.4 °C and maximums peaking at 34.4 °C (Sharma et al. 2012; Dhankar et al. 2024).

The district receives most of its annual rainfall, averaging 2073.3 mm, between June and September, predominantly driven by the southwest monsoon. This heavy monsoonal rainfall is critical in sustaining the district's agricultural activities and replenishing water resources. However, pre-monsoon showers and occasional winter rain contribute to the overall precipitation pattern. The varied climatic conditions and rainfall distribution significantly influence the region's ecosystem and hydrological dynamics. (Indian Meteorological Department.

12.3 Methodology

This study utilized Google Earth Engine (GEE) approach to assess Average Soil Loss (A) patterns using the RUSLE. This analysis incorporated multiple parameters: rainfall–run-off erosivity (R), soil erodibility (K), slope length (L), slope steepness (S), cropping management (C), and supporting conservation practices (P), which were integrated using thematic layers derived from satellite imagery and environmental datasets. Rainfall erosivity (R) was determined based on long-term precipitation records. Soil erodibility (K) was assessed through GEE, considering Dehradun's distinct soil characteristics.

The LS factor, cropping Management (C), and supporting conservation practices (P) were evaluated using GEE and satellite data. The study also incorporated the NDVI into the cover and management (C) factor calculation to reflect changes in

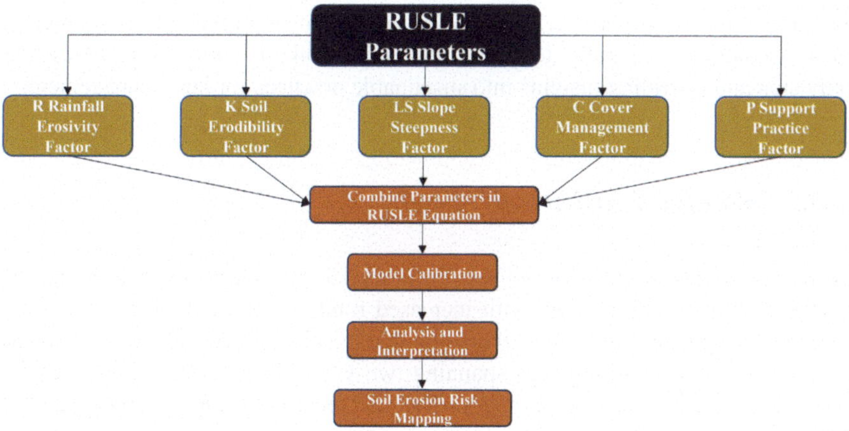

Fig. 12.1 A flowchart illustrating the methodology used in the study

seasonal plant cover. The slope of the terrain was considered a key factor. The GEE-based methodology provided detailed spatial insights into soil erosion risks across the Dehradun region, demonstrating scalability, efficiency, and accuracy in analyzing large-scale geospatial data for erosion modeling. The results contribute insights into sustainable land management strategies in Dehradun. The process of workflow is given in Fig. 12.1.

12.4 RUSLE Computation

12.4.1 Soil Loss in Unit Area (A)

To estimate the average annual A within the study area, the RUSLE was applied using the Google Earth Engine (GEE) framework (Lucà et al. 2018; Milodowski et al. 2020; Mitasova et al. 2013; Salvacion 2023). The equation, expressed in Eq. (12.1), is presented as follows.

$$A = R' \cdot K' \cdot L' \cdot S' \cdot C' \cdot P \tag{12.1}$$

In this equation, A denotes the average annual soil loss in tons per hectare per year. The components of the equation are defined as follows: R' represents the rainfall-runoff erosivity factor (MJ mm ha^{-1} h^{-1} yr^{-1}), K' signifies the soil erodibility factor (ton ha hr MJ^{-1} ha^{-1} mm^{-1}), L' is the slope length factor (dimensionless), S' is the slope gradient factor (dimensionless), C' indicates the cropping management factor (dimensionless, ranging from 0 to 0.5), and P reflects the supporting conservation practice factor (dimensionless, ranging from 0 to 1). Based on GEE, this approach integrates various geospatial layers derived from satellite imagery and environmental datasets, enabling a spatially detailed evaluation of soil erosion potential across the study area and providing insights into sustainable practices for land management.

12.4.2 Rainfall Erosivity (R') Factor

The R' is critical in determining soil erosion potential. The intensity of rainfall significantly influences soil erosion, with increased rainfall rates and larger drop sizes leading to heightened sheet and rill erosion due to enhanced run-off flow. To quantify R', the monthly rainfall data spanning two years (2020–2022), were used by employing a widely accepted equation (Eq. 12.2) that relates R' to annual rainfall (P):

$$R\prime = 79 + 0.363 \cdot P \tag{12.2}$$

The spatial distribution of the R' was derived using Kriging interpolation methods within the GEE framework. Satellite-based precipitation data at 1×1 KM gridded around the study area comprehensively represented regional R' patterns.

12.4.3 Soil Erodibility (K') Factor

The soil erodibility factor (K') is essential for evaluating the inherent vulnerability of soils to erosion caused by rainwater and run-off. This factor reflects various soil characteristics, including mineralogical, physical, chemical, and morphological properties. In the present analysis of the Dehradun district, a cloud-based geospatial analysis through GEE to estimate K' was used. By examining soil types and texture maps specific to the watershed, K' values were assigned that ranged from 0 to 1, where higher values indicate greater susceptibility to erosion.

12.4.4 Topographic (LS) Factor

The LS factor is vital in assessing soil erosion rates. It combines slope length (L′) and slope steepness (S′) into a unified index. The L′ factor accounts for how slope length affects erosion by considering the distance from where runoff initiates to where it deposits material. In contrast, the S' factor reflects how steepness influences erosion rates. The 30 m SRTM digital elevation model (DEM) from USGS was used to compute this LS factor. The calculation was performed using Eq. (12.3).

$$LS = \left(\frac{Flow\ accumulation \cdot Cell\ Size}{0.0896}\right)^{0.4} \cdot (sin(Slope)/0.0896)^{1.3} \quad (12.3)$$

This calculation-effectively showed the areas prone to increased run-off and erosion.

12.4.5 Land Cover (C′) Factor

The C' is influenced by land use and is particularly susceptible to anthropogenic activities aimed at mitigating erosion. Calculating this factor across a large watershed can be complex due to spatial variations in land cover patterns. To address this challenge, MODIS land use and land cover data at a resolution of 500 m within GEE were utilized using a supervised classification approach based on maximum likelihood algorithms for classification accuracy verification for each LULC class. The study area was categorized into five major classes: agricultural land, barren land, built-up areas, vegetation, and water bodies, achieving an overall accuracy rate of approximately 87.39%. Each class was assigned a C value between 0 and 1; lower values indicate minimal soil loss, while higher values suggest increased susceptibility.

12.4.6 Conservation Practice (P) Factor

The conservation practice (P) factor reflects the effectiveness of management practices in reducing soil erosion through elements such as vegetation cover and run-off control measures. The RUSLE integrates this P factor to represent the combined influence of land cover, support practices, land use, slope length, and custom adjustments on soil erosion dynamics. GEE's capabilities were used for scalable analyses to calculate this factor by extracting relevant properties related to cover (C'), support practices (M), land use (L'), slope length (S'), and custom adjustments.

12.4.7 NDVI

NDVI is a key pointer of vegetation health and cover, directly influencing the cover management factor in RUSLE calculations. By analyzing its values over different periods, seasonal variations in vegetation cover were captured using Eq. (12.4).

$$NDVI = \frac{NIR - Red}{NIR + Red} \quad (12.4)$$

Higher NDVI values correspond with denser vegetation cover, which leads to lower C factors.

12.4.8 Slope

Slope dynamics are fundamental in estimating soil loss as they directly impact water run-off's erosive potential; steeper slopes enhance rainfall force and accelerate surface run-off, contributing significantly to soil detachment and transport processes. The LS factor in RUSLE explicitly accounts for slope length and steepness when assessing erosion susceptibility across varying terrains. Through this comprehensive approach utilizing GEE's capabilities, notified policymaking in land management practices aimed at reducing soil loss while promoting sustainable strategies for land use within the study area.

12.5 Results

12.5.1 Soil Loss

Soil loss was quantified in tons per hectare per year (t/ha/year), revealing that a significant portion of the study area experiences high levels of soil erosion. The spatial distribution of soil loss is closely associated with regions characterized by steeper slopes. Soil loss was categorized into five classes: Slight (< 10 t/ha/year), Moderate (10–40 t/ha/year), High (40–70 t/ha/year), Very High (70–100 t/ha/year), and Severe (> 100 t/ha/year). The Severe category accounts for approximately 46% of the study area, covering 84,594.6 hectares. The Very High category represents around 12% of the area (20,678.68 ha), while 20.4% (38,349.55 ha) falls under the High category. The Moderate category constitutes 22.4% of the area (42,109.31 ha), and the Slight category accounts for only 0.9% (1691.89 ha). The distribution of soil loss throughout these categories is shown in Fig. 12.2.

12.5.2 R′

Various studies (Jain et al. 2001; Dabral et al. 2008) have indicated the rate of soil erosion in catchment areas exhibits heightened sensitivity to variations in rainfall. Daily rainfall measurements are more effective in reflecting changes in soil erosion rates, thereby aiding in characterizing the seasonal sediment yield distribution. In contrast, annual rainfall data offers benefits such as easy accessibility, straightforward calculations, and enhanced consistency of the exponent across regions (Shinde et al.

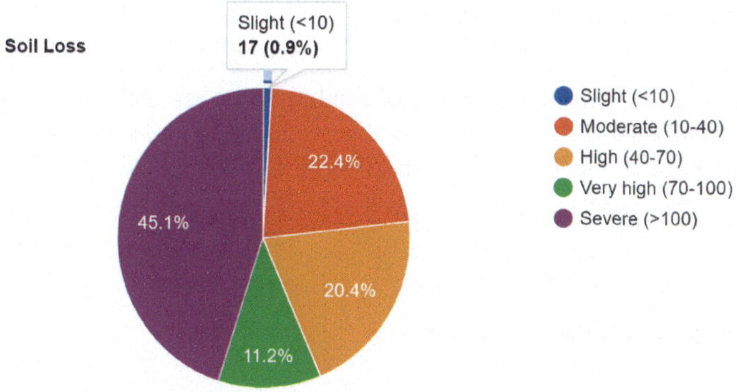

Fig. 12.2 Pie chart of soil loss class distribution

2010). Consequently, this study employed average annual rainfall—calculated by dividing total rainfall by the number of rainy days—for the computation of the R factor (Eq. 12.2). The estimated R factor values ranged from 1168.87 to 1642.16 MJ/mm·ha^{-1} h^{-1}/year, highlighting that the study region experiences significant rainfall as evidenced by the analysis results Fig. 12.3.

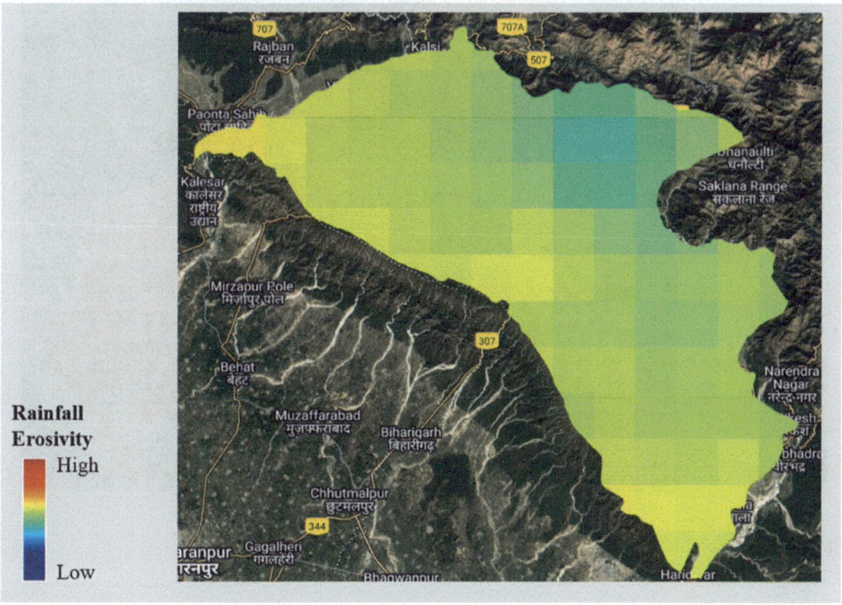

Fig. 12.3 The spatial distribution of R' in Dehradun Districts

12.5.3 Soil Erodibility Factor (K′)

K′ factor values were allocated to the corresponding soil types on the soil map to create a soil erodibility map. The *K′* factor was observed to range from 0.06 to 0.1. Lower *K′* factor values are typically linked with soils characterized by low permeability and reduced antecedent moisture content.

12.5.4 Topographic Factor (LS)

The LS reflects the impact of slope length and steepness on the erosion process. It was determined by incorporating flow accumulation and slope percentage as input parameters. The analysis indicates that the value of the topographic factor increases, ranging from 1.7 to 179.35, in response to rising flow accumulation and slope steepness. A map of the LS factor is presented in Fig. 12.4.

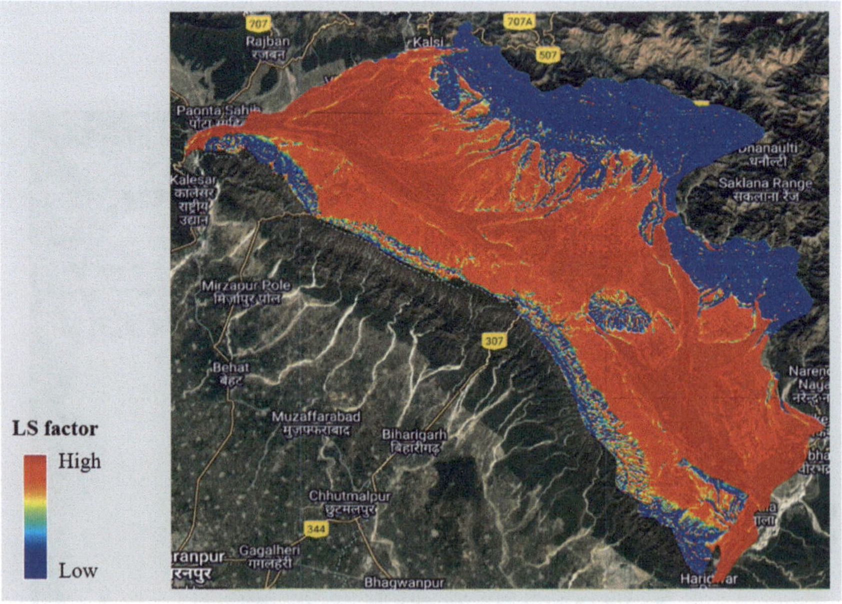

Fig. 12.4 Spatial distribution of LS factors in the study area

Fig. 12.5 A land-use (C) factor map showing the spatial distribution of erosion susceptibility

12.5.5 Land-Use Factor (C)

The analysis indicates that various land-use classes within the region of interest exhibit differing potential degrees of soil erosion. A higher C factor indicates more erosion. Notably, the built-up regions, riverbeds, and barren areas demonstrate the highest susceptibility to erosion—conversely, areas characterized by dense or sparse vegetation and cropland present relatively lower erosion risks. Comprehending land-use classes and their associated erosion risks is crucial for effective land-use planning and implementing sustainable agricultural practices to minimize soil erosion.

The C-factor map, illustrated in Fig. 12.5, is beneficial for identifying areas with varying erosion potentials. This approach allows for the creation of targeted soil conservation strategies, enhances the effectiveness of erosion control measures, and fosters long-term environmental sustainability by directing conservation efforts toward the areas that need them most.

12.6 Discussion

The Dehradun district is important for examining the dynamics of soil erosion. It is crucial to implement an integrated strategy for land-use planning and to create effective conservation policies. A detailed analysis of various key factors is vital for making well-informed decisions concerning the management of soil erosion and the encouragement of sustainable land use in this environmentally sensitive area. Among these factors, the topographic (LS) factor plays an important role, as it highlights the erosion vulnerability of specific locations by integrating data on slope length and steepness as it directs conservation efforts toward areas most at risk, emphasizing the importance of slope analysis (LS) in the strategic formulation of erosion management interventions. Regions with steeper slopes or longer lengths are recognized as particularly susceptible to erosion, warranting prioritized attention.

The approach is further enhanced by assessing the land cover (C) factor, which categorizes the district into distinct primary land-use classes. This classification reveals that areas with dense tree cover and built-up regions are less prone to erosion, while regions with sparse vegetation may experience higher erosion risks. This classification facilitates the development of targeted conservation strategies, allowing for concentrated interventions in the most at-risk land-use areas, which supports the aims of SDG 13 (Climate Action). Moreover, assessing the effectiveness of current conservation efforts through the conservation practice (P) factor is crucial for enhancing and optimizing erosion control methods. The analysis showed that 78% of the watershed, covering 142,000 ha, falls into the severe soil erosion category, which predominates the landscape and indicates critical soil loss. This highlights the pressing necessity for timely and well-organized erosion control initiatives to fulfill the objectives of SDG 15 (Life on Land). The findings offer valuable insights into land managers and policymakers to prioritize and tailor conservation strategies based on the distinct features of each soil class, ensuring sustainable land use and environmental protection in the Dehradun District. The study's outcomes are illustrated in Fig. 12.6, which reveals that the annual mean soil loss rate for the area is 488.58 t/ha/year. Such levels of soil loss can contribute to landslides in steep regions, increasing the area's susceptibility.

Furthermore, the soil erodibility (K) factor provides insights into the variability of soil characteristics across Dehradun. Understanding the diverse erosion susceptibilities of different soil types is essential for prioritizing and customizing protective measures, ensuring that interventions are specifically adapted to address particular susceptibilities.

Integrating NDVI data to this research offers important insights into the health of vegetation in the district. It underscores the crucial function of strong vegetative cover in preventing soil erosion. This combined approach allows for the creation of sustainable land-use planning strategies that are grounded in a comprehensive evaluation of both vegetation resilience and terrain characteristics. Additionally, it aids in promoting sustainable land-use planning, aligning with Sustainable Development Goal 15 (Life on Land).

Moreover, the R' showed a sharp spatial correlation with the spatial distribution of annual rainfall. The districts' north-eastern zone, which is in high slope zones, showed more susceptibility to rainfall erosivity.

Given the significant implications of soil erosion within the Dehradun district, well-designed erosion control measures and tailored conservation initiatives are necessary. These efforts align with SDG 6 (Clean Water and Sanitation) goals by

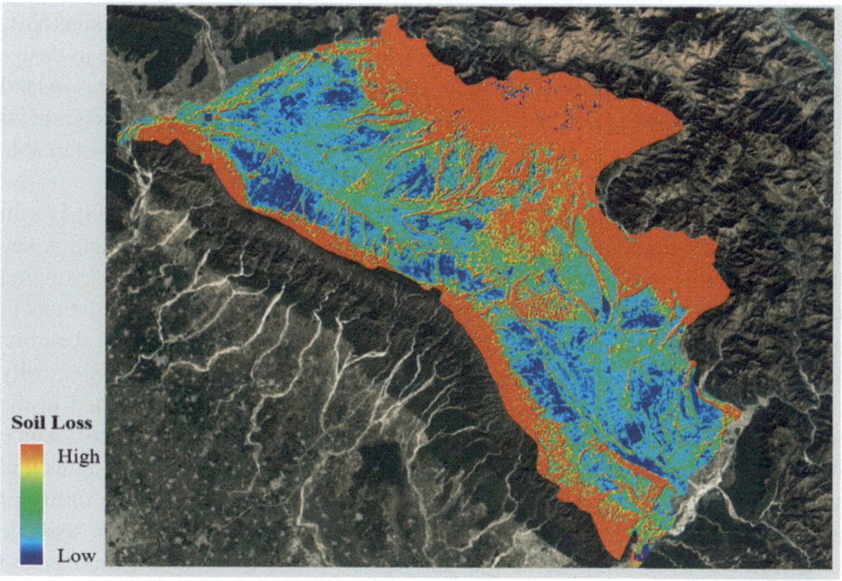

Fig. 12.6 Spatial distribution of soil loss in Dehradun Districts

reducing soil sedimentation in water bodies and SDG 13 by mitigating climate-induced risks.

The district and surrounding areas must necessitate comprehensive and sustainable land by using planning initiatives to safeguard the long-term vitality of ecosystems, human communities, agriculture, and water resources. This highlights the urgent need for a comprehensive perspective on the far-reaching impacts of soil erosion, calling for a unified strategy that harmonizes environmental conservation with community resilience and sustainable development goals.

To address these vulnerabilities, targeted conservation strategies must be implemented to focus on the most affected areas. For instance, forest covers can mitigate soil erosion by enhancing soil stability and reducing runoff. Additionally, integrating advanced geospatial methodologies, such as those utilizing cloud-based platforms like Google Earth Engine, can facilitate real-time monitoring and assessment of soil erosion dynamics.

Furthermore, community engagement and education are vital in promoting sustainable land management practices. By fostering awareness about the impacts of soil erosion and the importance of conservation efforts, local communities can contribute to enhancing ecosystem resilience. Ultimately, a concerted effort that combines scientific research with community involvement will be essential for developing effective strategies to combat soil erosion in Dehradun, thereby ensuring the sustainability of its natural resources for future generations.

Dehradun's unique geographical features, including its mountainous terrain and significant annual rainfall exceeding 3000 mm, further influence its soil erosion dynamics. The district has historically been known for its agricultural productivity, including litchi orchards and tea gardens; however, rapid urbanization since its designation as Uttarakhand's interim capital in 2000 has led to significant changes in land use and increased pressure on soil resources. Understanding these dynamics is crucial for developing effective strategies to mitigate soil erosion and promote sustainable land management practices in Dehradun.

Finally, this study enhances soil erosion assessment by integrating RUSLE with Google Earth Engine (GEE), creating a scalable and cloud-based approach. A key innovation of this research is the dynamic inclusion of the NDVI for the cropping management factor (C). This allows for the consideration of seasonal variations in vegetation and facilitates the efficient handling of large-scale geospatial datasets. These advancements make the methodology suitable for steep and ecologically sensitive terrains like Dehradun. In contrast to previous studies conducted in the Indian Himalayas and Nepal (e.g., George et al. 2021; Uddin et al. 2016), which often utilized coarse spatial resolutions, this research offers finer-scale insights that capture variations within grids. This increased granularity enhances the utility of the findings for localized conservation planning. On a global scale, this scalable framework serves as a replicable tool for regions prone to climatic extremes and rapid land-use changes, contributing to the Sustainable Development Goals (SDGs), particularly SDG 13 (Climate Action) and SDG 15 (Life on Land). However, the study has its limitations. The lack of extensive field validation, such as erosion plots or sediment deposition studies, highlights the need for future research to verify the model outputs empirically. Employing higher-resolution spatial data and conducting sub-watershed analyses could further refine the predictions. Explicitly modeling anthropogenic factors, including urbanization and deforestation, is another area that requires improvement. Expanding the study area or integrating machine learning for dynamic hotspot detection could enhance the methodology's applicability. Furthermore, combining RUSLE with hydrological and sediment transport models would provide a more comprehensive understanding of erosion dynamics. The findings also lead to relevant policy recommendations, such as implementing afforestation in high-erosion zones, promoting sustainable land-use planning, and establishing community-based conservation programs. Aligning these measures with national policies and SDGs will foster long-term environmental sustainability and resilience against challenges related to soil erosion.

12.7 Conclusion and Future Directions

This study offers valuable insights for local stakeholders in Dehradun, including urban planners, policymakers, conservationists, and agricultural managers. The study facilitates targeted interventions such as afforestation, soil stabilization, and sustainable farming practices by utilizing geospatial analysis to identify high-risk soil erosion zones. Integrating Google Earth Engine (GEE) provides a cost-effective, scalable tool for continuous soil monitoring, making it accessible for local agencies with limited resources. The immediate benefits of these efforts include improved land management decisions, reduced soil erosion, and minimized sediment deposition in rivers and reservoirs, essential for maintaining water quality and agricultural productivity.

Soil erosion is a significant challenge in Dehradun, a city characterized by its complex terrain, heavy rainfall, and rapid land-use changes. This research highlights the urgent need for a comprehensive strategy in land-use planning and soil conservation, considering crucial factors such as topography, land cover, soil erodibility, and rainfall erosivity. The study's findings reveal an alarming spatial distribution of soil loss in Dehradun, except in areas with rich vegetation and green cover, where soil erosion is considerably less. This extreme rate of soil depletion has far-reaching consequences, including heightened vulnerability to landslides and increased chances of floods, particularly in areas with steep gradients. Addressing soil erosion and its consequences in hilly regions such as Dehradun, prone to natural processes such as earthquakes, landslides, and extreme precipitation, can enhance resilience against climate-induced hazards by promoting adaptive stormwater management strategies that align with SDG 13. Thus, through this study, we suggest several key measures to improve soil management practices in Dehradun for sustainable land use management:

- Future work should leverage advanced technologies, such as real-time geospatial tools, machine learning, remote sensing, and cloud platforms, to monitor, predict, and map soil erosion hotspots more precisely.
- Detailed soil characterization is needed to develop site-specific interventions based on soil erodibility. The effectiveness of bioengineering solutions like vetiver grass and geotextiles for slope stabilization should also be explored.
- Participatory programs are crucial for involving local communities in sustainable land management, which addresses the goals of SDG 17. Public awareness campaigns are important to highlight the socio-economic impacts of soil erosion and the benefits of maintaining vegetative cover.
- Stronger policies are needed to regulate deforestation and promote afforestation in at-risk areas. By combating land degradation and conserving biodiversity, these policies will promote the life-on-land goals of SDG 15. Land-use planning should balance urban development and expansion with ecosystem conservation (SDG 11).
- Studies need to investigate the total soil loss in Dehradun due to climate change and changing rainfall patterns, which impact soil erosion through changing rainfall patterns and, hence, increased run-off. Therefore, proper channeling of stormwater management must be directed to prevent contamination of clean water (Dhakal and Chevalier, 2016), with strategies focusing on efficiently handling stormwater run-off.

Addressing these research areas will enhance resilience against soil erosion, promote sustainable development, and protect Dehradun's natural resources for the future while contributing holistically to SDG's goals for a sustainable future for Dehradun.

Given Dehradun's rapid urbanization and steep topography, strategic land-use planning is crucial for mitigating soil erosion. Recommendations include limiting construction in high-slope and erosion-prone areas and establishing buffer zones around rivers to lessen the impacts of sedimentation. Nature-based solutions, which incorporate green infrastructures such as rain gardens, bioswales, and permeable pavements, can effectively manage stormwater run-off. Successful initiatives inspire these solutions in the USA (e.g., Portland's Green Streets Program) and Europe (e.g., Copenhagen's Blue-Green Infrastructure). Also, promoting afforestation and sustainable agriculture is essential. This involves encouraging native vegetation in high-erosion areas and implementing practices such as contour farming, terracing, and cover crops for soil conservation. Lastly, community engagement is vital for raising awareness of soil erosion's socio-economic impacts and involving local communities in conservation efforts. These measures can balance urban development and environmental sustainability by aligning with the best global practices, ensuring Dehradun's long-term ecological resilience.

Competing Interests The authors declare no potential conflicts of interest.

References

Arnold CL, Gibbons CJ (1996) Impervious surface coverage: the emergence of a key environmental indicator. J Am Plann Assoc 62(2):243–258

Bansal N, Mukherjee M, Gairola A (2015) Causes and impact of urban flooding in Dehradun. Int J Curr Res 7(2):12615–12627

Behera M, Sena DR, Mandal U, Kashyap PS, Dash SS (2020) Integrated GIS-based RUSLE approach for quantification of potential soil erosion under future climate change scenarios. Environ Monit Assess 192(11):1–18

Booth DB, Jackson CR (1997) Urbanization of aquatic systems: degradation thresholds, stormwater detention, and the limits of watershed restoration. J Am Water Resour Assoc 33(5):1077–1090

Bouma J, Montanarella L (2016) Facing policy challenges with inter- and transdisciplinary soil research focused on the UN Sustainable Development Goals. SOIL 2(2):135–145

Bouma J, Montanarella L, Evanylo G (2019) The challenge for the soil science community to contribute to the implementation of the UN Sustainable Development Goals. Soil Use Manag 35(4):538–546

Burns M J et al (2012) Hydrologic shortcomings of conventional urban stormwater management and opportunities for reform. Landscape Urban Plan 105:230–240

Chang NB (2010) Hydrological connections between low-impact development, watershed best management practices, and sustainable development. J Hydrol Eng 15(6):384–385

Chanyal PC (2020) Research article applications of remote sensing and GIS for watershed characterization and soil loss assessment of tons watershed in Dehradun, Garhwal Himalaya. Int J Agric Appl Sci 1(1):56–67

Chen L, Wang J, Fu B, Qiu Y (2001) Land use change in a small catchment of northern Loess Plateau, China. Agric Ecosyst Environ 86:163–172

Chen L, Guo C, Yu Y, Zhou X, Fu Y, Wang S, Shen Z et al (2024) Optimization of green infrastructures for sustaining urban stormwater quality and quantity: an integrated resilience evaluation. J Hydrol 640:131682

Chitra V (2022) Remembering the river: flood, memory and infrastructural ecologies of stormwater drainage in Mumbai. Urban Stud 59(9):1855–1871

Costea M (2012) Using the Fournier indexes in estimating rainfall erosivity. Case study-the Secasul Mare Basin. Air Water Compon Environ 313–320

Dabral P, Baithuri N, Pandey A (2008) Soil Erosion Assessment in a Hilly Catchment of North Eastern India Using USLE, GIS and Remote Sensing. Water Resour Manag 22:1783–1798

Dapin IG, Ella VB (2023) GIS-based soil erosion risk assessment in the watersheds of Bukidnon, Philippines using the RUSLE model. Sustainability 15(4):3325

Davis AP, Hunt WF, Traver RG, Clar M (2013) Bioretention technology: overview of current practice and future needs. J Environ Eng 139(10):1304–1311

Devatha CP, Deshpande V, Renukaprasad MS (2015) Estimation of Soil loss Using USLE Model for Kulhan Watershed, Chattisgarh- A Case Study. Aquatic Procedia, 4, 1429–1436

Dhakal KP, Chevalier LR (2016) Urban stormwater governance: the need for a paradigm shift. Environ Manage 57:1112–24

Dhankar S, Mishra AK, Kumar K (2024) Satellite derived air pollution climatology over India and its neighboring regions: Spatio-temporal trends and insights. Physi Chem Earth A/B/C 136:103769

Dhruva NVV (1987) Downstream impacts of soil conservation in the Himalayan Region. Mt Res Dev 7(3):287–298

Dissanayake DMSLB, Morimoto T, Ranagalage M (2019) Accessing the soil erosion rate based on RUSLE model for sustainable land use management: a case study of the Kotmale watershed, Sri Lanka. Model Earth Syst Environ 5:291–306

Dogra VN (2011) CSWCRTI vision 2030. Central Soil & Water Conservation Research & Training Institute (CSWCRTI), 1–46. Equation RUSLE. U.S. Dept. of Agriculture, Agriculture Handbook No. 703, p 404

Eekhout JPC, Hunink JE, Terink W, de Vente J (2018) Why increased extreme precipitation under climate change negatively affects water security. Hydrol Earth Sys Sci 22(11):5935–5946

Elsen E, Hessel R, Liu B (2003) Discharge and sediment measurements at the outlet of a watershed on the Loess Plateau of China. CATENA 54:147–160

Faulkner SP (2004) Soils and sediment: understandingWetland biogeochemistry. In: Spray S, McGlothlin K (eds) Wetlands, pp 25–46. Rowman Littlefield, Inc, Lanham, MD, p 216

Fayas CM, Abeysingha NS, Nirmanee KGS, Samaratunga D, Mallawatantri A (2019) Soil loss estimation using the RUSLE model to zprioritize erosion control in Kelani River basin in Sri Lanka. Int Soil Water Conserv Res 7(2):130–137

Flanagan D (2002) Erosion encyclopedia of soil science (Lal R (ed). Marcel Dekker, New York, pp 395–398

Flanagan DC, Nearing MA (1995) USDA-water erosion prediction project: Hillslope profile and watershed model documentation. NSERL report no. 10, pp 1196–47097

Gao Y, Church SP, Peel S, Prokopy LS (2018) Public perception towards river and water conservation practices: opportunities for implementing urban stormwater management practices. J Environ Manage 223:478–488

Gardner R (2000) Soil erosion and nutrient loss in the middle hills of Nepal (1996–1998). ARS Lumle, Nepal Agricultural Research Council (NARC)

Garg V, Anand A (2022) Impact of city expansion on hydrological regime of Rispana Watershed, Dehradun, India. GeoJournal 87(Suppl 4):973–997

Gayen A, Saha S, Pourghasemi HR (2020) Soil erosion assessment using RUSLE model and its validation by FR probability model. Geocarto Int 35(15):1750–1768

George G, Merrill RK, Schillebeeckx SJD (2021) Digital sustainability and entrepreneurship: how digital innovations are helping tackle climate change and sustainable development. Entrep Theory Pract 45(5):999–1027

Ghosh, P., & Mukherji, S. (2022). Desorption kinetics of soil sorbed carbazole, fluorene, and dibenzothiophene by P. aeruginosa RS1 from single and multicomponent systems and elucidation of their interaction effects. Biochem Eng J 180:108367

Gilani H, Ahmad A, Younes I, Abbas S (2021) Estimation of annual soil erosion dynamics (2005–2015) in Pakistan using revised universal soil loss equation (RUSLE). Authorea. https://doi.org/10.22541/au.160946369.92099648/v1

Gogate NG, Rawal PM (2012) Sustainable stormwater management in developing and developed countries: a review. In: International conference on advances in design and construction of structures (ADCS 2012), Bangalore, India

Govers G, Everaert W, Poesen J, Rauws G, De Ploey J, Latridou JP (1990) A long flume study of the dynamic factors affecting the resistance of loamy soil to concentrated flow erosion. Earth Surf Proc Land 11:515–524

Government of India (2011) Census of India 2011. Office of the Registrar General & Census Commissioner, India

Gupta K (2007) Urban flood resilience planning and management and lessons for the future: a case study of Mumbai, India. Urban Water J 4(3):183–194

Gupta S, Kumar S (2017) Simulating climate change impact on soil erosion using RUSLE model—a case study in a watershed of the mid-Himalayan landscape. J Earth Syst Sci 126:43

Gyeltshen S, Adhikari R, Budha PB, Thapa G, Subedi KK, Singh BK (2022) Remote sensing and GIS-based soil loss estimation for Bhutan using the RUSLE model. Geocarto Int 37(21):6331–6350

Hovius N (1998) Controls on sediment supply by large rivers, the relative role of eustasy, climate, and tectonism in continental rocks. Soc Sed Geol 59:3–16 (Special Publication)

Indian Meteorological Department (2020) Climatological Data of Dehradun. IMD

Jacobson CR (2011) Identification and quantification of the hydrological impacts of imperviousness in urban catchments: a review. J Environ Manage 92(6):1438–1448

Jain SK, Kumar V (2012) Trend analysis of rainfall and temperature data for India. Curr Sci 102(1):37–49

Jain S, Kumar S, Varghese J (2001) Estimation of soil erosion for a Himalayan watershed using GIS technique. Water Resour Manag 15:41–54

Jamal S, Sen A (2024) Urban flooding as an emerging challenge: evidences from Chennai City. In: Making India disaster resilient: challenges and future perspectives. Springer International Publishing, Cham, pp 15–26

Jones P, Macdonald N (2007) Making space for unruly water: sustainable drainage systems and the disciplining of surface run-off. Geoforum 38(3):534–544

Justin GK, Suresh K, Hole Rani M (2021) Geospatial modelling of soil erosion and risk assessment in Indian Himalayan region—a study of Uttarakhand state. Environ Adv 4:100039

Kansal ML, Singh S (2022) Flood management issues in Hilly regions of Uttarakhand (India) under changing climatic conditions. Water 14(12):1879

Kalambukattu J, Kumar S (2017) Modelling soil erosion risk in a mountainous watershed of Mid-Himalaya by integrating RUSLE model with GIS. Eurasian J Soil Sci 6(2):92–105

Karvonen T, Koivusalo H, Jauhiainen M (1999) A hydrological model for predicting run-off from different land use areas. J Hydrol 217:253–265

Keesstra SD, Bouma J, Wallinga J, Tittonell P, Smith P, Cerdà A, Montanarella L, Quinton JN, Pachepsky Y, van der Putten WH, Bardgett RD, Moolenaar S, Mol G, Jansen B, Fresco LO (2016) The significance of soils and soil science towards realization of the United Nations Sustainable Development Goals. SOIL 2(2):111–128

Koirala P, Thakuri S, Joshi S, Chauhan R (2019) Estimation of soil erosion in Nepal using a RUSLE modelling and geospatial tool. Geosciences 9(4):147

Krishna SVV, Dikshit AK, Pandey K (2016) Flood modelling with global precipitation measurement (GPM) satellite rainfall data: a case study of Dehradun, Uttarakhand, India. In: Multispectral, hyperspectral, and ultraspectral remote sensing technology, techniques and applications VI, vol 9880. SPIE, pp 302–310

Kumar S and Kushwaha S (2013). Modelling soil erosion risk based on RUSLE-3D using GIS in a Shivalik sub-watershed. J Earth Syst Sci 122:389–398

Kumar A, Devi M, Deshmukh B (2014) Integrated remote sensing and geographic information system based RUSLE modelling for estimation of soil loss in Western Himalaya, India. Water Resour Manage 28:3307–3317
Kumar R, Devrani R, Dangwal A, Deshmukh B, Dutt S (2022) Extreme hydrological event-induced temporal variation in soil erosion of the Assiganga River Basin, NW Himalaya. Adv Remote Sens Technol Three Poles 230–246
Kumar S, Vishwakarma RK, Tyagi VK, Kumar V, Kazmi AA, Ghosh NC, Joshi H (2024a) Stormwater run-off characterization and adaptation of best management practices under urbanization and climate change scenarios. J Hydrol 635:131231
Kumar R, Khaira JK, Ahmed R, Devrani R, Deshmukh B (2024b) Land degradation vulnerability mapping using geospatial techniques: a case study of Nandakini River basin, NW Himalaya, India. Int J River Basin Manage 1–16
Lal R, Monger C, Nave L, Smith P (2021) The role of soil in regulation of climate. Philos Trans R Soc B, Biol Sci 376(1834):20210084
Lucà F, Buttafuoco G, Terranova O (2018) 2.03—GIS and soil. In: Comprehensive geographic information systems. Elsevier, pp 37–50
Mahapatra SK, Reddy GPO, Nagdev R, Yadav RP, Singh SK, Sharda VN (2018) Assessment of soil erosion in the fragile Himalayan ecosystem of Uttarakhand, India using USLE and GIS for sustainable productivity. Curr Sci 115(1):108–121
Mandal R, Nishant N, Chutia D, Aggarwal SP, Sharma B (2024) LoRa Enabled IoT sensor framework for monitoring urban flood in Guwahati City. In: NIELIT's international conference on communication, electronics and digital technologies. Springer Nature Singapore, Singapore, pp 55–76
McGrane SJ (2016) Impacts of zurbanization on hydrological and water quality dynamics, and urban water management: a review. Hydrol Sci J 61(13):2295–2311
Milodowski DT, Hancock S, Silvestri S, Mudd SM (2020) Chapter 5—linking life and landscape with remote sensing. In: Tarolli P, Mudd SM (eds) Developments in earth surface processes, vol 23. Elsevier, pp 129–182
Misra AK (2011) Impact of urbanization on the hydrology of Ganga Basin (India). Water Resour Manage 25:705–719
Mitasova H, Barton M, Ullah I, Hofierka J, Harmon RS (2013) 3.9 GIS-based soil erosion modeling. In: Treatise on geomorphology. Academic Press, pp 228–258
Mkhonta MM (2000) Use of remote sensing and geographical information system (GIS) on soil erosion assessment in the Gwayimane and Mahhuku catchment areas with special attention on soil erodobility (K-Factor). ITC, Enschede
Narasimhan B, Sreethu S, Modi K, Arun RS, Anelli R, Imani M, Bhallamudi SM (2023) Sustainable urban drainage systems. In: Technological Solutions for water sustainability: challenges and prospects, IWA Publishing, pp 255–264
National Environment Engineering Research Institute (NEERI) (2016) Report on stormwater management practices in Indian Cities. NEERI
Negese A, Fekadu E, Getnet H (2021) Potential soil loss estimation and erosion-prone area prioritization using RUSLE, GIS, and remote sensing in Chereti watershed, Northeastern Ethiopia. Air Soil Water Res 14:1–17
Oduro-Afriyie K (1996) Rainfall erosivity map for Ghana. Geoderma 74(1/2):161–166
Okorafor OOA, Adeyemo AJ, Egwuonwu CC (2017) Determination of rainfall erosivity index (R) for Imo State, Nigeria. Am J Eng Res 6(2):13–16
Olokeogun OS, Kumar M (2020) An indicator based approach for assessing the vulnerability of riparian ecosystem under the influence of urbanization in the Indian Himalayan city, Dehradun. Ecol Indic 119:106796
Pandey K, Vishwakarma DK (2019) Flash floods cause and remedial measures for their control in hilly regions. In: Applied agricultural practices for mitigating climate change, vol 2. CRC Press, pp 77–100

Pandey P, Gurung A (2022) Soil loss and erosion potential estimation of Jhimruk watershed, Nepal. Environ Earth Sci 81:496

Pathak S, Sharma S, Banerjee A, Kumar S (2024) A methodology to assess and evaluate sites with high potential for stormwater harvesting in Dehradun, India. Big Data Res 35:100415

Pimentel D, Kounang N (1998) Ecology of soil erosion in ecosystems. Ecosystems 1:416–426

Prasannakumar V, Vijith H, Abinod S, Geetha N (2012) Estimation of soil erosion risk within a small mountainous sub-watershed in Kerala, India, using Revised Universal Soil Loss Equation (RUSLE) and geo-information technology. Geosci Front 3(2):209–215

Raj A, Sharma LK, Naik R (2023) Spatial monitoring of soil health using remote sensing of distinct land cover in the Central Himalayan region using GEE platform. In: Soil carbon dynamics in Indian Himalayan region. Springer Nature Singapore, Singapore, pp 303–319

Rawat S, Jain MK, Rawat KS, Nikam BR, Mishra SK (2017) Vulnerability assessment of soil erosion/deposition in a Himalayan watershed using a remote sensing and GIS-based sediment yield model. Int J Curr Microbiol App Sci 6(3):40–56

Reinelt L, Whitaker J, Kazakou E, Bonnal L, Bastianelli D, Bullock JM, Ostle NJ (2023) Drought effects on root and shoot traits and their decomposability. Funct Ecol 37(4):1044–1054

Renard KG, Foster GR, Weesies GA, McCool DK, Yoder DC (1997) Predicting soil erosion by water: a guide to conservation planning with the revised universal soil loss

Rhodes G, Zebrowitz LA, Clark A, Kalick SM, Hightower A, McKay R (2001) Do facial averageness and symmetry signal health? Evol Human Behav 22(1):31–46

Rodak CM, Jayakaran AD, Moore TL, David R, Rhodes ER, Vogel JR (2020) Urban stormwater characterization, control, and treatment. Water Environ Res 92(10):1552–1586

Römkens MJM, Helming K, Prasad SN (2002) Soil erosion under different rainfall intensities, surface roughness, and soil water regime. CATENA 46(2–3):103–123

Saha A, Ghosh P, Mitra B (2018) GIS-based soil erosion estimation using RUSLE model: a case study of Upper Kangsabati, Watershed, West Bengal, India. Int J Environ Sci Nat Resour 13(5):001–008

Salvacion AR (2023) Chapter 14—soil erosion modeling under future climate change: a case study on Marinduque Island, Philippines. In: Water, land, and forest susceptibility and sustainability, vol 1. Elsevier, pp 381–398

Sati SP, Sundriyal YP, Rana N, Dangwal S (2011) Recent landslides in Uttarakhand: nature's fury or human folly. Curr Sci (Bangalore) 100(11):1617–1620

Senanayake I, Yeo I-Y, Robinson NJ, Dahlhaus PG, Hancock GR (2024) Identification of high-performing soil groups in grazing lands using a multivariate analysis method. Soil Security 16:100163

Sharma R and Malaviya P (2021) Management of stormwater pollution using green infrastructure: the role of rain gardens. Wiley Interdiscip Rev Water 8(2):e1507

Sharma P, Jha AB, Dubey RS, Pessarakli M (2012) Reactive Oxygen Species, Oxidative Damage, and Antioxidative Defense Mechanism in Plants under Stressful Conditions. J Bot 2012(1):217037

Shin E, Kim H (2015) Analyzing green roof effects in an urban environment: a case of Bangbae-dong, Seoul. J Asian Archit Build Eng 14(2):315–322

Shinde V, Tiwari K, Singh M (2010) Prioritization of micro watersheds on the basis of soil erosion hazard using remote sensing and geographic information system. Int J Water Resour Environ Eng 2:130–136

Shrestha DP (1997) Assessment of soil erosion in the Nepalese Himalaya: a case study in Likhu Khola Valley, Middle Mountain Region. Land Husbandry 2(1):59–80

Shukla AK, Ojha CSP, Garg RD, Shukla S, Pal L (2020) Influence of spatial urbanization on hydrological components of the Upper Ganga River Basin, India. J Hazard Toxic Radioactive Waste 24(4):04020028

Shuster WD, Bonta J, Thurston H, Warnemuende E, Smith DR (2005) Impacts of impervious surface on watershed hydrology: a review. Urban Water J 2(4):263–275

Singh N, Poonia T, Siwal SS, Srivastav AL, Sharma HK, Mittal SK (2022) Challenges of water contamination in urban areas. In: Current directions in water scarcity research, vol 6. Elsevier, pp 173–202

Singha R, Tiwarib KN, Malb BC (2006) Hydrological studies for the small watershed in India using the ANSWERS model. J Hydrol 318:184–199

Sourn T, Pok S, Chou P, Nut N, Theng D, Prasad PV (2022) Assessment of land use and land cover changes on soil erosion using remote sensing, GIS and RUSLE model: a case study of Battam-bang province, Cambodia. Sustainability 14(7):4066

Thomas M, Prakash A, Dhyani S, Pujari PR (2024) Governing green change to improve resilience by assessing urban risks for localizing nature based solutions in fast sprawling Dehradun, India. Int J Disaster Risk Reduction 111:104684

Tsihrintzis VA, Hamid R (1997) Modeling and management of urban stormwater run-off quality: a review. Water Resour Manage 11:137–164

Tyagi RS, Singh SK, Goyal PK (2024) Rejuvenation of water bodies with recycled water. Water Pract Technol 19(3):839–851

Uddin K, Murthy MSR, Wahid SM, Matin MA (2016) Estimation of soil erosion dynamics in the Koshi basin using GIS and remote sensing to assess priority areas for conservation. PLoS ONE 11(3):e0150494

Uddin K, Abdul Martin M, Maharajan S (2018) Assessment of land cover change and its impact on changes in soil erosion risk in Nepal. Sustainability 10(12):4715

Urbonas B (2000) Assessment of stormwater best management practice effectiveness. In: Field JPH (ed) Innovative urban wet-weather flow management systems. Michigan Technomic Publishing Company, Lancaster

Vanino S, Pirelli T, Di Bene C, Bøe F, Castanheira N, Chenu C, Cornu S, Feiza V, Fornara D, Heller O, Kasparinskis R, Keesstra S, Lasorella MV, Madenoglu S, Meurer K, O'Sullivan L, Peter N, Piccini C, Siebielec G, Farina R (2023) Barriers and opportunities of soil knowledge to address soil challenges: Stakeholders' perspectives across Europe. J Environ Manag 325:116581

Walsh CJ, Fletcher TD, Ladson AR (2005) Stream restoration in urban catchments through redesigning stormwater systems: looking to the catchment to save the stream. J N Am Benthol Soc 24(3):690–705

Walsh E, McDonnell KP (2012) The influence of added organic matter on soil physical, chemical, and biological properties: A small-scale and short-time experiment using straw. Arch Agronom Soil Sci 58(sup1):S201–S205

Walson A (1985) Soil erosion and vegetation damage near ski lifts at Cairn Gorm. Scotland. Biol Conserv 33(4):363–381

Winsemius H, Aerts J, van Beek L et al (2016) Global drivers of future river flood risk. Nature Clim Change 6:381–385

Wischmeier WH, Smith DD (1978) Predicting rainfall erosion losses. USDA Agricultural Handbook No. 537. Department of Agriculture, Science and Education Administration, USA

Wei W, Chen L, Yang L, Fu B, Sun R (2012) Spatial scale effects of water erosion dynamics: complexities, variabilities, and uncertainties. Chin Geogra Sci 22:127–143

Xiong L, Lu S, Tan J (2023) Optimized strategies of green and grey infrastructures for integrated control objectives of run-off, waterlogging and WWDP in old storm drainages. Sci Total Environ 901:165847

Yereseme AK, Surendra HJ, Kuntoji G (2022) Sustainable integrated urban flood management strategies for planning of smart cities: a review. Sustain Water Resour Manage 8(3):85

Zerihun M, Mohammed Yasin MS, Sewnet D, Adem AA, Lakew M (2018) Assessment of soil erosion using RUSLE, GIS and remote sensing in NW Ethiopia. Geoderma Reg 12:83–90

Chapter 13
Genus *Fissidens* Hedw. (Fissidentaceae) in the Eastern Ghats, India: Diversity, Distribution and Remote Sensing Prospects

Priyanshu Srivastava and Ashish Kumar Asthana

Abstract A study on the moss flora of the Eastern Ghats, has documented the occurrence of eleven taxa of the genus *Fissidens* Hedw., namely *Fissidens intromarginatulus* Bartr., *Fissidens pulchellus* Mitt., *Fissidens sylvaticus* Griffith., *Fissidens taxifolius* Hedw., *Fissidens crenulatus* Mitt., *Fissidens crenulatus* var. *tityalyanus* (Muell. Hal.) Gangulee, *Fissidens ceylonensis* Doz. and Molk., *Fissidens orishae* Gangulee, *Fissidens flaccidus* Mitt., *Fissidens bryoides* Hedw. and *Fissidens bryoides* Hedw. subsp. *schimidii* (Müll. Hal.) Nork. These taxa are distributed across ten localities of Odisha (Kalinga, Kantia, Berbera-Rajin, Buduli, Orchidarium, Kollah, Gurguria Research Station, Lanjighosra, Bureakhata, Joranda Falls), five in Andhra Pradesh (Katiki Waterfalls, Bhairav Kona, Sunkarimetta, Galikonda, Rajeev Gandhi Wildlife Sanctuary), seven in Tamil Nadu (Kolli Hills, Kalrayan Forest Reserve, Bodamalai Forest Reserve, Javadi Hills, Sirumalai Hills, Yercaud, Rose Garden and one locality in Karnataka region i.e. Malai Mahadeshwara Wildlife Sanctuary. Among these, *Fissidens pulchellus* is newly reported from the eastern Ghats. The species exhibit diverse ecological preferences, with eight being terrestrial, four growing on rocks, and three as epiphytes. This study underscores the importance of remote sensing technologies, such as high-resolution satellite imagery and vegetation indices, for mapping moss habitats and identifying regions with favorable microclimatic conditions. Such interactive approaches can significantly enhance the understanding and conservation of *Fissidens* species in the Eastern Ghats.

Keywords *Fissidens* · Moss · Eastern Ghats · Remote sensing

P. Srivastava (✉)
Department of Applied Science, Dr. K.N. Modi University, Newai, Rajasthan, India
e-mail: priyanshu.srivastava1@gmail.com

A. K. Asthana
CSIR-National Botanical Research Institute, Lucknow, Uttar Pradesh 226001, India

V. Saritha et al. (eds.), *Remote Sensing for Environmental Monitoring*,
https://doi.org/10.1007/978-981-96-5546-5_13

13.1 Introduction

The study region, known as the Eastern Ghats, is located between 11° 30′ and 22° N latitude and 76° 50′ and 86° 30′ E longitude, and includes parts of northern Odisha, Andhra Pradesh, Tamil Nadu, and parts of Karnataka. The study region has a tropical monsoon climate, receiving rainfall from both the South-West monsoon and the North-East receding monsoon. The average annual temperature in Odisha is from 24 to 34 °C, whereas in Andhra Pradesh it ranges from 13 to 45 °C and in Tamil Nadu it goes from 14 to 32 °C. Red soil, black soil, laterite soil and alluvial soil are the most frequent soil types in the Eastern Ghats. The altitude of the Eastern Ghats varies due to fragmented hills.

Family Fissidentaceae with 67 taxa was listed from India by Lal (2005) has isobilateral and vaginant lamina which represent true leaf (Brown 1819) belongs to order Dicranales. The family consists of a single genus with more than 450 species (Stone 2012), divided in to subgenera and sections by Müller (1848, 1900). Noteworthy contribution on Fissidentaceae from time to time was provided by Norkett (1969), Gangulee (1969, 1972), Iwatsuki (1969, 1980), Pursell (1982), Stone (1991), Bruggeman-Nannenga and Pursell (1995), Zhang et al. (1998), Dabhade (1998), Nair et al. (2005) etc.

Several studies have employed remote sensing techniques to analyze ecological changes in the Eastern Ghats. Jayakumar et al. (2002) mapped forest changes in Kolli Hills (1990–1999) using remote sensing and GIS, identifying significant forest losses and areas vulnerable to degradation, aiding conservation efforts. Sakthivel et al. (2010) assessed soil erosion in the Kalrayan Hills, Eastern Ghats, using similar techniques, pinpointing highly erosion prone areas like Karnelli, Uppur and Pattivalavu and suggesting targeted conservation measures. Ramachandran et al. (2016) highlighted discrepancies between satellite-based forest cover assessments and forest soil quality index analysis, revealing significantly higher forest degradation (42.4%) in the southern Eastern Ghats compared to the reported 1.8%.

Integrating remote sensing data with taxonomic approaches holds great potential for enhancing our understanding of moss distribution and ecology in the Eastern Ghats, offering a comprehensive framework for monitoring habitat quality and addressing the impacts of environmental change. This synthesis of advanced geospatial tools and ecological research could improve biodiversity assessments and support targeted conservation planning in this biodiversity rich region.

13.2 Materials and Methods

The specimens were collected during the years 2014, 2015, 2016 respectively from different localities of Eastern Ghats region of Odisha, Andhra Pradesh and Tamil Nadu covering an altitudinal range of 106 to 1406 m. The specimens were collected from terrestrial habitats such as rocks, soil etc. and from tree bark and have been

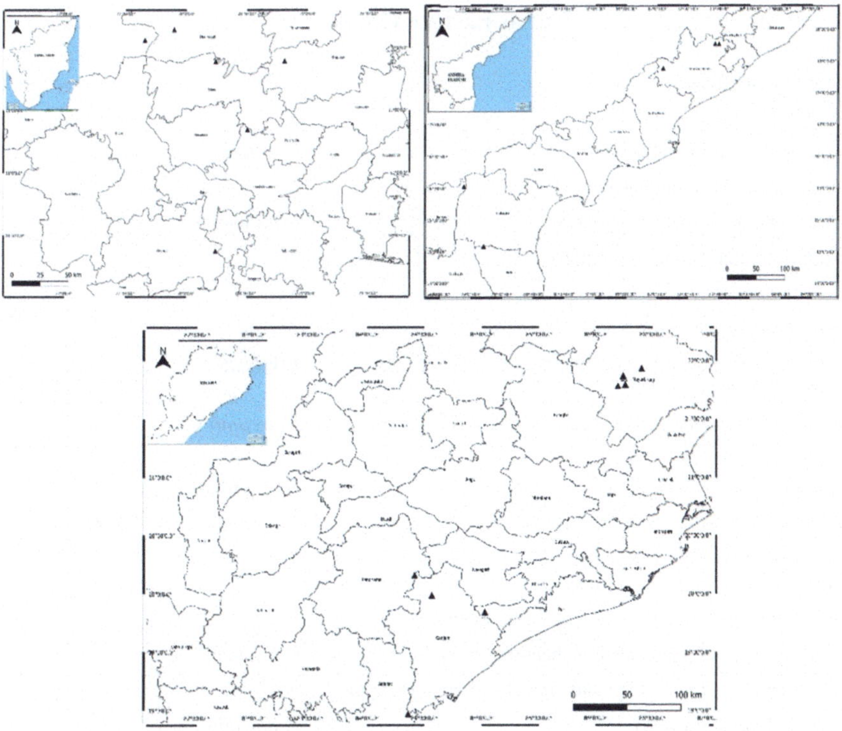

Fig. 13.1 Map showing study sites (Tamil Nadu, Andhra Pradesh and Odisha)

deposited in the Bryophyte Herbarium, NBRI, Lucknow (LWG), India. The area covered under eastern Ghats and the major sites of plant collection has been shown in the Map (Fig. 13.1). The samples were carefully examined, and the taxonomic findings were noted.

13.3 Results

Plant samples gathered from various ecological habitats across the states- Odisha, Andhra Pradesh and Tamil Nadu have been meticulously studied. Detailed analysis of their morphological and anatomical characteristics was conducted to accurately determine and validate the taxonomic classification of each species. This comprehensive investigation aimed to authenticate the identity and status of these taxa, contributing valuable insights into their diversity and systematics. Furthermore, integrating remote sensing technologies with this study offers promising future prospects for monitoring bryophyte habitats, assessing environmental changes, and enhancing conservation strategies.

Key to the species of *Fissidens* at Eastern Ghats

1. Limbidium present	2
Limbidium absent	3
2. Sub-marginal limbidium present at proximal end	***F. intromarginatulus***
Sub-marginal limbidium absent	4
3. Leaf margin dentate proximally	***F. pulchellus***
Leaf margin not dentate proximally	5
4. Costa excurrent	6
Costa ending below apex	7
5. Presence of glandular protuberance on stem, sheathing lamina closed type	***F. sylvaticus***
Absence of glandular protuberance on stem, sheathing lamina open	***F. taxifolius***
6. Limbidium covering entire sheathing lamina	***F. crenulatus***
Limbidium covering 1/2 length around vaginant lamina	***F. crenulatus*** var. ***tityalanus***
7. Limbidium breaks down few cells at proximal as well as distal end of Vaginant lamina	***F. ceylonensis***
Limbidium covering almost of Sheathing lamina	8
8. Limbidium covering 2/3rd of sheathing lamina	***F. orishae***
Limbidium covering all vaginant, apical and dorsal lamina	9
9. Limbidium 2–3 rowed	***F. flaciddus***
Limbidium 1–4 rowed	10
10. Leaves oblong-lingulate	***F. bryoides***
Leaves ovate-lanceolate	***F. bryoides* subsp. *schimidii***

1. ***Fissidens intromarginatulus*** Bartr. in *Rev. Bryol. Lichen.*, 23: 242. 1954.

Plants small sized, lax, gregarious, stem erect, about 6 mm, unbranched; axillary hyaline nodule absent. Leaves not curled when dry, yellowish-green, 8–12 pairs of leaves, aggregated toward apex, small and lax near base, equally broad from base to apex, oblong-lingulate, 1.4–2.0 × 0.3–0.4 mm; Dorsal lamina having rounded base ending on stem; sheathing lamina unequal, open type; B/L = 20/100, S/L = 60–100; limbidium incomplete of (2–3 layered) elongated cells restricted to sheathing lamina, presence of sub-marginal limbidium proximally; Leaf apical cells rounded, 7.5–10 × 3.7–6.2 μm, median cells rounded-hexagonal, 12.5–15 × 7.5–8 μm, basal cells irregular hexagonal, 10–25 × 7.5 μm. Sporophyte not seen.

Range of Distribution: Asia [Myanmar, East Nepal, India].

Distribution: INDIA: Central India; Eastern Ghats [Odisha]; Eastern Himalaya [West Bengal (Darjeeling)]; Western Himalaya [Uttarakhand (Dehradun)]; Western Ghats.

Habitat: Soil.

Additional Specimen Examined: Odisha, Ganjam Dist., Kalinga, alt ca. 610 m, on soil, 24.10.1959, leg.?, Det. H.C. Gangulee, (3/1) 9-3 (CAL).

2. ***Fissidens pulchellus*** Mitt. in *Musc. Ind. Or.*: 140. 1859.

Plants small sized, caespitose, pale green, stem erect, about 5 × ± 2 mm, unbranched; axillary hyaline nodules not present. Leaves strongly curled when dry, 12–15 pairs of leaves, oblong-lanceolate, ± 1.2 × ± 0.30 mm, apiculate; Dorsal lamina slightly decurrent ending on nerve; sheathing lamina usually unequal (open); B/L = 25/100, S/L = 50/100; Limbidium not present; margin denticulate with sharp cells on border; costa may be percurrent, ending below apex or excurrent in a short apiculus,; Laminar apical cells of leaves irregularly quadrate-hexagonal, ± 11.4 × 5.7–7.6 μm, median cells quadrate-hexagonal, 7.6–9.5 × 7.6 μm, basal cells quadrate- hexagonal, 11.4–15.2 × 7.6 μm. Sporophyte not seen.

Range of Distribution: Asia [East Nepal, India].

Distribution: INDIA: Central India [Amarkantak]; Eastern Ghats [Andhra Pradesh, Karnataka, Odisha, Tamil Nadu]; Eastern Himalaya [West Bengal (Darjeeling)]; Western Ghats [Kerala, Kanyakumari].

Habitat: Tree barks, Soil, Rocks.

Specimens Examined: Tamil Nadu, Namakkal, Kolli Hills, alt ca. 1182 m, Epiphytic, 14.4.2014, leg. A.K. Asthana and Party, 256065C (LWG); Salem, Kalrayan Forest Reserve, Karmandurai, Giant Orchard, alt ca. 789 m, on bricks, 15.4.2014, 256142 (LWG); Bodamalai Forest Reserve, alt ca. 270 m, on rocks, 16.4.2014, 256165 (LWG); soil covered rocks, 16.4.2014, 256168 (LWG); 256169 (LWG); on soil, 16.4.2014, 256171 (LWG); Tiruvannamalai, Javadi Hills, alt ca. 792 m, on soil, 17.4.2014, 257171B (LWG); on logs, 17.4.2014, 257177 (LWG); on soil, 17.4.2014, 257187B (LWG); on rocks, 17.4.2014, 257192 (LWG); 257193A (LWG); 257200 (LWG); Dindigul, Sirumalai Hills, alt ca. 1046 m, Epiphytic, 18.4.2014, 257234C (LWG); alt ca. 1070 m, on soil, 18.4.2014, 257280 (LWG); Epiphytic, 18.4.2014, leg A.K. Asthana and Party, 257281 (LWG); Karnataka, Chamrajanagar, Kollegal, Malai Mahadeshwara Wildlife Sanctuary, alt ca. 312 m, on soil, 16.4.2014, leg A.K. Asthana and Party, 256184B (LWG); 256187A (LWG); Andhra Pradesh, Prakasam, Bhairavkona, alt ca. 305 m, on rocks, 12.3.2015, leg A.K. Asthana and Party, 258230 (LWG); alt ca. 361 m, on soil covered rocks, 12.3.2015, 258234 (LWG); Epiphytic, 12.3.2015, 258235 (LWG); rocks, 12.3.2015, 258238 (LWG); Visakhapatnam, Araku Valley, Sunkarimetta, alt ca. 1161 m, Epiphytic, 16.3.2015, 258327B (LWG); alt ca. 1164 m, Epiphytic, 16.3.2015, 258345A (LWG); rocks, 16.3.2015, 258348 (LWG); Galikonda, alt ca. 1246 m, on wet rocks, 16.3.2015, 258361B (LWG); 258365C (LWG); alt ca. 1181 m, on rocks, 17.3.2015, 300045B (LWG); Odisha, Mayurbhanj, Similipal Tiger and Biosphere Reserve, near Orchidarium, alt ca. 593 m, on soil, 1.4.2016, leg A.K. Asthana and Party, 300663B (LWG); Epiphytic, 1.4.2016, 300671C (LWG); 300674B (LWG); on rocks, 1.4.2016, 300675 (LWG); 300676B (LWG); on soil covered rocks, 1.4.2016, 300677C (LWG); Kollah, alt ca. 793 m, Epiphytic, 2.4.2016, 300687A (LWG); Orchidarium, alt ca. 593 m, on rocks, 3.4.2016, 300812 (LWG); Gurguria Research Station, alt ca., Epiphytic, 3.4.2016, 300846C (LWG); Lanjighosra, alt ca. 642 m, on rocks, 3.4.2016, 300861B

(LWG); Gajapati, Mahendragiri Forest Reserve, Bureakhata, alt ca. 1051 m, on rocks, 05.04.2016, 300935 (LWG).

3. ***Fissidens sylvaticus*** Griffith. in *Cal. J. Nat. Hist.*, 2: 507. 1842.

Plants minute, gregarious, yellowish-brown, stem simple, erect, ± 10 mm, unbranched; presence of glandular protuberance of 6–7 turgid cells. Leaves curled when dry, 12–14 pairs of leaves, oblong-lanceolate to broader oblong-lingulate, 1.5–1.8 × 0.2–0.27 mm, acuminate apex, curled distally; Dorsal lamina ending on nerve base; sheathing lamina equal, closed; B/L = 15/100, S/L = 55.5/100; Limbidium not present; margin crenulate due to projection of cell; costa ends slightly below tip to slightly percurrent; leaf apical cells quadrate-hexagonal, 5–7.5 × ± 3.7 μm, median cells quadrate, hexagonal, 7.5–12.5 × 5–7.5 μm, basal cells irregular, sub-quadrate to hexagonal, 12.5–15 × 5–7.5 μm. Sporophyte not seen.

Range of Distribution: Africa [Algeria, Madagascar]; Asia [Borneo, East Nepal, Celebes, Ceylon, Hongkong, Japan, Java, Myanmar, Philippines, Sumatra, Thailand, Vietnam,]; Oceania [New Guinea, New Zealand, Samoa].

Distribution: INDIA: Andaman Is.; Central India [Madhya Pradesh (Pachmarhi)]; Eastern Ghats [Odisha, Tamil Nadu]; Eastern Himalaya [West Bengal (Darjeeling)]; North-East India [Assam, Meghalaya (Khasi Hills)]; Gangetic plains [Bihar, Chottanagpur, West Bengal]; Western Ghats [Maharashtra (Bombay, Khandala), Tamil Nadu (Nilgiri, Palni, Coorg)]; Western Himalaya [Uttarakhand (Kathgodam, Kumaon)].

Habitat: Soil.

Additional Specimen Examined: Tamil Nadu, Tinnevelly Dist., Kathalaimalai, alt ca. 914 m, habitat.?,?.05.1927, leg.?, Det. A.H. Norkett, (3/1) 33-5 (CAL); Odisha, Cuttack, Jajpur, Vill. Kantia, alt ca.? on soil, 28.10.1958, leg.?, Det by. E.B. Bartram, 33(3/1)-1 (CAL).

4. ***Fissidens taxifolius*** Hedw. in *Sp. Musc.*:155. 1801.

Plants small, gregarious, yellow-green, stem erect, up to 5 × 2 mm, unbranched; axillary nodules not present. Leaves erectopatent, curled circinately at top when dry, 10–13 pairs of leaves, oblong-lingulate, broadly acuminate, 0.80–1.50 × 0.20–0.50 mm; Dorsal Lamina ending on nerve; sheathing lamina usually of open type (unequal); B/L = ± 33.3/100, S/L = 53.3/100; Limbidium not present; margin crenulate or somewhat entire; costa prominent, ending below apex; Leaf apical cells rounded-quadrate, tumescent, ± 15.2 × ± 7.6 μm, median cells quadrate-hexagonal, ± 7.6 × ± 5.7 μm, basal cells quadrate-hexagonal, about 11.4 × up to 7.6 μm. Sporophyte not seen.

Range of Distribution: Africa [Canary Island, Madeira, North Africa]; Asia [Central Asia, Caucasus, East Nepal, Persia, Korea, Sakhalin, Japan, Ryukyus]; Europe [Azores]; North, central and south America.

Distribution: INDIA: Andaman Is.; Eastern Ghats [Tamil Nadu]; Eastern Himalaya [West Bengal (Darjeeling)]; Gangetic Plains; Western Himalaya [Uttarakhand (Ranikhet, Nainital), Himachal Pradesh (Shimla), Kashmir].

Habitat: Rocks.

Specimen Examined: Tamil Nadu, Salem, Yercaud, alt ca. 1216 m, on soil covered rocks, 13.4.2014, leg A.K. Asthana and Party, 254940A (LWG).

5. ***Fissidens crenulatus*** Mitt. in *Musc. Ind. Or.*: 140. 1859.

Plants minute, gregarious, yellowish green, stem erect, ± 5 × 1.2 mm, unbranched; axillary hyaline nodule absent. Leaves curled when dry, up to 10 pairs of leaves, oblong-lanceolate to oblong-lingulate, ± 1 × ± 0.35 mm; Dorsal lamina ending on nerve, narrower at base; sheathing lamina usually unequal (open type); B/L = 35/100, S/L = 60/100; Semi-limbidium of 2–3 layered bordered over sheathing lamina; margin serrulate; costa excurrent in a short apiculus; leaf apical cells rounded-hexagonal, multipapillate up to 7.6 × 5.7 μm, median cells irregular, hexagonal, about 7.6 × 4.5 μm, basal cells sub-rectangular, up to 9.5 × about 7.6 μm. Sporophyte not seen.

Range of Distribution: Asia [East Nepal, India, Upper Myanmar].

Distribution: INDIA: Central India; Eastern Ghats [Andhra Pradesh, Odisha]; Gangetic Plains.

Habitat: Tree barks, Rocks.

Specimen Examined: Andhra Pradesh, Visakhapatnam, Araku Valley, Katiki waterfalls, alt ca. 914 m, on rocks, 17.3.2015, leg A.K. Asthana and Party, 300068 (LWG); alt ca. 954 m, wet rocks, 17.3.2015, leg A.K. Asthana and Party, 300074 (LWG); Odisha, Mayurbhanj, Similipal Tiger and Biosphere Reserve, Joranda Falls, alt ca. 761 m, Epiphytic, 2.4.2016, leg A.K. Asthana and Party, 300776 (LWG); Orchidarium, alt ca. 593 m, Epiphytic, 3.4.2016, 300814A (LWG).

6. ***Fissidens crenulatus* var. *tityalyanus*** (Muell. Hal.) Gangulee in *Mosses E. India* 2:506. 1971.

Plants small sized, yellowish-green, stem erect, 4 mm long, unbranched; axillary hyaline nodules absent. Leaves with 8–10 pairs, oblong-lingulate, 1.1–1.2 × ± 0.25 mm; Dorsal lamina ending on stem, decurrent; sheathing lamina equal (closed), 0.6–0.75 mm; B/L = 20.8/100, S/L = 62.5/100; semilimbidium (1–3 row) present in sheathing lamina region leaving sheathing lamina apex; margin crenulate due to mamillae; costa excurrent to ± percurrent; Leaf apical cells rounded-hexagonal, conical mamillae, 7.5–12.5 × 6.2–7.5 μm, median cells hexagonal, 10–12.5 × 5 μm, basal cells elongated irregular, quadrate, 10–20 × ± 5 μm. Seta reddish in color; capsule elongate, cylindric.

Range of Distribution: Asia [India, Nepal].

Distribution: INDIA: Central India; Eastern Ghats [Odisha]; Gangetic plains [North Bengal, West Bengal].

Habitat: Soil.

Additional Specimens Examined: Odisha, Puri Dist., Berbera-Rajin Road, alt ca. 244 m, on red laterite soil, 20.10.1959, Leg.?, Det. H.C. Gangulee, (3/1)4-4 (CAL); (3/1)4-5 (CAL); Ganjam Dist., Buduli, Near Bhanhanagar, alt ca. 122 m, on soil, 24.10.1959, leg.?, Det. H.C. Gangulee; (3/1)4-6 (CAL).

7. ***Fissidens ceylonensis*** Doz. and Molk. in *Ann. Sci. Nat. Bot.* Ser.3, 2:304. 1844.

Plants small to medium sized, gregarious, yellowish green, stem erect, ± 9 × 1.2 mm, unbranched; axillary hyaline nodules absent. Leaves curled when dry, 12–16 pairs of leaves, oblong-lingulate, sudden apiculate from a broader base, 1–1.2 × 0.30 mm; Dorsal lamina base rounded, decurrent below; sheathing lamina unequal (open); B/L = 30/100, S/L = 60/100; semilimbidium 1 to 3 rows wide below and narrow above of elongated cells, cover most of sheathing lamina of all leaves but it breaks down to a few parenchymatous at tip and at base; margin crenulate; costa strong, excurrent; Apical cells of leaves sub-obscure, rounded-hexagonal, multipapillate (3–4 per cell), 7.6–15.2 × 7.6 μm, median cells hexagonal multipapillate (4–5 papillae) in cell wall, 7.6 × 7.6 μm, basal cells irregularly hexagonal, smooth, 7.6–11.4 × 7.6 μm. Sporophyte not seen.

Range of Distribution: Asia: [Borneo, East Nepal, Ceylon, Java, Malay, Philippines, Sumatra, Thailand, Vietnam, Yunnan]; Oceania [New Zealand].

Distribution: INDIA: Eastern Ghats [Odisha, Tamil Nadu]; Eastern Himalaya [West Bengal (Darjeeling), Sikkim]; Gangetic Plains [West Bengal]; Western Ghats [Tamil Nadu (Nilgiri, Palni), Kerala (Trivandrum)]; Western Himalaya.

Habitat: Soil.

Specimen Examined: Odisha, Mayurbhanj, Similipal Tiger and Biosphere Reserve, Joranda Falls, alt ca. 761 m, on soil, 2.4.2016, leg A.K. Asthana and Party, 300787 (LWG); 300788 (LWG).

8. ***Fissidens orishae*** Gangulee in *Nova Hedwigia*, 8:140. 1964.

Plants small sized, in loose tufts, yellowish-brown, stem erect, 3 mm long, unbranched; axillary hyaline nodules absent. Leaves not curled when dry, 6–8 pairs of leaves, oblong-lingulate, broader in sheathing region, 0.8–1.1 × 0.25–0.42 mm; Dorsal lamina narrowing down at base and vanishing in nerve; sheathing lamina unequal or open type; B/L = ± 38.1/100, S/L = ± 63.6/100; imperfect limbidium (covering 2/3rd of S.L.) of 2 rows of elongated pellucid cells at sheathing lamina base; margin smooth to crenulate; costa ending few cells before apex; leaf apical cells mamillose, rounded-quadrate, 10–17.5 × 5–7.5 μm, median cells quadrate, 7.5–10 × 7.5–8.7 μm, basal cells hexagonal, 10–20 × 7.5–8.7 μm. Seta short, apical, brown, 1 mm; capsule erect, ovate-cylindrical, ± 0.75 × 0.3 mm in diameter; operculum conic-rostrate.

Distribution: Eastern Ghats [Odisha]; Gangetic plains [West Bengal].

Habitat: Soil.

Additional Specimen Examined: Odisha, Ganjam Dist., Gallery R.F., Buduli, alt ca. 122 m, on soil, 24.10.1959, leg.?, Det. H.C. Gangulee, (3/1) 45-1 (CAL) (TYPE).

9. ***Fissidens flaccidus*** Mitt. in *Trans. Linn. Soc. London* 23:56. 1860.

Synonym: ***Fissidens splachnobryoides*** Broth in Fl. Schutzgeb. Südsee 81. 1900.

Plants small sized, light green, stem erect, 2.4 × 1.25 μm, unbranched; axillary nodules not present. Leaves curled when dry, up to 6 pairs of leaves, oblong-lanceolate, 0.66–0.83 × 0.14–0.21 mm, acuminate; dorsal lamina narrowing down at proximal end, wedge shaped; sheathing lamina equal (closed type); limbidium 2–3 layered, becoming indistinct near base, present all around leaf; margin entire; costa thin, ending far below apex; Leaf apical cells rhombic to elliptic ovate, 6.6–10 × ± 5 μm, median cells rhombic-quadrate, up to 6.6 × 4 μm, basal cells rectangular, 10–16.6 × ± 6.6 μm. Sporophyte not seen.

Range of Distribution: Asia [East Nepal, Ceylon, Upper Myanmar, Java, North Borneo, Ryukyus, Japan, Philippines, Sri Lanka]; Oceania [New Guinea].

Distribution: INDIA: Central India [Madhya Pradesh (Pachmarhi)]; Eastern Ghats [Andhra Pradesh]; Eastern Himalaya; Gangetic plains [Lower Bengal], Punjab & West Rajasthan [Punjab (Kalka)]; Western Ghats [Maharashtra (Bombay, Khandala)].

Habitat: Rocks.

Specimen Examined: Andhra Pradesh, Prakasam, Nemaligundla Ranganayaka Swamigundam, alt ca. 300 m, soil covered rocks, 13.3.2015, leg A.K. Asthana and Party, 258254 (LWG).

10. ***Fissidens bryoides*** Hedw. in *Sp. Musc.*:153. 1801.

Plants small sized, gregarious, yellowish-green, stem erect, up to 4 × 2 mm, unbranched; axillary hyaline nodules absent. Leaves not much curled when dry, up to 7 pairs of leaves, oblong-lingulate, 1.2 × 0.50 mm, acuminate; Dorsal Lamina ending on nerve base; sheathing lamina unequal (open type); B/L = 41.6/100, S/L = 50/100; Limbidium present, yellowish, cartilaginous elongated all around leaf, on dorsal lamina Limbidium is single row at tip, 2–3 rowed at base, on sheathing lamina Limbidium is 3–4 rowed at base and 2 rowed at apex; margin entire; costa yellow–brown, percurrent to slightly excurrent; Leaf apical cells smooth, quadrate, chlorophyllose, 7.6–22.8 × 3.8–7.6 μm, median cells smooth, quadrate, 11.4 × 7.6 μm, basal cells smooth, elongated, 15.2–19 × 7.6–11.4 μm. Sporophyte not seen.

Range of Distribution: Asia [Ceylon, China, East Nepal, India, Japan, Java, Malay, Philippines, Siberia, Taiwan]; Europe [Caucasus]; North and Central Africa; North and South America.

Distribution: INDIA: Punjab and West Rajasthan [Rajasthan]; Eastern Ghats [Odisha, Tamil Nadu]; Eastern Himalaya; Gangetic Plains [Lower Bengal]; North-East India [Meghalaya (Khasi Hills), NEFA]; Western Himalayas [Uttarakhand (Ranikhet), Himachal Pradesh (Shimla)]; Western Ghats [Tamil Nadu (Nilgiris, Coonoor)].

Habitat: Tree barks, Soil.

Specimen Examined: Tamil Nadu, Salem, Yercaud, Rose Garden, alt ca. 1428 m, Epiphytic, 13.4.2014, leg A.K. Asthana and Party, 254991D (LWG); Odisha, Mayurbhanj, Similipal Tiger and Biosphere Reserve, alt ca. 642 m, on soil, 3.4.2016, leg A.K. Asthana and Party, 300873 (LWG).

11. ***Fissidens bryoides*** Hedw. subsp. ***schimidii*** (Müll. Hal.) Nork. in *Mosses. E. India.* 2:471. 1971.

Plants small sized, caespitose, yellowish-green, stem erect, ± 5 mm long, procumbent; axillary hyaline nodules not present. Leaves contorted when dry, 7–8 pairs of leaves, ovate-lanceolate, tapering down from a wider base, 1–1.8 × 0.27–0.37 mm, shortly acuminate; dorsal lamina ending on nerve; sheathing lamina equal (closed type); B/L = 20.5/100, S/L = 41.6/100; Limbidium of elongated cells all around leaf with 2 rows in dorsal lamina, 1 row near apex, 3–4 rows near sheathing lamina and devoid at distal end; margin smooth or entire; costa percurrent to ending below apex; leaf apical cells smooth, irregularly quadrate, about 10 × 5–6.2 μm, median cells shape, 7.5–8 × ± 7.5 μm, basal cells quadrate to sub-quadrate, 12.5–20 × 7.5–12.5. Capsule inclined, arcuate, clavate, 0.8 × 0.3 mm.

Range of Distribution: Africa; Asia [Ceylon, East Nepal, India, Japan, Java, Malay, Philippines].

Distribution: INDIA: Central India [Jharkhand (Chottanagpur)]; Eastern Ghats [Odisha]; Eastern Himalaya [West Bengal (Darjeeling)]; Western Ghats [Karnataka (Mangalore), Tamil Nadu (Nilgiris, Palni)]; Western Himalaya [Uttarakhand (Ranikhet)].

Habitat: Soil.

Additional Specimen Examined: Odisha, Ganjam Dist., Kalinga, alt ca. 610 m, on soil, 24.10.1959, leg.?, Det. H.C. Gangulee, (3/1) 7-4 (CAL).

13.4 Discussion

The genus *Fissidens* Hedw. (Fissidentaceae) is represented by diverse species in the Eastern Ghats, India, showcasing unique morphological and ecological characteristics. Among these, five taxa- *Fissidens intromarginatulus, Fissidens sylvaticus, Fissidens crenulatus* var. *tityalyanus, Fissidens orishae* and *Fissidens bryoides* Hedw. subsp. *schimidii* were examined from specimens housed at the Central National Herbarium, BSI, Kolkata. Notable distinctions include the sub-marginal limbidium at the proximal end in *F. intromarginatulus,* and prominent dentate leaves in *F. pulchellus. F. sylvaticus* was characterized by a glandular protuberance on the stem and a closed sheathing lamina type, distinguishing it from *F. taxifolius.* The complete limbidium along the sheathing lamina separated *F. crenulatus* from its variant *F. crenulatus* var *tityalanus.* Limbidium pattern also varied between *F. ceylonensis,* where it breaks down at the proximal and distal ends of the vaginant lamina, and *F. orishae,* where it spans the entire sheathing lamina.

Additionally, advancements in climate science and geospatial technology offer new avenues to explore moss ecology in the eastern Ghats. Climate datasets, encompassing variations in temperature and rainfall, can illuminate the relationship between environmental factors and moss growth patterns. Coupling species distribution models with remote sensing data provides a robust framework to predict shifts in moss habitats under future climate scenarios, enabling the identification of vulnerable zones and ecological refugia.

Remote sensing techniques, previously utilized to study forest changes in the Eastern Ghats, highlights its potential for moss distribution research. For instance, studies in Kolli and Kalrayan Hills mapped forest losses and erosion-prone areas using GIS, while discrepancies between soil quality indices and satellite-based forest assessments revealed greater forest degradation than earlier reported. Applying such tools to moss studies can monitor habitat quality, restoration efforts and biodiversity conservation. Integrating taxonomic insights with remote sensing and climate data provides a comprehensive approach to understanding *Fissidens* distribution, predicting its response to climate change, and guiding conservation strategies in the ecologically rich Eastern Ghats.

Acknowledgements The authors are grateful to the Director, CSIR-National Botanical Research Institute, Lucknow, India for providing the laboratory facilities and the Ministry of Environment, Forests and Climate Change, New Delhi for financial assistance. Thanks, are also due to the Dr. K.N. Modi University, Newai, Rajasthan for invaluable support.

Competing Interests The authors declare no potential conflicts of interest.

References

Brown R (1819) Characterization and description of *Lyellia*, etc. Trans Linn Soc Lond 12:560–583

Bruggeman-Nannenga MA, Pursell RA (1995) Notes on *Fissidens* V. Lindbergia 20:49–55

Dabhade GT (1998) Mosses of Khandala and Mahabaleshwar in Western Ghats, Classic Graphics, Thane, India

Gangulee HC (1969–1972) Mosses of Eastern India and Adjacent Regions. I. Books and Allied (P) Ltd., Calcutta, India

Iwatsuki Z (1969) Notes on Japanese *Fissidens*. J Hattori Bot Lab 32:311–318

Iwatsuki Z (1980) A preliminary study of *Fissidens* in China. J Hattori Bot Lab 48:171–186

Jayakumar S, Arockiasamy DI, John Britto S (2002) Conserving forest in the eastern Ghats through remote sensing and GIS—a case study in Kolli Hills. Curr Sci 82(10):1259–1267

Lal J (2005) A checklist of Indian mosses. Bishen Singh Mahendra Pal Singh, Dehradun, India

Nair MC, Rajesh KP, Madhusoodanan PV (2005) Bryophytes of Wayanad in western Ghats. Malabar Natural History Society, Kozhikode, India

Norkett AH (1969) Some problems in the monographic revision of the genus *Fissidens* Hedw. With special reference to Indian species. Bull Bot Soc Bengal 23:75–82

Ramachandran A, Radhapriya P, Jayakumar S, Dhanya P, Geetha R (2016) Critical analysis of forest degradation in the southern eastern Ghats of India: comparison of satellite imagery and soil quality index. PLoS ONE 11(1):1–19

Sakthivel R, Manivel M, Jawahar Raj N, Pugalanthi V, Ravichandran N, Anand VD (2010) Remote sensing and GIS bases forest cover change detection study in Kalrayan Hills, Tamil Nadu. J Environ Biol 31(5):737–747

Stone IG and Catcheside (2012) Australian mosses online. 64. Fissidentaceae. http://www.anbg.gov.au/abrs/Mosses_Online/Fissidentaceae.pdf

Stone IG (1991) *Fissidens linearis* Brid. and its synonyms. J Bryol 18:159–167

Zhang L, Lin P-J, Chau L, Lee W (1998) Taxonomic studies on the bryophytes from Hong Kong I. Fissidentaceae. J Hattori Bot Lab 84:1–10

Chapter 14
Bird Diversity, Habitat Degradation, and Ecosystem Services Evaluation of Bisalpur Wetland, Rajasthan

Raj Singh, Vara Saritha, and Sachchidanand Singh

Abstract A wetland ecosystem in a place with high temperatures all year, notably in Rajasthan's arid and semi-arid areas, is critical for preventing water-related challenges and maintaining ecological balance. Therefore, as an essential home for both resident and migratory bird species, Rajasthan's Bisalpur Wetland contributes significantly to the biodiversity of birds. This study observed that the majority of bird species at Bisalpur wetland are classified as Least Concern, the existence of Near Threatened and Vulnerable species emphasizes the need for focused conservation efforts, according to the IUCN Red List assessment. Migratory species like the Bar-headed Goose, Common Pochard, and Northern Pintail highlight the wetland's biological importance as a stopover and wintering location on major migration routes. However, it has been identified that such vital ecosystems are constantly being destroyed as a result of local human involvement. People in the surrounding areas commonly use forest wood for cooking and shelter. Moreover, originating from the construction of a dam on the Banas River, this wetland supports a variety of ecosystem services, including irrigation, fisheries, flood control, and biodiversity conservation. Surrounding villages rely heavily on the wetland for fuelwood, water, and agricultural sustenance, while the site also holds significant cultural and religious value due to its historic Bisaldeo temple. The wetland fosters biodiversity, with a variety of fish, plankton, and avian species enriching its ecosystem. Government environmental bodies are required to take action regarding resource depletion and unauthorized residents in the wetland catchment areas for sustainable management of these precious resources.

Keywords Wetland birds · Vegetation loss · Habitat degradation · Sustainable management

---

R. Singh · V. Saritha (✉)
Department of Life Sciences, Environmental Science Division, GITAM Deemed to be University, Visakhapatnam, India
e-mail: svara@gitam.edu

S. Singh
National Institute of Hydrology, Western Himalayan Regional Centre, Jammu, India

V. Saritha et al. (eds.), *Remote Sensing for Environmental Monitoring*,
https://doi.org/10.1007/978-981-96-5546-5_14

14.1 Introduction

Wetlands are productive ecosystems representing a transition zone between aquatic and terrestrial regions (Scholte et al. 2016; Zhou et al. 2020). It plays an important role in sustaining diverse biodiversity, conservation, and nutrient recycling. Wetland works as a natural barrier against high tides, drought, and flooding and helps in climate change adaptation (Blackwell and Pilgrim 2011; Costa et al. 2024; Londe et al. 2024; Rahman et al. 2018).

However, wetlands are suitable habitats for both local and migratory birds. Birds use this ecosystem generally for feeding and breeding purposes (Marasinghe et al. 2022; Panda et al. 2021). They move from different countries to locate suitable areas during distinct seasons. The shallow vegetative region of the wetland supports rich molluscans, crustaceans, and fish diversity, which are the primary source of nutritious food for water birds (Bhendekar et al. 2024; Qu et al. 2023; Singh and Mishra 2019).

India's varied climatic and geographic factors support diverse wetlands, including high-altitude lakes in the Himalayas and floodplains in the Gangetic region (Naik and Sharma 2022). These diverse environmental conditions support a wide variety of wetlands, ranging from small ponds to large lakes, with variations in water quality, including fresh, saline, and brackish characteristics (Bachheti et al. 2023; Mishra et al. 2024). Thus, Indian wetlands attract a significant number of migratory birds each winter (Rajpar et al. 2022). These beautiful birds attract tourists, adding substantially to the nation's economic value. Also, the rich diversity of birds is an indicator of a healthy ecosystem and good water quality (Aarif et al. 2023; Mishra et al. 2023; Yadav and Rai 2024). SAC ISRO classified wetland services into four major categories (Table 14.1), adopted from the International Lake Environment Committee Foundation (ILEC), 2007 (12th World Lake Conference (Taal 2007) | International Lake Environment Committee | ILEC).

Moreover, these precious resources have continuously been destroyed due to rapid urbanization and industrialization. Furthermore, drought and other anthropogenic activities pose severe challenges to wetland biodiversity (Let and Pal 2023). Water pollution from agricultural runoff, industrial discharge, and untreated sewage lowers water quality, endangering aquatic life and disturbing the ecological balance (Fluet-Chouinard et al. 2023; Xi et al. 2021, 2022). Excessive extraction of marsh vegetation, frequently utilized for cooking fuel or cattle grazing, causes habitat degradation and reduces food and shelter for many species (Kadlec 2020; Sekey et al. 2023). Furthermore, excessive irrigation water usage, particularly in agriculturally intensive regions, depletes wetland water levels, putting these ecosystems under further stress (Li et al. 2021; Mattson et al. 2024; Pereira et al. 2024). Together, these processes contribute to biodiversity loss, affecting both local species and the critical ecosystem services that wetlands provide, such as water purification, flood regulation, and carbon sequestration (Ahmad et al. 2024; Hu et al. 2020; Singh et al. 2025).

However, this study focused on bird monitoring and assessment of habitat loss along interpretation of wetland goods and services at the Bisalpur wetland, Rajasthan,

Table 14.1 Ecosystem services provided by wetlands

Regulating services	Resource provision services	Supporting services	Cultural service
Aquatic habitat	Fuel	Primary production	Scenic and aesthetic values
Health provision	Hydropower	Geological formation	Spiritual and religious value
Capacity	Fiber and wood	Nutrient cycling	Educational resources
Drought and flood mitigation	Irrigation and drinking water	Heat energy	Historic sites
Fertile lands	Fish	Physical structure	
Self-purification capacity			
Navigation routes			
Climate mediation			
Diverse food-chain			

Source Data has been compiled from Wetland Atlas (Gupta et al. 2024)

India. This study has the potential to fill key knowledge gaps about the ecological dynamics of Bisalpur Wetland. The findings will not only help to develop conservation plans for bird habitats, but will also provide significant insights into sustainable wetland management.

Objectives of the Study:

- To assess the diversity of bird species in the Bisalpur Wetland.
- To evaluate the extent of habitat degradation (impact of human activities on wetland).
- To analyze the goods and services provided by the Bisalpur Wetland.

14.2 Material and Methods

14.2.1 Study Area

The Bisalpur Wetland (bird monitoring site) is located in Tonk district, Rajasthan, India, at latitude 25° 55′ 22.14″ and longitude 75° 27′ 21.16″ (Fig. 14.1). Bisalpur Wetland is a man-made wetland formed due to the construction of the Bisalpur Dam in the Khamnor hills of the Aravalli range. Construction of the Dam was started in 1985 and completed in 1996 (World Bank Document). The dam's water is commonly used for irrigation and drinking purposes (Singh et al. 2024). The availability of water, along with the beautiful mountainous slopes and greenery, makes this geographical region a favorable habitat for both migratory and local birds.

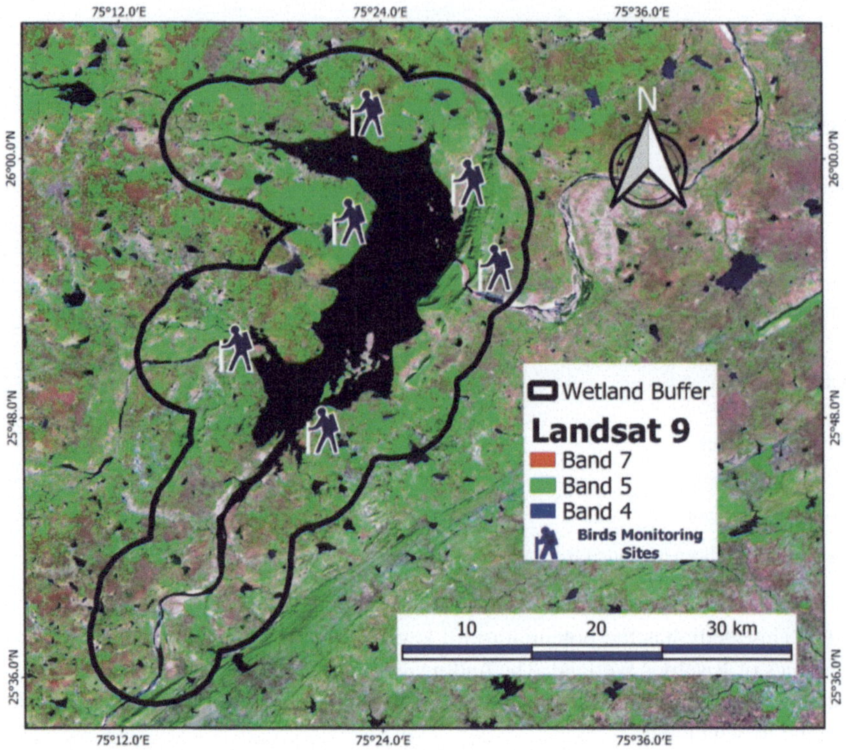

Fig. 14.1 Study area

14.2.2 Survey Methods of Birds

From October to March, avian surveys were conducted in the mornings (6.30–11.00 AM) and evenings (3.00–5.00 PM). Actual head counts were conducted for bird species with very small populations. The study area is categorized into six different zones based on habitat preferences by migratory and local birds, as well as the consideration of surrounding vegetation, water availability, and other ecological factors. The categorized zones were analyzed throughout the research period, and each zone was revisited the following month using a 20 × 50 prismatic field binocular. Regular field surveys were conducted at 10-day intervals throughout the study period, with surveys avoided on foggy and wet days. Bird species were identified using a field reference book called The Book of Indian Birds. Moreover, the eBird checklist has been used to compile the common birds' data (https://ebird.org/home). In addition, information on avian status and anthropogenic activity was gathered from the adjacent residential area of the marsh.

14.2.3 Monitoring of Major Goods and Services of Bisalpur Wetland

The wetland goods and services data were analyzed and compiled from both secondary sources and field surveys. Data collecting for this investigation with the help of site visits in different sessions between 2022 and 2024 at many places in and around the wetland. Most of the general information, such as the use of fodder, fuel wood, use of wetland water, fish food, and other recreational activity data, was compiled with local people's opportunistic interviews. Moreover, people around the wetland hesitate to interact with stranger people. Also, we have observed that they communicate in their native Rajsthan language, and if they don't understand our question or reply, they simply deny further talk. Therefore, we have mostly taken opportunistic interviews. At a field visit on February 11, 2024, we recorded several responses at the wetland site through group discussions with interested people (Fig. 14.2a). These respondent people were assembled that day in a Mahapanchayat (Fig. 14.2b). The Mahapanchayat (gathering for discussion on regional or community matters) was organized by local leaders and villagers. The participating people in Mahapachayat raised their demands regarding solving the issue related to irrigation water and profitable value for agricultural products in front of the state government and Dam management authority.

Fig. 14.2 **a** Local respondent, **b** gathering of local people for Mahapanchayat

14.3 Results and Discussion

The bird diversity survey conducted at Bisalpur Wetland provided a detailed overview of the bird species in their catchment area. The species were classified into groups like waterfowl, pigeons, shorebirds, and others, with reference to the eBird field checklist (Bisalpur Dam Water Catchment Area, Tonk, Rajasthan, India, eBird Hotspot). This method helped categorize the birds and understand their conservation status and any potential threats to their populations.

14.3.1 Water Birds Status

A wide variety of waterfowl species was observed at the Bisalpur Wetland (Table 14.2). It highlights the abundance of fauna in the Bisalpur catchment area. However, the recorded species, such as the Bar-headed Goose, is classified as "Least Concern" on the IUCN Red List, but its population is declining. This decline raises concerns about the species' long-term survival. The Common Pochard sees a similar pattern of declining populations, likewise classified as "Vulnerable," highlighting the necessity for concentrated conservation initiatives. On the other hand, the Common Shelduck seems to be doing very well in its wetland home, as its population trend IUCN status shows positive. Additionally, groups of species like the Cotton Pygmy-Goose are observed in the study area, showing that the wetland is good for many types of waterfowl.

14.3.2 Pigeons and Doves

The wetland supports common species like the Eurasian Collared Dove (*Streptopelia decaocto*) and Rock Pigeon (*Columba livia*), both of which are classified as "Least Concern" (Table 14.3). However, these birds face falling population trends globally, which could be attributed to habitat changes or other environmental stresses. Monitoring these species is critical for learning how they adapt to changing environments.

14.3.3 Cuckoos

Several cuckoo species, including the Greater Coucal (*Centropus sinensis*), were seen (Table 14.4). All cuckoo species recorded are rated as "Least Concern" and show stable populations in the IUCN list, indicating that they are not currently endangered.

Table 14.2 Diversity of waterfowl at Bisalpur Wetland

S. No.	Common name	Scientific name	Diversity status	IUCN red list status	IUCN population trend
1	Bar-Headed Goose	*Anser indicus*	Occ	Least concern	Decreasing
2	Common Pochard	*Aythya ferina*	C	Vulnerable	Decreasing
3	Common Shelduck	*Tadorna tadorna*	Lc	Least concern	Increasing
4	Cotton Pygmy-Goose	*Nettapus coromandelianus*	Lc	Least concern	Stable
5	Eurasian Wigeon	*Mareca penelope*	Occ	Least concern	Decreasing
6	Ferruginous Duck	*Aythya nyroca*	Lc	Near threatened	Decreasing
7	Gadwall	*Mareca strepera*	C	Least concern	Increasing
8	Garganey	*Spatula querquedula*	Occ	Least concern	Decreasing
9	Graylag Goose	*Anser anser*	Occ	Least concern	Increasing
10	Green-Winged Teal	*Anas crecca*	Lc	Least concern	Unknown
11	Indian Spot-billed Duck	*Anas poecilorhyncha*	Vc	Least concern	Decreasing
12	Knob-Billed Duck	*Sarkidiornis melanotos*	Lc	Least concern	Decreasing
13	Lesser Whistling—Duck	*Dendrocygna javanica*	Lc	Least concern	Decreasing
14	Mallard	*Anas platyrhynchos*	C	Least concern	Increasing
15	Northern Pintail	*Anas acuta*	A	Least concern	Decreasing
16	Northern Shoveler	*Spatula clypeata*	Lc	Least concern	Decreasing
17	Red-Crested Pochard	*Netta rufina*	C	Least concern	Unknown
18	Ruddy Shelduck	*Tadorna ferruginea*	Lc	Least concern	Unknown
19	Tufted Duck	*Aythya fuligula*	Lc	Least concern	Stable

Note A = Abundant (100 birds), Vc = Very common (51–100), C = Common (11–50), Lc = Less common (1–10), Occ = Occasional (one or more stray birds spotted once in a while)

This is a good indicator since it indicates that the wetland and other habitat regions maintain a balance for species requiring certain ecological conditions.

Table 14.3 Diversity of Pigeons and Doves at Bisalpur Wetland

S. No.	Common name	Scientific name	Diversity status	IUCN red list status	IUCN population trend
1	Eurasian Collared-Dove	*Streptopelia decaocto*	Lc	Least concern	Decreasing
2	Red Collared-Dove	*Streptopelia tranquebarica*	Lc	Least concern	Decreasing
3	Rock Pigeon	*Columba livia*	Lc	Least concern	Decreasing
4	Spotted Dove	*Spilopelia chinensis*	Lc	Least concern	Increasing

Table 14.4 Diversity of Cuckoos at Bisalpur Wetland

S. No.	Common name	Scientific name	Diversity status	IUCN red list status	IUCN population trend
1	Common Hawk-Cuckoo	*Hierococcyx varius*	Lc	Least concern	Stable
2	Greater Coucal	*Centropus sinensis*	Lc	Least concern	Stable
3	Pied Cuckoo	*Clamator jacobinus*	Lc	Least concern	Stable
4	Sirkeer Malkoha	*Taccocua leschenaultii*	Lc	Least concern	Stable

14.3.4 Rails, Gallinules, and Allies

The wetland is home to birds such as the White-breasted Waterhen (*Amaurornis phoenicurus*) and the Eurasian Coot (*Fulica atra*). Although the populations of these species are stable, additional information is required to comprehend the population patterns of others, like the White-breasted Waterhen (Table 14.5). If further conservation measures are required, this will be ascertained with the aid of ongoing monitoring.

14.3.5 Shorebirds

Table 14.6 showing the diversity of shorebirds at Bisalpur Wetland gives information about the various species found in the area, their conservation status, and global population trends. Most shorebird species, including the Black-winged Stilt, Common Greenshank, and Green Sandpiper, are classified as "Least Concern" on the IUCN

Table 14.5 Diversity of Rails, Gallinules, and Allies at Bisalpur Wetland

S. No.	Common name	Scientific name	Diversity status	IUCN red list status	IUCN population trend
1	Brown Crake	*Zapornia akool*	Lc	Least concern	Unknown
2	Eurasian Coot	*Fulica atra*	Lc	Least concern	Stable
3	Eurasian Moorhen	*Gallinula chloropus*	Lc	Least concern	Stable
4	White-breasted Waterhen	*Amaurornis phoenicurus*	Lc	Least concern	Unknown

Red List, indicating that they are not in imminent danger of extinction. The population patterns for these species differ, with some, such as the Black-winged Stilt and Green Sandpiper, exhibiting an increase in numbers, which is a good indicator of the wetland's ecology. However, certain species, such as the Common Sandpiper, Common Snipe, and Indian Thick-knee, are suffering population decreases, indicating potential challenges to their survival, probably due to habitat loss or environmental stress. The Great Thick-knee, which is classified as "Near Threatened," likewise has a declining population, underlining the need for conservation efforts to prevent further loss. Population trends remain unknown for several species, such as the Common Redshank, Red-wattled Lapwing, and Small Pratincole, indicating a data gap that necessitates additional research to fully estimate their conservation requirements.

14.3.6 Storks, Cormorants, and Anhingas

Species like the Painted Stork (*Mycteria leucocephala*) and the Great Cormorant (*Phalacrocorax carbo*) have shown increasing populations trend in IUCN status and are listed as "Least Concern" with all other observed species (Table 14.7). These birds are important for maintaining the ecological balance of the wetland, as they are top predators or play key roles in nutrient cycling.

14.3.7 Herons, Ibis, and Allies

The recorded bird species, Black-headed Ibis (*Threskiornis melanocephalus*), classified as "Near Threatened," was documented with a decreasing population trend in IUCN Red list status (Table 14.8). This calls for conservation measures to ensure its survival in the region. Other species, such as the Great White Egret (*Ardea alba*),

Table 14.6 Diversity of Shorebirds at Bisalpur Wetland

S. No.	Common name	Scientific name	Diversity status	IUCN red list status	IUCN population trend
1	Black-Winged Stilt	*Himantopus himantopus*	Lc	Least concern	Increasing
2	Common Greenshank	*Tringa nebularia*	Lc	Least concern	Stable
3	Common Redshank	*Tringa totanus*	Lc	Least concern	Unknown
4	Common Sandpiper	*Actitis hypoleucos*	Lc	Least concern	Decreasing
5	Common Snipe	*Gallinago gallinago*	Lc	Least concern	Decreasing
6	Great Thick-Knee	*Esacus recurvirostris*	C	Near threatened	Decreasing
7	Green Sandpiper	*Tringa ochropus*	Lc	Least concern	Increasing
8	Indian Thick-Knee	*Burhinus indicus*	C	Least concern	Decreasing
9	Little Ringed Plover	*Charadrius dubius*	Lc	Least concern	Stable
10	Red-wattled Lapwing	*Vanellus indicus*	Lc	Least concern	Unknown
11	Small Pratincole	*Glareola lactea*	Lc	Least concern	Unknown
12	Spotted Redshank	*Tringa erythropus*	Lc	Least concern	Stable
13	Yellow-wattled Lapwing	*Vanellus malabaricus*	Lc	Least concern	Stable

Table 14.7 Diversity of Storks, Cormorants, and Anhingas at Bisalpur Wetland

S. No.	Common name	Scientific name	Diversity status	IUCN red list status	IUCN population trend
1	Asian Openbill	Anastomus oscitans	C	Least concern	Unknown
2	Great Cormorant	*Phalacrocorax carbo*	Lc	Least concern	Increasing
3	Indian Cormorant	*Phalacrocorax fuscicollis*	Lc	Least concern	Unknown
4	Little Cormorant	*Microcarbo niger*	Lc	Least concern	Unknown
5	Painted Stork	*Mycteria leucocephala*	C	Least concern	Increasing

Table 14.8 Diversity of Herons, Ibis, and Allies at Bisalpur Wetland

S. No.	Common name	Scientific name	Diversity status	IUCN red list status	IUCN population trend
1	Black Bittern	*Ixobrychus flavicollis*	Lc	Least concern	Decreasing
2	Black-Crowned Night Heron	*Nycticorax nycticorax*	Lc	Least concern	Decreasing
3	Black-Headed Ibis	*Threskiornis melanocephalus*	Lc	Near threatened	Decreasing
4	Gray Heron	*Ardea cinerea*	Lc	Least concern	Unknown
5	Great White Egret	*Ardea alba*	C	Least concern	Unknown
6	Indian Pond-Heron	*Ardeola grayii*	C	Least concern	Unknown
7	Little Egret	*Egretta garzetta*	A	Least concern	Increasing
8	Medium Egret	*Ardea intermedia*	C	Least concern	Decreasing

have unknown population trends, highlighting the need for more detailed studies to assess their conservation status.

14.3.8 Owls, Hornbills, Kingfisher, Vultures, Hawks, and Allies

Table 14.9, highlights various birds of prey and related species. Most of the species listed, such as the Black-winged Kite (*Elanus caeruleus*), Crested Serpent-Eagle (*Spilornis cheela*), and White-eyed Buzzard (*Butastur teesa*), are classified as "Least Concern" by the IUCN and have stable population trends. These species are vital for controlling prey populations and maintaining ecological balance in the wetland. However, the Brown Fish-Owl (*Ketupa zeylonensis*) has a decreasing population trend, which may require monitoring and conservation efforts to prevent further decline. For the Common Kingfisher (*Alcedo atthis*), the population trend remains unknown (as per the IUCN record), indicating a gap in data and the need for further study at the global and local levels to assess its conservation status.

14.3.9 Bulbuls, Crows, Starlings, Mynas, and Ravens

Some of the most common passerine species found near the marsh. The majority of the species, such as the Common Babbler (*Argya caudata*) and House Crow (*Corvus*

Table 14.9 Recorded diversity of Owls, Hornbills, Kingfisher, Vultures, Hawks, and Allies at Bisalpur Wetland

S. No.	Common name	Scientific name	Diversity status	IUCN red list status	IUCN population trend
1	Black-Winged Kite	*Elanus caeruleus*	Lc	Least concern	Stable
2	Brown Fish-Owl	*Ketupa zeylonensis*	Lc	Least concern	Decreasing
3	Common Kingfisher	*Alcedo atthis*	C	Least concern	Unknown
4	Crested Serpent-Eagle	*Spilornis cheela*	Lc	Least concern	Stable
5	Indian Gray Hornbill	*Ocyceros birostris*	C	Least concern	Stable
6	Long-Legged Buzzard	*Buteo rufinus*	Lc	Least concern	Stable
7	Short-Toed Snake-Eagle	*Circaetus gallicus*	Lc	Least concern	Stable
8	White-Eyed Buzzard	*Butastur teesa*	Lc	Least concern	Stable

splendens), are classified as "Least Concern" and have stable population trends, indicating that they are thriving in the area (Table 14.10). Some species, such as the Common Myna (*Acridotheres tristis*) and Indian Pied Starling (*Gracupica contra*), have growing populations.

Table 14.10 Recorded diversity of Bulbuls, Crows, Starling, Mynas and Ravens at Bisalpur Wetland

S. No.	Common name	Scientific name	Diversity status	IUCN red list status	IUCN population trend
1	Common Babbler	*Argya caudata*	Lc	Least concern	Stable
2	Common Myna	*Acridotheres tristis*	C	Least concern	Increasing
3	House Crow	*Corvus splendens*	C	Least concern	Stable
4	Indian Pied Starling	*Gracupica contra*	C	Least concern	Increasing
5	Red-Vented Bulbul	*Pycnonotus cafer*	Lc	Least concern	Increasing
6	Rosy Starling	*Pastor roseus*	C	Least concern	Unknown

14.3.10 Diversity of Old-World Flycatchers

Table 14.11 focuses on flycatchers and other small birds. All of the mentioned species, including the Black Redstart (*Phoenicurus ochruros*), Blue Rock-Thrush (*Monticola solitarius*), and Indian Robin (*Copsychus fulicatus*), are classed as "Least Concern" by the IUCN, with stable populations. These little insect-eating birds help to keep the ecology balanced by regulating insect populations.

The image depicting the IUCN Red List status of bird species at Bisalpur Wetland shows a visual breakdown of the conservation categories for the observed species. Most birds are designated as "Least Concern," which means they are not currently in danger of extinction. However, a few species are listed as "Near Threatened" or "Vulnerable," indicating that they are more likely to become endangered if population reductions continue Fig. 14.3. The image helps to highlight which species deserve immediate conservation attention and acts as a guide for prioritizing future wetland protection efforts.

Table 14.11 Recorded diversity of Old-World Flycatchers at Bisalpur Wetland

S. No.	Common name	Scientific name	Diversity status	IUCN red list status	IUCN population trend
1	Black Redstart	*Phoenicurus ochruros*	Lc	Least concern	Stable
2	Blue Rock-Thrush	*Monticola solitarius*	Lc	Least concern	Stable
3	Indian Robin	*Copsychus fulicatus*	Lc	Least concern	Stable
4	Oriental Magpie-Robin	*Copsychus saularis*	Lc	Least concern	Stable

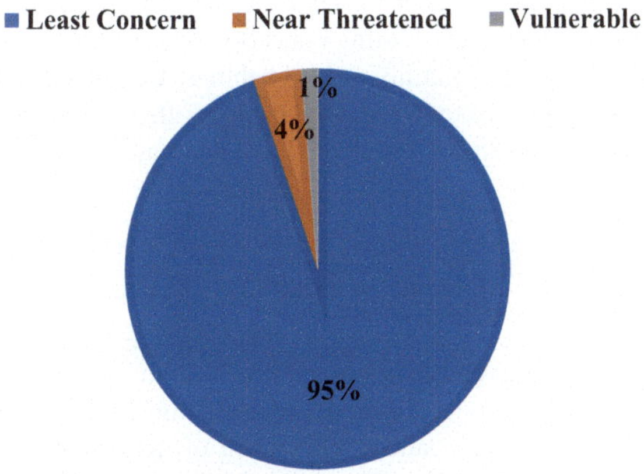

Fig. 14.3 IUCN red list status of total bird species at Bisalpur wetland

Several migrating bird species were spotted at Rajasthan's Bisalpur Wetland, highlighting the wetland's importance as a critical stopover and wintering destination for these birds. Migratory birds are important indicators of wetland ecological health, as they contribute to biodiversity and signal environmental changes (Liu et al. 2015; Webb et al. 2010; Yao et al. 2020). Among the recorded species, Bar-headed Goose (*Anser indicus*), Common Pochard (*Aythya ferina*), Garganey (*Spatula querquedula*), Northern Pintail (*Anas acuta*), and Northern Shoveler (*Spatula clypeata*) are notable migratory waterfowl that travel to the wetland during winter from regions as far as Central Asia and Siberia. These birds rely on wetlands like Bisalpur for food and shelter during their long migratory journeys.

Additionally, shorebirds such as the Black-winged Stilt (*Himantopus himantopus*) and Common Sandpiper (*Actitis hypoleucos*) are also migratory species, contributing to the wetland's diversity. The presence of these migratory species highlights the significance of the wetland as part of global bird migration networks, underlining the need for conservation efforts to protect these habitats, which support both resident and migratory birds.

Thus, this study underlines the need for ongoing monitoring and habitat protection to ensure the long-term survival of bird species at Bisalpur Wetlands. Effective conservation efforts should be designed to help declining and at-risk species while simultaneously preserving circumstances that sustain healthy populations.

14.3.11 Habitat Degradation

Industrialization, human contamination, and climate change are the primary causes of wetlands habitat degradation (Cao et al. 2020; Casazza et al. 2021; Donnelly et al. 2022). The Bisalpur Wetland has experienced significant habitat degradation in recent years, as illustrated by various environmental changes and human activities that directly impact bird populations and other wildlife. An analysis of land cover changes and direct observations from 2015 to 2023 reveals the scale and specifics of this decline, underscoring concerns for the wetland ecosystem's sustainability and the long-term viability of bird habitats. Compared to habitat fragmentation's more varied and weaker effects, habitat loss often significantly negatively impacts species distribution and abundance (Gutzwiller and Flather 2011; Quesnelle et al. 2013).

14.3.12 Vegetation Loss

In 2015, the vegetation around the Bisalpur Wetland was dense, providing important cover and resources for bird species (Fig. 14.4). By 2023, this area had undergone significant degradation, with much of the original vegetation replaced by barren ground or built areas (Fig. 14.5). The loss of greenery is essential because birds rely on vegetated regions for nesting, shelter, and food. The decline of native plant species disturbs food webs, reducing insect populations that feed many birds. Forested regions surrounding wetlands act as crucial buffer zones, providing shade, lowering runoff, and filtering pollutants before they enter the wetland (Nair et al. 2024; Walton et al. 2020). Trees and native plants in these locations serve as habitat and food for a variety of bird species. Reduced forest cover, as seen in many wetland areas, causes increased erosion, sedimentation, and nutrient runoff, disrupting the wetland's ecological balance and reducing habitat quality for birds (Irvine et al. 2022; Wu et al. 2024; Xu et al. 2023).

14.3.13 Construction and Land Use Changes

Construction in the Bisalpur basin has also contributed significantly to habitat degradation (Fig. 14.6). Informal building complexes have extended across previously undeveloped areas, diminishing available habitat for wildlife (Fig. 14.7). Furthermore, the increase in unplanned structures, many of which lack effective waste management, introduces pollution, degrading the quality of the wetland ecosystem (Danso et al. 2021; Sarkar and Maji 2022).

14.3.14 Deforestation and Pollution

The local community's reliance on firewood for fuel has accelerated degradation around the wetland (Figs. 14.8 and 14.9). Tree felling for firewood directly harms the habitat by removing bird roosting spots and native vegetation, which maintains a larger ecological balance. Increased human activity depletes resources and causes disruptions in avian activities such as breeding and foraging (Cao et al. 2024; Duan et al. 2022). Solid trash accumulation and algae blooms pose additional risks in the Bisalpur Wetland. Human activities, such as garbage disposal in and around wetlands, contribute to pollution, degrade water quality, and introduce poisons into the environment. Algal blooms, which are most likely caused by nutrient runoff, deplete oxygen levels in the water, endangering fish populations and other aquatic species that provide food for birds. These blooms can also release toxic poisons to birds and humans, reducing the wetland's ecological health.

However, the degradation of the Bisalpur Wetland poses severe threats to resident and migratory bird species, including the following major concerns:

- **Habitat Loss and Fragmentation**: As vegetation cover decreases and built-up areas expand, bird habitats become fragmented, reducing the available space and making it difficult for birds to find food, shelter, and mates.
- **Food Source Decline**: Reduced plant cover disrupts the food chain, impacting insect populations and aquatic life.

Fig. 14.4 Status of wetland surrounding vegetation in 2015 (full greenery in the highlighted zone)

Fig. 14.5 Status of wetland surrounding vegetation in 2023 (total degradation in the highlighted zone)

Fig. 14.6 Land degradation and construction work in Bisalpur catchment

Fig. 14.7 Constructed informal housing and firewood collection

Fig. 14.8 Firewood collection for cooking fuel by wetland surrounding people

- **Increased Human Disturbance**: The rise in human presence and activities around the wetland disturbs the natural behavior of birds, potentially reducing breeding success and survival rates.
- **Pollution and Toxins**: Solid waste and algal blooms introduce toxins.

14.3.15 Wetland Ecosystem Services

14.3.15.1 General, Social and Demographic Environment

Rajasthan, India's largest state, faces significant water scarcity due to its harsh climate and geographical conditions. With 85 out of 142 desert blocks, it struggles to meet water demand and agricultural and non-agricultural needs (Goyal and Gaur 2022; Gummagolmath and Anand 2023; Pointet 2022). Dams have played a

Fig. 14.9 Human disturbance, algal bloom, solid waste, and land degradation at wetland

crucial role in addressing these challenges by providing water for drinking, irrigation, urban needs, hydroelectric power generation, and wildlife conservation, delivering substantial social and economic benefits (Hoque et al. 2022; Laxmi and Goyal 2023; Saad and Gamatié 2020). However, the Bisalpur wetland was created in the 1990s after constructing a dam on Banas River. In compliance with the authority conferred by Rule 7 of the Wetland (Conservation and Management) Rules, 2017, and the stipulations of the Environment (Protection) Act, 1986 (Central Act No. 29 of 1986), the State Government formally designates the Bisalpur wetland (http://rmsc.health.rajasthan.gov.in/content/dam/environment/env-swa/FinalWetlandGazetteNotification/Bisalpur%20wetland%20Tonk.pdf). It is located in the Toda Raisingh tehsil of Tonk district. Four nearby villages, Rajmahal, Rampura, Botunda, and Tharoli, are situated within a 5-km radius downstream from the dam. Figures 14.10, 14.11, 14.12 and 14.13 show the demographic details of interconnected villages in Bisalpur wetland; the figures' statistical data were compiled

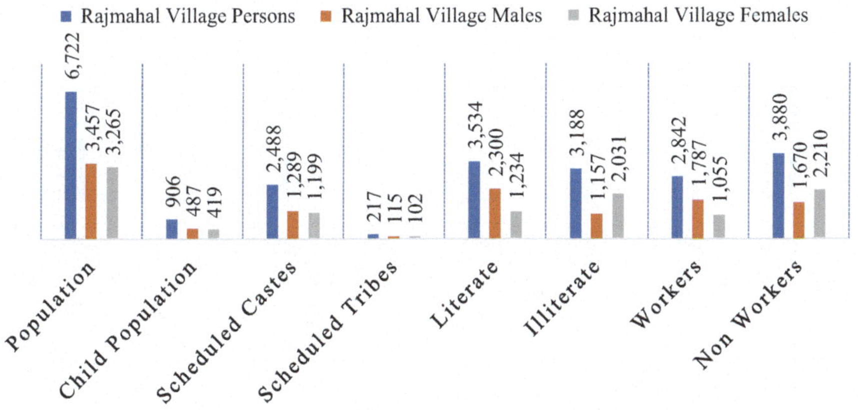

Fig. 14.10 Census status of the village Rajamahal

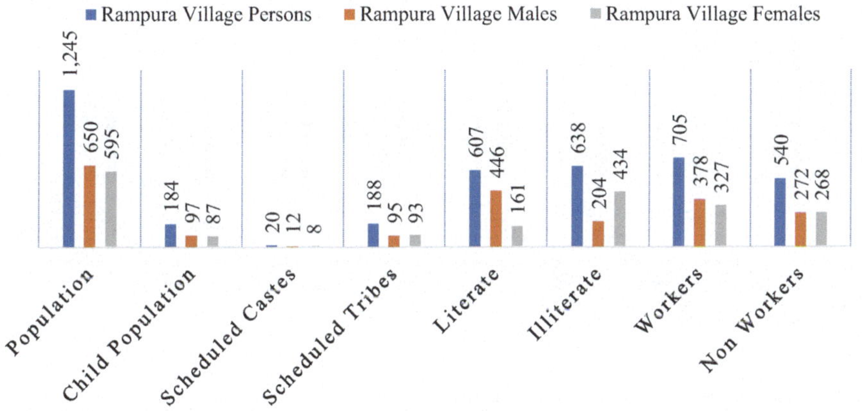

Fig. 14.11 Census status of the village Rampura

from India Primary Census Abstract 2011 (Visualizations | Government of India). However, the investigation of the Bisalpur wetland identified major social activities carried out by locals, including as visiting religious places, performing traditional rites on the marsh, collecting wood for fuel, and using water for household purposes. These activities highlight the close relationship between the wetland and the cultural practices of the surrounding populations.

14.3.15.2 Interpretation of Various Goods and Services in and Around Wetland

Wetlands are key ecosystems that provide many goods and services, which play an essential role in contributing to local and global socio-economic values (Ahmad et al.

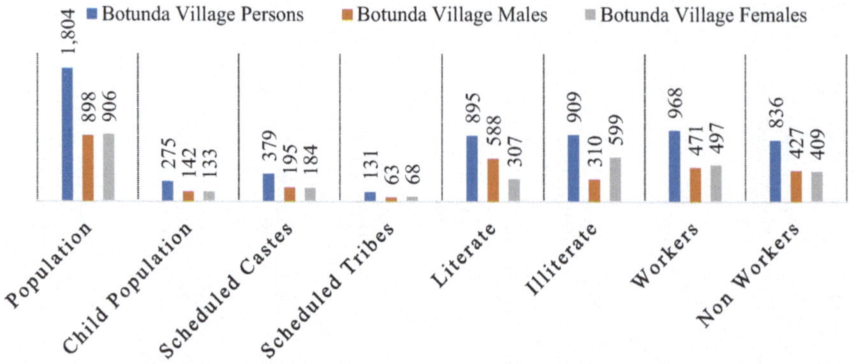

Fig. 14.12 Census status of the village Botunda

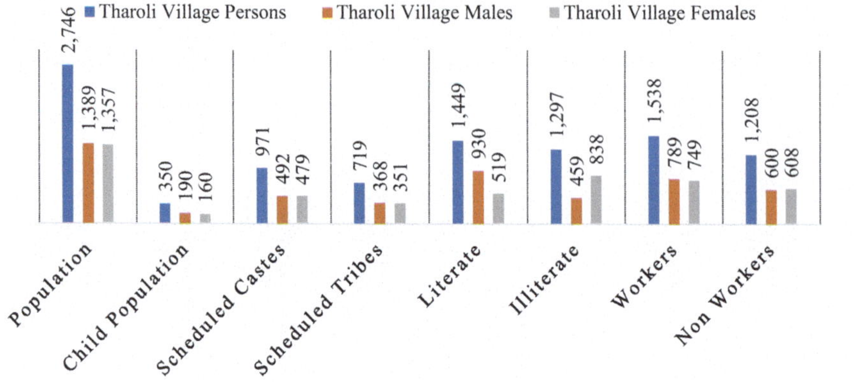

Fig. 14.13 Census status of the village Tharoli

2024; Singh et al. 2023; Slagter et al. 2020; Smalling et al. 2021). During the study period, it was observed that the Bisalpur Wetland supports numerous use and non-use values. These include providing food, wood, irrigation water, wildlife habitat, drinking water, flood and drought mitigation in the desert region, aquatic habitats, fertile agricultural land, and fulfilling religious and spiritual needs. Additionally, it holds historical significance and meets many primary needs of the local households. Moreover, the wetland values data was compiled from literature, secondary sources, and field surveys. The various services offered by the Bisalpur Wetland are detailed below.

14.3.15.3 Use Value

a. **Direct Use Value**

Fish

The Bisalpur wetland, which is supported by a constant supply of fresh water from the Banas River and extensive submerged vegetation across varied habitats, sustains a good fishery and contributes significantly to Rajasthan's economy. The previous study recorded a total of 21 species of fish on the Bisalpur wetland (Banyal and Kumar 2015). People can easily interpret the fish's movement in transparent shallow water. Moreover, the Rajasthan State Wetland Authority prohibited the public from feeding and capturing fish. Where registered commercial fishing activities are permitted on the Bisalpur wetland. However, the local communities that live near the wetlands rely significantly on fish as a staple of their diet, considering it their principal source of nutrition. Local people are directly or indirectly involved such as working as boat operators or engaging in tasks like setting up and retrieving fishing nets in fishing activity with the Bisalpur Wetland Fishery Authority (RR Fisheries, Figs. 14.14 and 14.15). Fishing is an essential part of their livelihood and sustenance because it is ingrained in their everyday lives and cultural customs.

Agriculture

The Bisalpur wetland plays a crucial role in supporting a large population by providing irrigation water to the downstream regions of the Tonk district. A survey

Fig. 14.14 Fish collection, storage, and supply facilities at the Bisalpur wetland

Fig. 14.15 Fisheries and fish catching boat near wetland

revealed that the wetland consistently supplies sufficient water for various agricultural practices, particularly in areas receiving dam water discharge (Figs. 14.16 and 14.17). Agriculture remains the primary occupation for people living in the villages surrounding the wetland (Fig. 14.18). This is because, unlike other arid and semi-arid regions of Rajasthan, where access to adequate irrigation water is a persistent challenge, the Bisalpur wetland ensures a reliable water supply. Additionally, employment opportunities in these areas are limited, as most families depend on agriculture, and there are few alternative income sources due to the scarcity of industries in resettled areas.

Moreover, a report of the District Environment Plan (DEP_Tonk.pdf) highlights that Tonk district (in which Bisalpur wetland is located) is surrounded by five major districts of Rajasthan (Jaipur, Ajmer, Bhilwara, Bundi, Swai Madhopur) and animal husbandry and agriculture are the primary professions of the people living there (Desai 2009; Mahala et al. 2023; Sharma and Chakraborty 2021).

Fuel Wood and Fodder

The Bisalpur wetland provides significant ecosystem services, such as fuelwood and fooder. Figure 14.19, the wetlands support local populations' harvesting of wood, which is largely utilized for cooking fuel. Furthermore, the nearby pastureland provides valuable food for livestock. These services are critical to the local population's survival (Adhya and Banerjee 2022), especially in rural locations where other resources are few. This dependency emphasizes the wetland's role in meeting both energy and agricultural needs.

Fig. 14.16 Agriculture field within wetland boundary and their surrounding

Fig. 14.17 Agriculture activity in downwards, water discharge region of wetland

Fig. 14.18 Agriculture cropping in the downstream region of wetland

Fig. 14.19 Collected wood for cooking fuel by local people and surrounding pasture land

Recreation and Tourism

The recreation value of the wetland is defined by various activities such as eco-tourism, boating, fishing, cultural, aesthetic, and spiritual benefits, which contribute to the local economy (Xu and He 2022; Yu et al. 2018). However, the Rajasthan government forest department has started the LavKush Vaatika near the transition zone of wetland water and land (Fig. 14.20). Moreover, it is under construction, and some primary child gaming equipment has been installed, including trails that allow people to explore and appreciate the beauty of this wetland (Fig. 14.21). The Bisalpur wetland, with its beautiful mountainous landscape, the ancient Lord Shiva temple, and a fish aquarium (Fig. 14.22), attracts a number of visitors from various regions of Rajasthan state as well as the Nation, particularly during the rainy season (July to September). Aquariums have a variety of fish species such as L. cichlid, Green terror, Flowerhorn cichlid, Asian arowana, Red tail, Stingray, Oscar, Bala Shark, Kais fish, Angelfish, Gourami fish, Tinfoil barb, Pirate perch, and many of it is not in the living stage due to poor maintenance of Aquarium (Fig. 14.23).

b. **Indirect Use Value**

In-Filtration, Nutrient Recycling, Flood Control and Micro Climate

Wetlands offer excellent natural filtering capacity due to their unique ecological properties (Bam and Ireson 2019; Xu et al. 2024). The Bisalpur wetland benefits from the continuous inflow of water from the Banas River and supports a sufficient growth of vegetation, much of which is submerged in shallow water (Fig. 14.24). This environment creates an ideal habitat for various fish and mollusk species. During the survey, we also observed lichens and algae thriving on stones and dam pillars. The submerged plants play an important role in preventing eutrophication; they trap the sediment and absorb the nutrients from the water. These factors collectively contribute to nutrient recycling and demonstrate the wetland's natural self-purification capabilities. However, this wetland is located in the Aravalli range, surrounded by beautiful

Fig. 14.20 Allocation of LavKush Vatika and their entrance gate

Fig. 14.21 Gaming zone, hiking, and nature walks

Fig. 14.22 Fish aquarium and ancient Lord Shiva temple

Khamnor Hill and 48.31 km square of Bisalpur Conservation Reserve. It creates a micro-climate to sustain various biodiversity habitats and resting sites for tourists and local people. Moreover, this hill is under moderate slopy geographical conditions, especially the discharge region at the Dam side, making possible threats of flood hazard. Hence, the Bisalpur wetland plays an essential role in trapping and slowing down the flow of surface water and reducing the possibility of flooding in

Fig. 14.23 Inner portion of the aquarium

Fig. 14.24 Submerges, stones in shallow water, and surrounding vegetation at wetland

the downstream region of the wetland, especially in extreme weather conditions in the rainy season.

Fig. 14.25 Bisaldev Temple at wetland

Religious

When religious and historic structures are combined with nature, the result is a stunning location that draws tourists from all around. The Bisalpur wetland has a beautiful ancient Temple on the bank of the Dam (Fig. 14.25). The name of the temple is Bisaldeo temple. The temple was built using ancient Hindu architectural design. The temple is dedicated to Lord Shiva and is one of the ancient famous temples in Rajasthan state. This temple has significant social and religious importance. The temple was built in the twelfth century by the Chahamana king Vigraharaja IV, also known as Bisal Deo, and is listed by the Archaeological Survey of India as a "Monument of National Importance" Vigraharaja IV, a devout follower of Gokarna.

The internal wall of the temple is full of stone carving art and the ancient structure (Fig. 14.26). There are four small modern temples on the opposite side of the Dam gate (water discharge region), along with one Suny Dev Temple at the side of the Dam. However, due to the availability of a resting room, a small guest house, a parking zone, and an open walking street on Dam, this area is the primary attraction for visitors (Fig. 14.27).

However, these religious places located on or near Bisalpur wetland attract local visitors, especially during festivals, rituals, or pilgrimages (Fig. 14.28). This number of visitors can provide economic benefits to the surrounding communities in several ways such as boosting of selling in local shop, cultural tourisem and not only support local livelihoods but also contribute to preserving the cultural and spiritual significance of these areas.

Fig. 14.26 Internal view of Bisaldev Temple

Fig. 14.27 Temples at the opposite side of the Dam

Fig. 14.28 Road side shops and visitors

(II) Non-use Value:

Biodiversity

The Bisalpur wetland catchment area is rich in vegetation cover, especially the hilly patch of Bisalpur Conservation Reserve. The Bisalpur Dam's construction, rehabilitation, and improvement, funded by the World Bank, have led to the classification of this conservation reserve as a "Natural Habitat" under Environmental and Social Standards (World Bank Document). Moreover, the conservation has declared a protective landscape for the conservation and management of flora and fauna habitat. Also, the rights of the people who are living within the protected area are not affected and they benefit from some local primary requirements for conservation reserve resources, including wood, fodder, and habitat for domestic animals. The *Prosopis juliflora* predominates in this region and is typical of a dry deciduous forest. One of the most common hyper-accumulating plants is *P. juliflora*, *P. juliflora* is widely distributed in the fluoride-endemic desert and semiarid regions of Rajasthan, India. The floodplain region of this area soil has good fertility due to sediment deposition by the flow of the Banas River. Hence, a good soil cover sustains a variety of agricultural crops. An aquatic environment's productivity is directly proportional to its plankton density. The plankton species is an important source of food for small fishes and plays an important role in the food chain of aquatic ecosystems (Ariadi et al. 2022; Mo et al. 2014; Rasdi et al. 2020). Previous study recorded a good number of plankton and fish diversity in Bisalpur wetland water (Banyal and Kumar 2015; Summarwar 2012). In the current study, all over 75 species have been listed in the Bisalpur wetland, which shows a good diversity of avian species and attracts visitors in the winter session.

14.4 Conclusion

This study discovered that the Bisalpur Wetland provides a unique home for various birds during winter. According to the IUCN status, it has 95% least concern, 4% near threatened, and 1% vulnerable bird species. Migratory species like the Bar-headed Goose, Common Pochard, and Northern Pintail highlight the wetland's biological importance as a stopover and wintering location on major migration routes. These species rely on Bisalpur Wetland for basic resources such as food and shelter, demonstrating the wetland's critical role in maintaining global bird migration networks. The ecological health of the Bisalpur Wetland is inextricably tied to the well-being of these migratory birds, who serve as indicators of the general environmental status. Bisalpur Wetland degradation, primarily caused by human activity, has resulted in severe habitat loss, vegetation decline, and pollution, all affecting wetland bird populations. Quick action is required to protect and restore this habitat. Moreover, the Bisalpur wetland serves as a lifeline for the local population and ecosystem, offering essential goods and services. Its contribution to agriculture, fisheries, and water availability directly supports the livelihoods of surrounding communities, while its biodiversity and cultural significance enhance its value as a conservation site. However, effective management is required to address issues such as facility maintenance and flood risk, ensuring long-term sustainability. Preserving this wetland is vital for balancing ecological health, cultural heritage, and socio-economic benefits, making it an indispensable resource for the region.

Competing Interests The authors declare no potential conflicts of interest.

References

Aarif KM, Rubeena KA, Nefla A, Musilova Z, Musil P, Shaju SS, Muzaffar SB et al (2023) Heavy metals in wetlands of southwestern India: from sediments through invertebrates to migratory shorebirds. Chemosphere 345:140445

Adhya T, Banerjee S (2022) Impact of wetland development and degradation on the livelihoods of wetland-dependent communities: a case study from the lower gangetic floodplains. Wetlands 42(7):65

Ahmad M, Ahmad WS, Ahmad SN, Jamal S, Saqib M (2024) Tracing the roots of wetland degradation in India: a systematic review of anthropogenic drivers, ecological consequences and conservation strategies. GeoJournal 89(1):24

Ariadi H, Khristanto A, Soeprapto H, Kumalasari D, Sihombing JL (2022) Plankton and its potential utilization for climate resilient fish culture. Aquac Aquarium Conserv Legislation 15(4):2041–2051

Bachheti A, Arya AK, Tyagi S, Deepti (2023) An assessment of wetland and water-dependent avifaunal diversity in selected wetlands of Uttarakhand, Western Himalayas, India. Biol Bull 50(4):729–735

Bam EKP, Ireson AM (2019) Quantifying the wetland water balance: a new isotope-based approach that includes precipitation and infiltration. J Hydrol 570:185–200

Banyal HS, Kumar S (2015) Study on fish faunal diversity in the Bisalpur reservoir near Deoli city, Rajasthan. J Env Bio-Sci 29(2):499–501

Bhendekar G, Shinde A, Wankhade V (2024) Seasonal bird diversity of an urban lake with the history of eutrophication and restoration. Trop Ecol 65:592–608

Blackwell MSA, Pilgrim ES (2011) Ecosystem services delivered by small-scale wetlands. Hydrol Sci J 56(8):1467–1484

Cao B, Bai C, Xue Y, Yang J, Gao P, Liang H, Xu J et al (2020) Wetlands rise and fall: six endangered wetland species showed different patterns of habitat shift under future climate change. Sci Total Environ 731:138518

Cao Y, Ge Y, Wang S, Fan B, Liu C, You H, Zhang S et al (2024) Reconstructing the rapid transitions of ecosystems during the mid-late Holocene: a pollen record from Haixing wetland in Bohai Bay, North China. Quat Sci Rev 344:108973

Casazza ML, McDuie F, Jones S, Lorenz AA, Overton CT, Yee J, Thorne KM et al (2021) Waterfowl use of wetland habitats informs wetland restoration designs for multi-species benefits. J Appl Ecol 58(9):1910–1920

Costa MDP, Wartman M, Macreadie PI, Ferns LW, Holden RL, Ierodiaconou D, Nicholson E et al (2024) Spatially explicit ecosystem accounts for coastal wetland restoration. Ecosyst Serv 65:101574

Danso GK, Takyi SA, Amponsah O, Yeboah AS, Owusu RO (2021) Exploring the effects of rapid urbanization on wetlands: insights from the Greater Accra Metropolitan Area, Ghana. SN Soc Sci 1:1–21

Desai MD (2009) Contribution of livestock production system to farmer's livelihood in Western region of Maharashtra, M. Sc.(Agri.) thesis. University of Agricultural Sciences, Dharwad, India

Donnelly JP, Moore JN, Casazza ML, Coons SP (2022) Functional wetland loss drives emerging risks to waterbird migration networks. Front Ecol Evol 10:844278

Duan H, Yu X, Zhang L, Xia S, Liu Y, Mao D, Zhang G (2022) An evaluating system for wetland ecological risk: case study in coastal mainland China. Sci Total Environ 828:154535

Fluet-Chouinard E, Stocker BD, Zhang Z, Malhotra A, Melton JR, Poulter B, McIntyre PB (2023) Extensive global wetland loss over the past three centuries. Nature 614(7947):281–286

Goyal RK, Gaur MK (2022) The implications of climate change on water resources of Rajasthan. In: Hydro-meteorological extremes and disasters. Springer, pp 265–278

Gummagolmath KC, Anand A (2023) Potential agri-business in the Thar Desert. In: Natural resource management in the Thar Desert Region of Rajasthan. Springer, pp 273–311

Gupta PK, Patel JG, Bhanderi RJ, Parmar M, Rathore BP, Mishra I, Dhar D et al (2024) Indian wetlands: a high resolution remote sensing assessment and analysis. Space Applications Centre, ISRO Ahmedabad, India, pp 1–189. https://vedas.sac.gov.in/static/downloads/atlas/Wetlands/wetland_L4_atlas_12March2024.pdf

Gutzwiller KJ, Flather CH (2011) Wetland features and landscape context predict the risk of wetland habitat loss. Ecol Appl 21(3):968–982

Hoque MM, Islam A, Ghosh S (2022) Environmental flow in the context of dams and development with special reference to the Damodar Valley Project, India: a review. Sustain Water Resour Manage 8(3):62

Hu T, Liu J, Zheng G, Zhang D, Huang K (2020) Evaluation of historical and future wetland degradation using remote sensing imagery and land use modeling. Land Degrad Dev 31(1):65–80

Irvine K, Dickens C, Castello L, Bredin I, Finlayson CM (2022) Vegetated wetlands: from ecology to conservation management. In: Fundamentals of tropical freshwater wetlands. Elsevier, pp 589–639

Kadlec RH (2020) Hydrologic factors in wetland water treatment. In: Constructed wetlands for wastewater treatment. CRC Press, pp 21–40

Laxmi S, Goyal R (2023) Analysis of water resources of Bisalpur dam using time series forecasting models. In: Geospatial and Soft Computing Techniques. HYDRO 2021. Lecture Notes in Civil Engineering, vol 339. Springer, Singapore

Let M, Pal S (2023) Socio-ecological well-being perspectives of wetland loss scenario: a review. J Environ Manage 326(Pt B):116692
Li D, Tian P, Luo Y, Dong B, Cui Y, Khan S (2021) Importance of stopping groundwater irrigation for balancing agriculture and wetland ecosystem. Ecol Ind 127:107747
Liu J, Liang J, Yuan X, Zeng G, Yuan Y, Wu H, Li X (2015) An integrated model for assessing heavy metal exposure risk to migratory birds in wetland ecosystem: a case study in Dongting Lake Wetland, China. Chemosphere 135:14–19
Londe DW, Davis CA, Loss SR, Robertson EP, Haukos DA, Hovick TJ (2024) Climate change causes declines and greater extremes in wetland inundation in a region important for wetland birds. Ecol Appl 34(2):e2930
Mahala A, Mallick PK, Sharma RC, Kumar A, Kumari S (2023) Health management practices of Malpura sheep flock in Tonk district of Rajasthan under mega sheep seed project. Pharma Innov J 12(2):3068–3072
Marasinghe S, Perera P, Newsome D, Kotagama S, Jayasinghe C (2022) Understanding the impact of recreational disturbance caused by motor vehicles on waterbirds: a case study from the Bundala Wetland, Sri Lanka. J Coast Conserv 26(2):6
Mattson M, Sousa D, Quandt A, Ganster P, Biggs T (2024) Mapping multi-decadal wetland loss: comparative analysis of linear and nonlinear spatiotemporal characterization. Remote Sens Environ 302:113969
Mishra AP, Kumar S, Patra R, Kumar A, Sahu H, Chandra N, Alshehri F et al (2023) Physico-chemical parameters of water and its implications on avifauna and habitat quality. Sustainability 15(12):9494
Mishra H, Bano F, Prateek, Mishra A, Kumar A, Kumar V (2024) Elucidation of diversity and habitat utilization of waterbirds in Khajuha Wetland, Northern India. Biol Bull 51(3):800–812
Mo WY, Cheng Z, Choi WM, Man YB, Liu Y, Wong MH (2014) Application of food waste based diets in polyculture of low trophic level fish: effects on fish growth, water quality and plankton density. Mar Pollut Bull 85(2):803–809
Naik R, Sharma LK (2022) Monitoring migratory birds of India's largest shallow saline Ramsar site (Sambhar Lake) using geospatial data for wetland restoration. Wetl Ecol Manage 30(3):477–496
Nair JV, Verma J, Shinde VR (2024) Enriching and maintaining the riparian buffer zone. In: Managing urban rivers. Elsevier, pp 161–180
Panda BP, Das AK, Jena SK, Mahapatra B, Dash AK, Pradhan A, Parida SP (2021) Habitat heterogeneity and seasonal variations influencing avian community structure in wetlands. J Asia Pac Biodivers 14(1):23–32
Pereira LS, Paredes P, Espírito-Santo D (2024) Crop coefficients of natural wetlands and riparian vegetation to compute ecosystem evapotranspiration and the water balance. Irrig Sci 42:1171–1197
Pointet T (2022) The United Nations world water development report 2022 on groundwater, a synthesis. LHB 108(1):2090867
Qu F, Wang S, Wang W, Liu S, Li S, Liu H, Zhang Z (2023) Macrobenthic community structure of Rudong coastal wetland, China: the impact of invasive Spartina alterniflora and its implication for migratory bird conservation. Wetlands Ecol Manage 31(1):159–168
Quesnelle PE, Fahrig L, Lindsay KE (2013) Effects of habitat loss, habitat configuration and matrix composition on declining wetland species. Biol Conserv 160:200–208
Rahman MM, Jiang Y, Irvine K (2018) Assessing wetland services for improved development decision-making: a case study of mangroves in coastal Bangladesh. Wetlands Ecol Manage 26(4):563–580
Rajpar MN, Ahmad S, Zakaria M, Ahmad A, Guo X, Nabi G, Wanghe K (2022) Artificial wetlands as alternative habitat for a wide range of waterbird species. Ecol Ind 138:108855
Rasdi NW, Arshad A, Ikhwanuddin M, Hagiwara A, Yusoff FM, Azani N (2020) A review on the improvement of cladocera (Moina) nutrition as live food for aquaculture: using valuable plankton fisheries resources. J Environ Biol 41:1239–1248

Saad A, Gamatié A (2020) Water management in agriculture: a survey on current challenges and technological solutions. IEEE Access 8:38082–38097

Sarkar D, Maji N (2022) Status and threats of wetland change in land use pattern and planning: impact of land use patterns and urbanization. In: Handbook of research on monitoring and evaluating the ecological health of wetlands. IGI Global, pp 106–127

Scholte SSK, Todorova M, Van Teeffelen AJA, Verburg PH (2016) Public support for wetland restoration: what is the link with ecosystem service values? Wetlands 36:467–481

Sekey W, Obirikorang KA, Boadu KB, Gyampoh BA, Nantwi-Mensah A, Israel EY, Adjei-Boateng D (2023) Mangrove plantation and fuelwood supply chain dynamics in the Keta Lagoon Complex Ramsar Site, Ghana. Wetlands Ecol Manage 31(1):143–157

Sharma S, Chakraborty D (2021) Traditional medicinal plants used by tribal communities in Tonk district, Rajasthan. Plant Sci Today 8(1):225–228

Singh R, Mishra A (2019) Physico-chemical characteristics of Asan wetland with reference to Avian and Molluscan diversity, Doon Valley (Uttarakhand), India. Int Res J Environ Sci 8(3):1–11

Singh R, Saritha V, Pande CB (2023) Monitoring of wetland turbidity using multi-temporal Landsat-8 and Landsat-9 satellite imagery in the Bisalpur wetland, Rajasthan, India. Environ Res 241:117638

Singh R, Saritha V, Pande CB (2024) Dynamics of LULC changes, LST, vegetation health and climate interactions in Wetland buffer zone: a remote sensing perspective. Phys Chem Earth Parts a/b/c 135:103660

Singh R, Saritha V, Mishra AP, Pande CB, Sahu H (2025) A comprehensive analysis of water quality index in a wetland ecosystem supporting drinking water to major cities in Rajasthan, India. J Cleaner Prod 487:144593

Slagter B, Tsendbazar N-E, Vollrath A, Reiche J (2020) Mapping wetland characteristics using temporally dense Sentinel-1 and Sentinel-2 data: a case study in the St. Lucia wetlands, South Africa. Int J Appl Earth Obs Geoinf 86:102009

Smalling KL, Rowe JC, Pearl CA, Iwanowicz LR, Givens CE, Anderson CW, Adams MJ et al (2021) Monitoring wetland water quality related to livestock grazing in amphibian habitats. Environ Monit Assess 193(2):58

Summarwar S (2012) Studies on plankton diversity in Bisalpur reservoir. Int J Life Sci Bot Pharm Res 1(4):65–72

Walton CR, Zak D, Audet J, Petersen RJ, Lange J, Oehmke C, Hoffmann CC et al (2020) Wetland buffer zones for nitrogen and phosphorus retention: impacts of soil type, hydrology and vegetation. Sci Total Environ 727:138709

Webb EB, Smith LM, Vrtiska MP, Lagrange TG (2010) Community structure of wetland birds during spring migration through the Rainwater Basin. J Wildl Manage 74(4):765–777

Wu Q, Wang L, Wang T, Ruan Z, Du P (2024) Spatial–temporal evolution analysis of multi-scenario land use and carbon storage based on PLUS-InVEST model: a case study in Dalian, China. Ecol Indic 166:112448

Xi Y, Peng S, Ciais P, Chen Y (2021) Future impacts of climate change on inland Ramsar wetlands. Nat Clim Change 11(1):45–51

Xi Y, Peng S, Liu G, Ducharne A, Ciais P, Prigent C, Tang X et al (2022) Trade-off between tree planting and wetland conservation in China. Nat Commun 13(1):1967

Xu S, He X (2022) Estimating the recreational value of a coastal wetland park: application of the choice experiment method and travel cost interval analysis. J Environ Manage 304:114225

Xu Z, Dong B, Wang C, Gao X, Xu H, Wei Z, Liu X et al (2023) Construction of international important wetland White-headed crane ecological corridor in Chongming Dongtan, China. Ecol Indic 149:110156

Xu D, Bisht G, Tan Z, Sinha E, Di Vittorio AV, Zhou T, Leung LR et al (2024) Climate change will reduce North American inland wetland areas and disrupt their seasonal regimes. Nat Commun 15(1):2438

Yadav A, Rai D (2024) Diversity and IUCN status of Winter Migratory Avifauna of Mandothi Wetlands, Haryana (India). Biol Bull 51:1197–1205

Yao S, Li X, Liu C, Zhang J, Li Y, Gan T, Kuang W et al (2020) New assessment indicator of habitat suitability for migratory bird in wetland based on hydrodynamic model and vegetation growth threshold. Ecol Indic 117:106556

Yu X, Mingju E, Sun M, Xue Z, Lu X, Jiang M, Zou Y (2018) Wetland recreational agriculture: balancing wetland conservation and agro-development. Environ Sci Policy 87:11–17

Zhou L, Guan D, Huang X, Yuan X, Zhang M (2020) Evaluation of the cultural ecosystem services of wetland park. Ecol Ind 114:106286

The manufacturer's authorised representative in the EU is Springer Nature Customer Service Centre GmbH, Europaplatz 3, 69115 Heidelberg, Germany. If you have any concerns regarding our products, please contact ProductSafety@springernature.com

Printed and bound by CPI Group (UK) Ltd, Croydon, CR0 4YY
01/10/2026
02225922-0009